A

TREATISE

ON

SURVEYING AND NAVIGATION:

UNITING

THE THEORETICAL, PRACTICAL, AND

EDUCATIONAL FEATURES OF THESE SUBJECTS

BY HORATIO N. ROBINSON, A. M.

FORMERLY PROFESSOR OF MATHEMATICS IN THE UNITED STATES NAVY; AUTHOR OF MATHEMATICAL, PHILOSOPHICAL, AND ASTRONOMICAL WORKS.

THIRD EDITION.

CINCINNATI:
JACOB ERNST, 112 MAIN STREET.
BOSTON:
B. B. MUSSEY & CO.; ROBERT S. DAVIS & CO.; W. J. REYNOLDS & CO.
NEW YORK:
MASON BROTHERS; D. BURGESS & CO.; A. S. BARNES & CO.; NEWMAN & IVISON;
PRATT, WOODFORD & CO.
PHILADELPHIA:
LIPPINCOTT, GRAMBO & CO.; THOMAS, COWPERTHWAITE & CO.;
E. H. BUTLER & CO.; URIAH HUNT & SON.
1853.

Entered, according to act of Congress, in the year 1852,
By H. N. ROBINSON,
in the Clerk's Office of the District Court for the Northern District of New York.

A. C. JAMES, STEREOTYPER,
167 WALNUT ST., CINCINNATI.

PREFACE.

This book is more than its title page proclaims it to be: it is the practical application of the Mathematical Sciences to Mensuration, to Land Surveying, to Leveling, and to Navigation.

Nor is the work merely practical. Elementary principles are here and there brought before the mind in a new light; and original investigations will be found in many parts of the work. To show the reader *how* a thing is to be done, is but a small part of the object sought to be obtained: the great stress is put upon the *reasons* for so doing, which gives true discipline to the mind, and adds greatly to the educational value of any book.

We have illustrated the subject of logarithms, and their practical uses, the same in this book as is common to be found in other books, and this is sufficient for the common pupil, or the ordinary practical man, whether surveyor or navigator; but in addition to this, we have carried logarithms much further in this work than in any I have seen. I do not mean by this that we have more voluminous tables than others. Such is not the fact.

Voluminous tables are not necessary for those who really understand the nature of logarithms, and such are mainly intended for those who are not expected to understand principles. To give a more practical illustration of logarithms, and to suggest artifices in using logarithms generally, we have given Table III and its auxilliaries, on page 70 of tables, showing logarithms to *twelve places* of decimals, a degree of accuracy which *practice never* demands. By the help of this table combined with a true knowledge of the subject, the logarithm of *any number* may be readily found true to *ten places* of decimals, or, conversely, the number corresponding to any given logarithm may be found to almost any degree of accuracy.

Our Traverse Table is not so full as in some other books, but it is full enough to answer every purpose; and latitude and departure, corresponding to any course and distance, can be found by it, provided the operator's good judgment is awake. Indeed a contracted table, in an educational point of view, is better than a full one; for the former calls forth and cultivates tact in the student, but the latter is best for the unanimated plodder.

In running lines, and computing the areas of surveys, we have endeavored to present the subject in such a manner that the reader must constantly keep Elementary Geometry in view, and the whole is so clear and simple, that many will think it unworthy of the rank that it seems to hold in the public estimation, but there are other reasons for this.

The chapter on surveys and surveyors will be found to be a little peculiar, but the information there given, will be highly useful to all those who are inclined to look upon a survey as a mathematical problem only.

On the compass, and the declination of the needle, we have been very full: the subject embraces meridians and astronomical lines drawn on the earth.

The manner in which we should proceed to make a survey, provided no such instrument as the compass existed, and there were no such thing as a magnetic needle, is taken up and illustrated in this work.

The subject of dividing lands is fully discussed and illustrated, and if any one has occasion to complain of mathematical abstrusity in this work, it will be found in this connection; yet there is nothing here above elementary algebra and geometry.

The method of taking levels and making a profile of the vertical section of a line for rail roads, is set forth in this work. The profile shows the necessary excavation or embankment, which it is necessary to cut down or build up at any particular point, to conform to any proposed grade that may be contemplated.

To determine the elevation of any place above the level of the sea, by means of the barometer, has been, and now is, a very interesting problem to all philosophical students, yet very few of them have been able to comprehend it beyond its first great principle, the variation of atmospheric pressure. To trace, or rather to discover the mathematical law which connects the elevation of any locality with the mean hight of the barometer at the same place, has been an obscure problem, and we have taken hold of it with a determination to break open some avenue of light (if such were possible) by which the simplicity of the problem might be brought to the comprehension of the every-day mathematical student, and we believe that we have succeeded in the undertaking.

The part on Navigation, might be regarded, at first view, an abridgment of that subject, and in one sense it is, for we have studied to be as brief as possible, but we would never let brevity stand in the place of perspicuity; and however it may appear, we have given all the mathematical essentials of the subject, and whoever acquires what is here given, will find very little necessity of looking elsewhere for the continuation of the study, unless it is for *sea terms* and seamanship; but these have nothing to do with Navigation as a science. Our method of working *lunars* is more brief than any other, where auxilliary tables and methods of approximation are not resorted to, but to attain this brevity, we have been compelled to use Natural Sines in part of the operation; but on the other hand, this should be no objection, for it gives us a clearer view of the unity and harmony of the mathematical sciences.

CONTENTS.

INTRODUCTION.

CHAPTER I.

CHAPTER II.

CHAPTER III.

SURVEYING.

CHAPTER I.

CHAPTER II.

CHAPTER II.

CHAPTER III.

CELESTIAL OBSERVATIONS.

CHAPTER I.

CHAPTER II.

CHAPTER III.

APPENDIX.

INTRODUCTION.

CHAPTER I.

Mensuration, Surveying, and Navigation, are but branches of the same science, and should be regarded as the application of geometry and trigonometry, and in this light we shall present them to our readers.

In this volume we shall not demonstrate geometrical truths unless we wish to present them in some new form, or unless the demonstration is not readily to be found in the proper places, in the elementary books.

It is expected that all readers of a work of this kind, have previously made themselves more or less acquainted with Algebra and Geometry, and where this is the case the reader will have no difficulty ; and readers who are not thus prepared should be careful not to charge *imaginary* defects to the book : in no work of this kind would it be proper to demonstrate every elementary principle. These remarks apply only to the *educational character* of the book.

Preparatory to a course of practical mathematics, it is proper to give such descriptions of the instruments to be used as will enable the operator to understand their use. But some of these instruments can never be understood from a book, it must be from the instrument itself ; we might as well attempt to give a person an idea of color by the means of language, as to give a person a correct idea of the sextant and theodolite by a mere book description. It is true we can do something by drawings and descriptions, and that something we intend to do.

To represent plane surfaces and tracts of land on paper, no other instruments are necessary than the scale, and dividers, and a protractor to measure angles. In fact, every thing can be done with the scale and dividers only — other instruments, as the protractor, sector, and parallel rulers,only add to our convenience ; at the same time they could be dispensed with.

THE PLANE SCALE.

The plane scale, or the plane diagonal scale of equal parts, as here represented, is the most common and useful of all the instruments used in drawing. It is also a ruler, and if wide and well made will serve as a square also, by which right angles may be drawn.

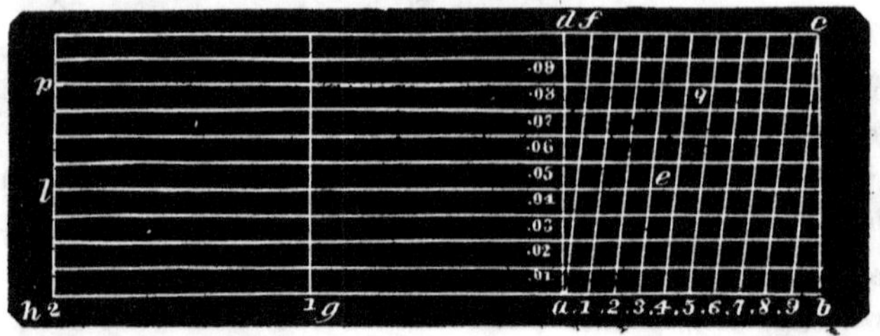

The very appearance of this scale will show its construction, the side of the square *a b* may be of any length whatever, it is generally taken an inch, but this is not imperative.

By means of the 10 parallel lines running along the length of the scale, and the 10 diagonal lines parallel to each other in the square *a b c d*, we have 100 *intersections* in the square, by which we are enabled to find any and every hundredth part of the division of *a b*.

For example, I wish to find 27 hundredths of the line *a b*. I go to the division 2 on *a b*, and then run up that diagonal line to the 7th parallel, and from that intersection to the line *a d* is 27 hundredths of *a b*.

The distances *a b*, *a g*, *g h*, &c., may be taken to represent 1, 10, 100, or in fact any number we please. Suppose we take any one of the equal divisions *a b*, *a g*, &c., to represent 100, and then require 234. From *l* to *e* represents that distance.

If the base *a b* were 10, from *l* to *e* would be 23.4 ; if 1, then from *l* to *e* would be 2.34 ; and so on proportionally for any other change of *base*, or change of the *unit.*

To transfer distances from the scale (as *l e*, *p q*, &c.) to paper, we require

DIVIDERS.

Dividers are nothing more than a delicate pair of compasses — two bars turning on a joint. They are too well known to require representation by a figure.

They are also used for describing circles and parts of circles.

THE PROTRACTOR.

The following diagram accurately represents this instrument. It consists of a semicircle of brass *ABC*, divided into degrees.

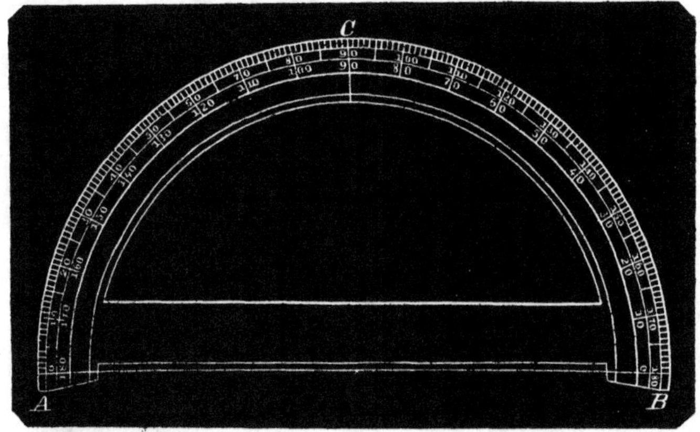

The degrees are numbered both ways, from *A* to *B* and from *B* to *A*. There is a small notch in the middle of *AB*, to indicate the center.

To lay off an angle. Place the diameter *AB* on the line, so that the center shall fall on the angular point.

Then at the degree required, at the edge of the semicircle make a point with a pin. Then remove the protractor and draw a line through the point so marked and the angular point; this line, with the given line, will make the required angle.

The reader will observe a great similarity between this instrument and the circumferentor, which is described in a subsequent portion of this work.

This instrument is designed merely to draw angles on paper, that to draw lines marking given angles, with other lines, in the field.

In addition to this, both the protractor and circumferentor may be used in taking levels, and measuring angles of altitude, when *no better instruments* for such purposes are at hand.

For instance, if a delicate plumb should be suspended from the center of the protractor, and the thread rest at the point *C*, while the instrument is held in a frame, then *A* and *B* would be as a level, and as many degrees as the plumb line rested from *C* so many degrees would be the inclination of *A* and *B* from a horizontal level.

Levels and angles of altitudes were formerly taken in this way.

With the instruments previously described, solve the following problems. The references are to Robinson's Geometry. Thus, (th. 15, b. 1, cor. 1,) indicates theorem 15, book 1, corrolary 1, where the demonstrations of the problem referred to will be found.

PROBLEM 1.

To bisect a given finite straight line.

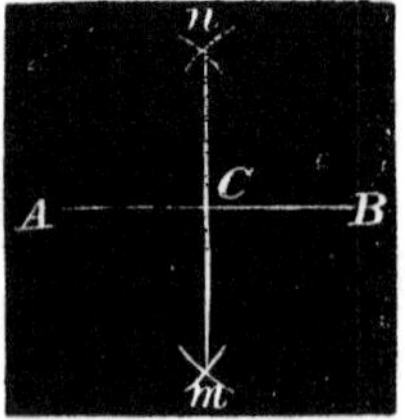

Let *AB* be the given line, and from its extremities, *A* and *B*, with any radius greater than the half of *AB* (Post. 3), describe arcs, cutting each other in *n* and *m*. Join *n* and *m;* and *C*, where it cuts *AB*, will be the middle of the line required.

Proof, (th. 15, b, 1, cor. 1).

PROBLEM 2.

To bisect a given angle.

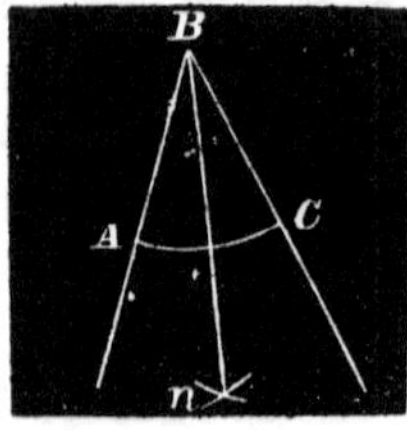

Let *ABC* be the given angle. With any radius, from the center *B*, describe the arc *AC*. From *A* and *C*, as centers, with a radius greater than the half of *AC*, describe arcs, intersecting in *n;* and join *Bn*, it will bisect the given angle.

Proof, (th. 19, b. 1).

PROBLEM 3.

From a given point, in a given line, to draw a perpendicular to that line.

Let AB be the given line, and C the given point. Take n and m equal distances on opposite sides of C; and from the points m and n, as centers, with any radius greater than nC or or mC, describe arcs cutting each other in S. Join SC, and it will be the perpendicular required. Proof, (th. 15, b. 1, cor.).

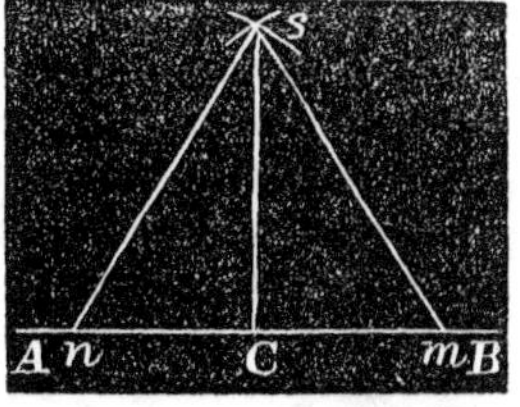

The following is another method, which is preferable, when the given point, C, is at or near the end of the line.

Take any point, O, which is manifestly one side of the perpendicular, and join OC; and with OC, as a radius, describe an arc, cutting AB in m and C. Join mO, and produce it to meet the arc, again, in n; mn is then a diameter to the circle. Join Cn, and it will be the perpendicular required. Proof, (th. 9, b. 3).

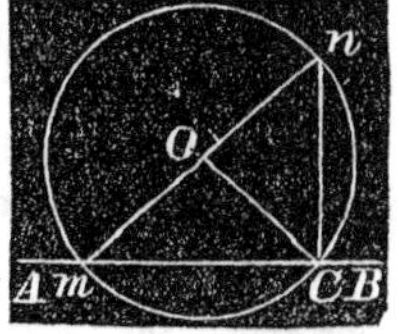

PROBLEM 4.

From a given point without a line, to draw a perpendicular to that line.

Let AB be the given line, and C the given point. From C, draw any oblique line, as Cn. Find the middle point of Cn by (problem 1), and from that point, as a center, describe a semicircle, having Cn as a diameter. From the point m, where this semicircle cuts AB, draw Cm, and it will be the perpendicular required. Proof, (th. 9, b. 3).

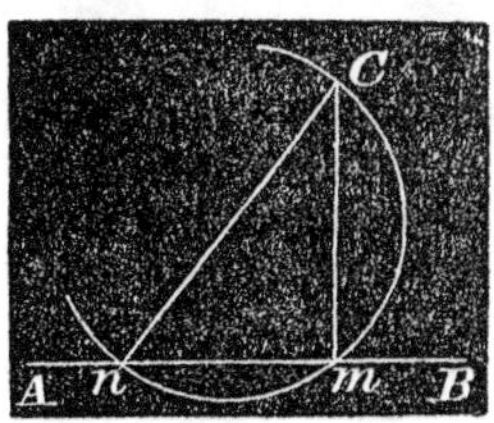

PROBLEM 5.

At a given point in a line, to make an angle equal to another given angle.

Let A be the given point in the line AB, and DCE the given angle.

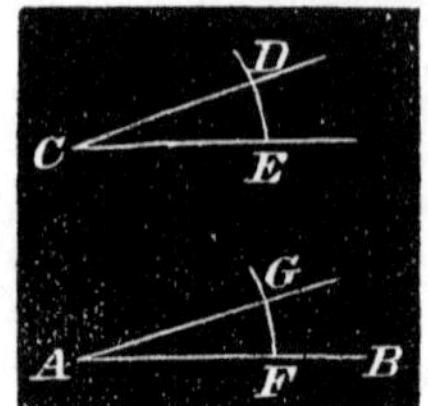

From C as a center, with any radius, CE, draw the arc ED.

From A, as a center, with the radius $AF = CE$, describe an indefinite arc; and from F, as a center, with FG as a radius, equal to ED, describe an arc, cutting the other arc in G, and join AG; GAF will be the angle required. Proof, (th. 5, b. 3).

PROBLEM 6.

From a given point, to draw a line parallel to a given line.

Let A be the given point, and CB the given line. Draw AB, making an angle, ABC; and from the given point, A, in the line AB, draw the angle $BAD = ABC$, by the last problem.

AD and CB make the same angle with AB; they are, therefore, parallel. (Definition of parallel lines).

PROBLEM 7.

To divide a given line into any number of equal parts.

Let AB represent the given line, and let it be required to divide it into any number of equal parts, say five. From one end of the line A, draw AD, indefinite in both length and position. Take any convenient distance in the dividers, as Aa, and set it off on the line AD; thus making the parts Aa, ab, bc, &c., equal. Through the last point, e, draw EB, and through the points a, b, c, and d, draw parallels to eB (problem 6.); these parallels will divide the line as required Proof (th. 17, b. 2).

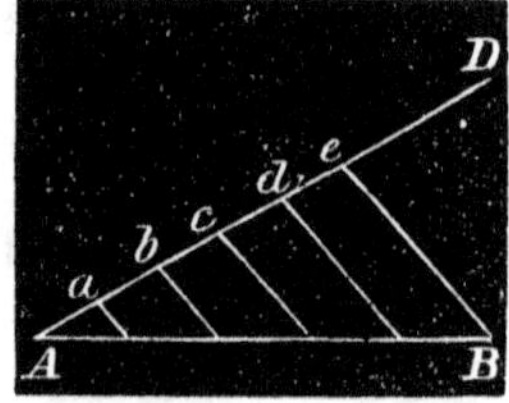

PROBLEM 8.

To find a third proportional to two given lines.

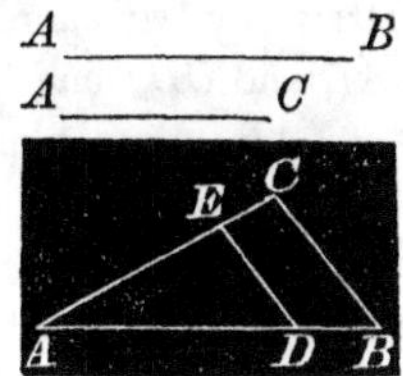

Let AB and AC be any two lines. Place them at any angle, and join CB. On the greater line, AB, take $AD=AC$, and through D, draw DE parallel to BC; AE is the third proportional required.

Proof, (th. 17, b. 2).

PROBLEM 9.

To find a fourth proportional to three given lines.

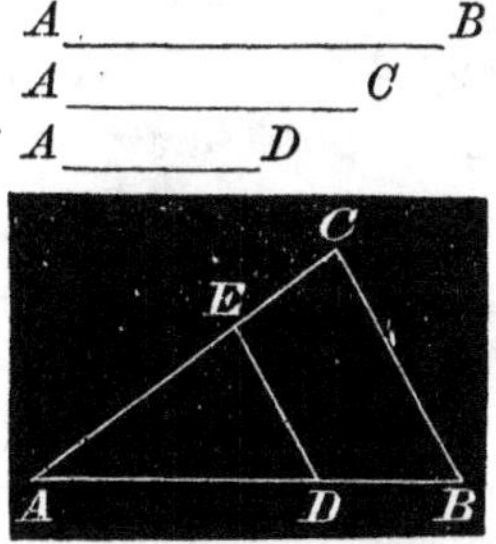

Let AB, AC, AD, represent the three given lines. Place the first two together, at a point forming any angle, as BAC, and join BC. On AB place AD, and from the point D, draw (problem 6) DE parallel to BC; AE will be the fourth proportional required.

Proof, (th. 17, b. 2).

PROBLEM 10.

To find the middle, or mean proportional, between two given lines.

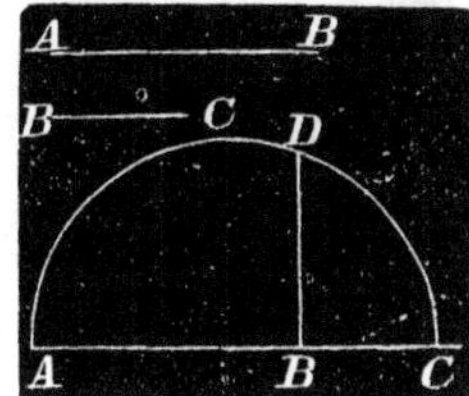

Place AB and BC in one right line, and, on AC, as a diameter, describe a semicircle (postulate 3), and from the point B, draw BD at right angles to AC (problem 3); BD is the mean proportional required.

Proof, (scholium to th. 17, b. 3).

PROBLEM 11.

To find the center of a given circle.

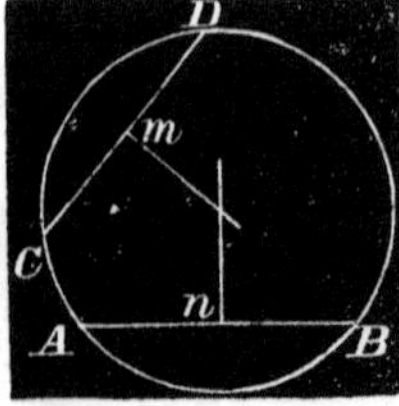

Draw any two chords in the given circle, as *AB* and *CD;* and from the middle point, *n*, of *AB*, draw a perpendicular to *AB;* and from the middle point, *m*, draw a perpendicular to *CD;* and where these two perpendiculars intersect will be the center of the circle. Proof, (th. 1, b. 3).

PROBLEM 12.

To draw a tangent to a given circle, from a given point, either in or without the circumference of the circle.

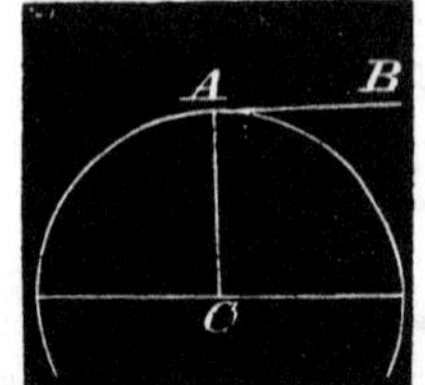

When the given point is in the circumference, as *A*, draw *AC* the radius, and from the point *A*, draw *AB* perpendicular to *AC; AB* is the tangent required.

Proof, (th. 4, b. 3).

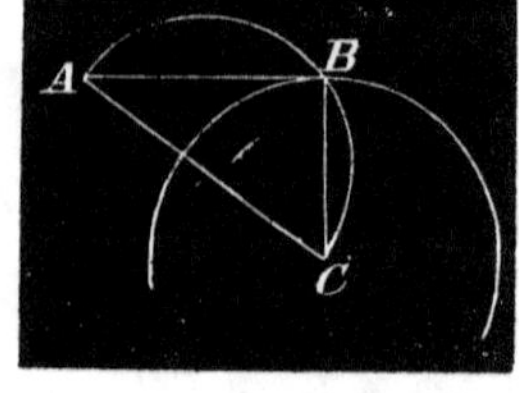

When *A* is without the circle, draw *AC* to the center of the circle; and on *AC*, as a diameter, describe a semicircle; and from the point *B*, where this semicircle intersects the given circle, draw *AB*, and it will be tangent to the circle.

Proof, (th. 9, b. 3), and (th. 4, b. 3).

PROBLEM 13.

On a given line, to describe a segment of a circle, that shall contain an angle equal to a given angle.

Let AB be the given line, and C the given angle. At the ends of the given line, make angles DAB, DBA, each equal to the given angle, C. Then draw AE, BE, perpendiculars to AD, BD; and from the center, E, with radius, EA or EB, describe a circle; then AFB will be the segment required, as any angle F, made in it, will be equal to the given angle, C.

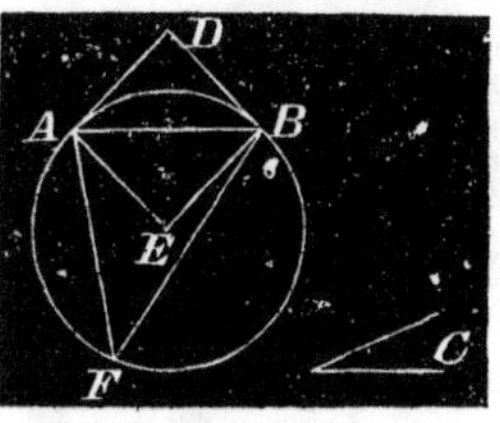

Proof, (th 11. b. 3), and (th. 8, b. 3).

PROBLEM 14.

To cut a segment from any given circle, that shall contain a given angle.

Let C be the given angle. Take any point, as A, in the circumference, and from that point draw the tangent AB; and from the point A, in the line AB, make the angle $BAD=C$ (problem 5), and AED is the segment required.

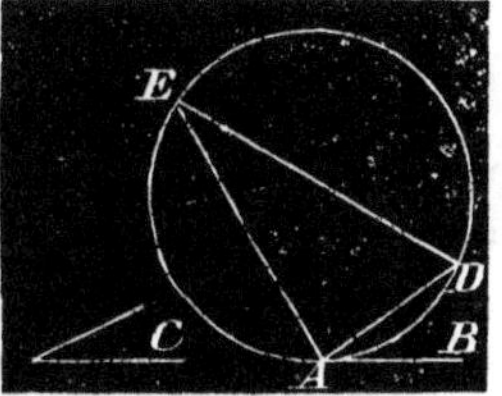

Proof, (th. 11, b. 3), and (th. 8, b. 3).

PROBLEM 15.

To construct an equilateral triangle on a given finite straight line.

Let AB be the given line, and from one extremity, A, as a center, with a radius equal to AB, describe an arc. At the other extremity, B, with the same radius, describe another arc. From C, where these two arcs intersect, draw CA and CB; ABC will be the triangle required.

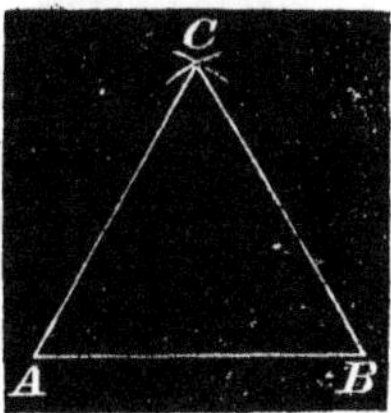

The construction is a sufficient demonstration. Or, (ax. 1).

PROBLEM 16.

To construct a triangle, having its three sides equal to three given lines, any two of which shall be greater than the third.

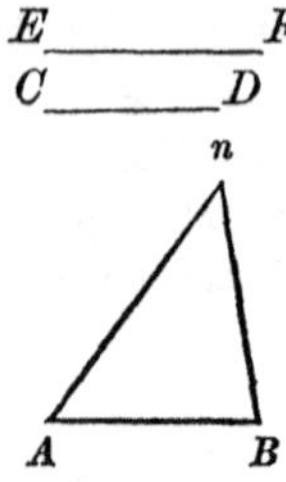

Let *AB*, *CD*, and *EF* represent the three lines. Take any one of them, as *AB*, to be one side of the triangle. From *A*, as a center, with a radius equal to *CD*, describe an arc; and from *B*, as a center, with a radius equal to *EF*, describe another arc, cutting the former in *n*. Join *An* and *Bn*, and *AnB* will be the △ required. Proof, (ax. 1).

PROBLEM 17.

To describe a square on a given line.

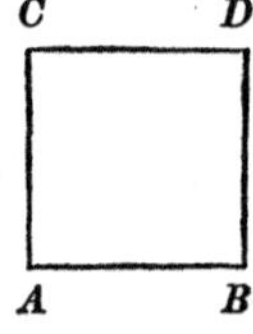

Let *AB* be the given line, and from the extremities, *A* and *B*, draw *AC* and *BD* perpendicular to *AB*. (Problem 3.)

From *A*, as a center, with *AB* as radius, strike an arc across the perpendicular at *C*; and from *C*, draw *CD* parallel to *AB*; *ACDB* is the square required. Proof, (th. 21, b. 1.)

PROBLEM 18.

To construct a rectangle, or a parallelogram, whose adjacent sides are equal to two given lines.

A ________ C

A ____________ B

Let *AB* and *AC* be the two given lines. From the extremities of one line, draw perpendiculars to that line, as in the last problem; and from these perpendiculars, cut off portions equal to the other line; and by a parallel, complete the figure.

When the figure is to be a parallelogram, with oblique angles, describe the angles by problem 5. Proof, (th. 21, b. 1).

PROBLEM 19.

To describe a rectangle that shall be equal to a given square, and have a side equal to a given line.

Let AB be a side of the given square, and CD one side of the required rectangle.

C________D
A__________B
E____________F

Find the third proportional, EF, to CD and AB (problem 8). Then we shall have,

$$CD : AB :: AB : EF$$

Construct a rectangle with the two given lines, CD and EF (problem 18), and it will be equal to the given square, (th. 13, b. 2).

PROBLEM 20.

To construct a square that shall be equal to the difference of two given squares.

Let A represent a side of the greater of two given squares, and B a side of the lesser square.

On A, as a diameter, describe a semicircle, and from one extremity, m, as a center, with a radius equal to B, describe an arc, n, and, from the point where it cuts the circumference, draw mn and np; np is the side of a square, which, when constructed, (problem 17), will be equal to the difference of the two given squares. Proof, (th. 9, b. 3, and 36, b. 1.)

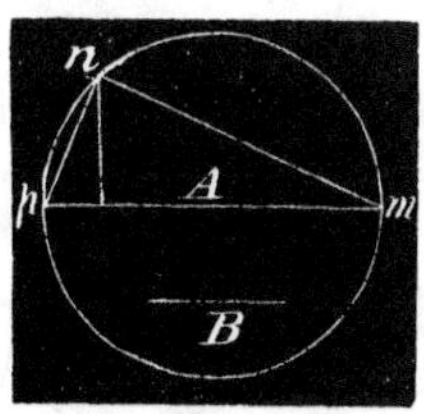

PROBLEM 21.

To construct a square, that shall be to a given square, as a line, M, *to a line,* N.

Place M and N in a line, and on the sum describe a semicircle From the point where they join, draw a perpendicular to meet the circumference in A. Join Am and An, and produce them indefinitely. On Am or An, produced, take AB= to the side of the given square; and from B, draw BC parallel to mn; AC is a side of the required square.

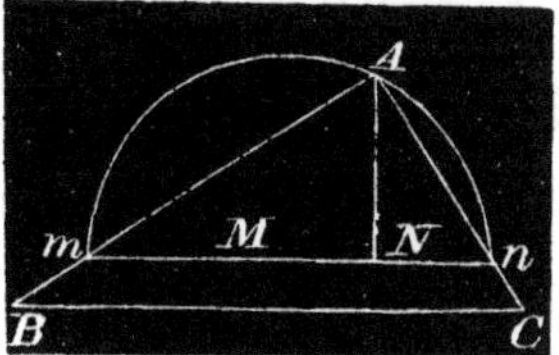

Besides the numerical scale of equal parts, we have scales of chords, sines, and tangents, which can be constructed corresponding to any radius.

Such scales of course *are not scales of equal parts.*

Such scales are constructed in the following manner.

Take CA any radius, and describe a semicircle. Draw CD at right angles to AB, and draw a tangent line from A. Divide the arc AD into equal parts 10, 20, 30, &c., beginning at D, and subdivide them as much as required.

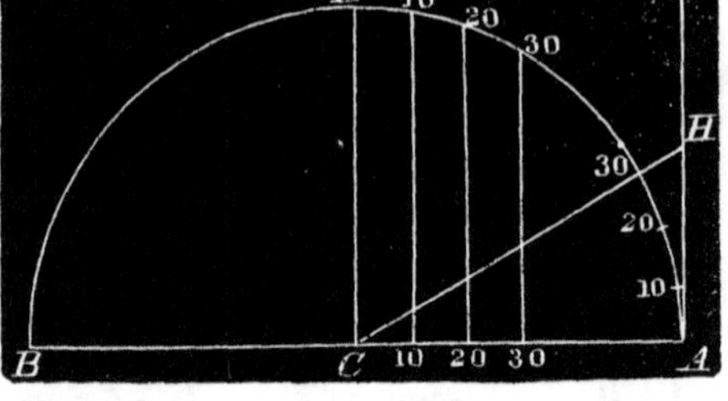

Draw 10 10,—20 20,—30 30, &c., all parallel to CD.

From C to 10 on the line CA, is the sine of 10°. From C to 20 is the sine of 20° &c. &c.

The line 10 10 is the sine of 80°, and CD or CA is the sine of 90°.

The distance from A to D is the chord of 90°, from A to 10 is the chord of 80°, and from A to 20 is the chord of 70°, and so on down. Thus we perceive that we can take off any sine or chord and lay it down on a ruler; and chords and sines thus laid off constitute the scale of chords, sines, &c.

Lines drawn from C through any division of the *arc, commencing at A* to strike the tangent line, will mark off the tangent corresponding to that arc. Thus, if the angle ACH is 30°, then the line AH placed on a scale, will represent the tangent of 30° to the radius CA, and thus any other tangent can be laid down on the same scale.

The scale of chords and sines, as well as the scale of equal parts, are to be found on the

SECTOR.

The sector is commonly made of ivory, and consists of two arms which open and turn round a joint at their common extremity.

For some operations, particularly the projection of solar eclipses, the sector is a very useful instrument.

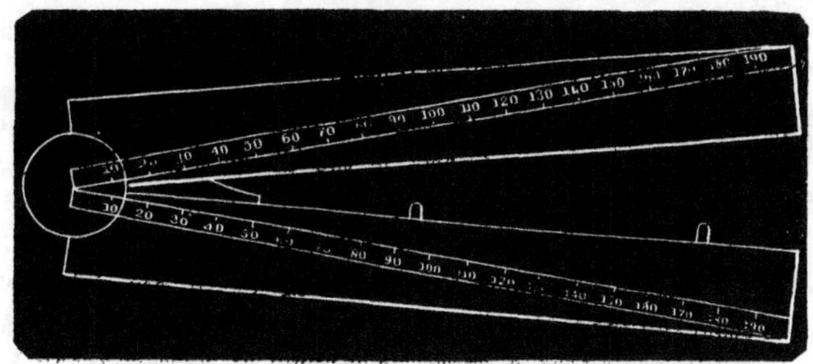

The figure before us represents one side of a sector with the plane scale only upon it. More than one scale can be put on to a side, but we represent but one to avoid confusion.

The scale must be alike on both arms — and it must commence exactly at the joint — hence when near the center the different scales *crowd* each other.

The two arms of the sector always form two sides of a triangle, and by opening and closing them we vary the angle, yet the distance across from one arm to the other is always proportional to the sides of the triangle.

The advantage of the sector will appear from the following problem.

A map is before me, its scale is 20 miles to an inch; I wish to find the distance in a right line between two points laid down on it.

1st. I take one inch in the dividers and open the sector, so that the distance between 20 and 20 on the two arms, shall just correspond to the measure in the dividers, that is, shall be one inch. Let the sector lie on the table thus opened.

2nd. Now take the distance you wish to measure. in the dividers; place one foot on one arm of the sector, and the other foot on the other arm; so that the feet of the dividers shall fall on the same number on both arms of the sector.

The number thus marked by the dividers will be the distance required. The distance between any other two points may be measured on the same map, without any computation whatever.

For another illustration of the utility of the sector, let us suppose, that the sine of 20° is required corresponding to a radius of 6 inches.

Take 6 inches in the dividers, and open the sector so that the sine of 90° from arm to arm shall be 6 inches.

The sector being thus open, take the distance from 20 to 20, on the line of sines from arm to arm, in the dividers, and that is the distance required.

GUNTER'S SCALE.

Gunter's scale is commonly two feet in length, containing the plane scale and the scale of sines, chords, and tangents on one side of it, and the scale for the *logarithms* of numbers, sines, and tangents on the other.

This scale is very ingenious, but it is not so much used nor considered so important as formerly.

CHAPTER II.

LOGARITHMS.

ART 1. *Logarithms are exponents.*

Thus, if $a^2=9$

and $a^3=27$

Then $a^5=243$; by multiplying the two equations together term by term.

The exponent 2 of the first equation may be considered the logarithm of 9; the exponent 3 the logarithm of 27, and the exponent 5 the logarithm of the number 243.

In these equations $a=3$ *the base of the system.*

By the preceding operation it is obvious that adding the exponents 2 and 3, corresponds to multiplying the numbers 9 and 27.

If we take the equation $a^5=243$, and divide it by $a^2=9$ member by member we shall have

$$a^{5-2}=a^3=27.$$

Hence adding exponents (logarithms) corresponds to the multiplica-

tion of their corresponding numbers and subtracting the exponents (logarithms) corresponds to the division of their numbers.

It is this property of logarithms that gives them their utility and importance.

ART. **2.** The base of our common system of logarithms is 10, and in any equation in the form $10^x=n$, x is the logarithm of the number n whatever number n may represent. If $n=10$, then the equation becomes $10^x=10$. Whence $x=1$ because $10^1=10$. Therefore in our common system of logarithms the logarithm of 10 must be 1.

Now because
$$10^0=1$$
$$10^1=10$$
$$10^2=100$$
$$10^3=1000$$
$$10^4=10000$$
&c. &c.; it is plain that the logarithm of 1 is 0, of 10 is 1, of 100 is 2, &c., *every power of* 10 *increasing the logarithm by* 1.

It is also obvious, that every number between 0 and 10 must have a fractional or *decimal* number for its logarithm, and every number between 10 and 100 must have *one, and some decimal* for its logarithm.

In the equation $10^x=3$, x is the logarithm of 3, and if we multiply this by $10^1=10$, member by member, we shall have

$$10^{1+x}=30$$

Multiply this by $\quad 10^1 \quad =10$

and we have $\quad 10^{2+x}=300.$

These results show that 3, 30, 300, have logarithms containing the same *decimal* number; x differs from each exponent only by whole numbers, and thus, generally; *any number multiplied or divided by* 10, *or any power of* 10, *will have logarithms containing the same decimal part.*

ART. **3.** For the general computation of logarithms we refer to algebra, and in a work like this we shall only attend to such portions of theory as to enable the student to use them understandingly and with as much practical facility as possible.

Let it be observed that the logarithm

of	10000	is	4,00000
of	1000	is	3,00000
of	100	is	2,00000
of	10	is	1,00000
of	1	is	0,00000
	$\frac{1}{10}$	is	—1,00000
	$\frac{1}{100}=10^{-2}$	is	—2,00000

For every division of the number by 10 we subtract 1 from its logarithm, and when the number comes down to 1, and its logarithm of course to 0, if we again divide by 10, making it $\frac{1}{10}$ or 10^{-1}, we must subtract *one* from the logarithm, making it —1.

The decimal portion of a logarithm *is always positive*, but the *index or whole number part of it, becomes minus when the value of the number is less than* 1.

ART. 4. The whole number belonging to any logarithm is called its *index*, a very appropriate term, because it *indicates*, it points out where to place the decimal point between whole numbers and decimals.

The *index*, or (as some call it) the *characteristic*, is never put in the tables (except from 1 to 100), because we always know what it is. It is always *one less* than the number of digits in the whole number. This is obvious from Art. 3.

Thus, the number 3754 has 3 for the index of its logarithm, because the number consists of 4 digits; that is, the logarithm is 3, *and some decimal.*

The number 34.785 has 1 for the index of its logarithm, because the number is between 34 and 35, and 1 is the index for all numbers between 1 and 100.

All numbers consisting of the same figures, whether integral, fractional, or mixed, have logarithms consisting of the same decimal part. (Art. 2.) The logarithms differ only in their *indices*.

Thus,	the number	7956.	has	3.900695	for its log.
	the number	795.6	has	2.900695	"
	the number	79.56	has	1.900695	"
	the number	7.956	has	0.900695	"

the number .7956 has —1.900695 for its log.
the number .07956 has —2.900695 "
&c., &c.

For every division by 10, we diminish the index by 1. When the index is *minus* it indicates a decimal number; but let the learner remember that the index only is minus; the decimal part is *always positive.*

ART. **5.** To take out the logarithm of any number from the tables, we only consider the digits; for the logarithms of 7956, or of 7.956, or of .007956, have the same *decimal part;* and when that decimal part is found we then consider the value of the number to prefix the index.

To prefix the index to a decimal, count the decimal point 1, and each cipher 1, up to the first significant figure, and this is the negative index.

For example, find the logarithm of the decimal .00085. To accomplish this we must look for the logarithm of the whole number 85, and we find its decimal part to be .929419; and now, to determine the index, we count *one* for the decimal point and three ciphers, making 4; hence, we have Num. .00085 - log. —4.929419.

The smaller the decimal, the greater the negative index; and when the decimal becomes 0, the logarithm becomes *negatively infinite.*

ART. **6.** The logarithm of any number consisting of *four digits or less,* can be taken out of the table directly and without the least difficulty.

Thus, to find the logarithm of the number 3725, we find the number 372 at the side, and over the top we find 5, and opposite the former and under the latter we find .571126 for the decimal part of the logarithm. The 57, the first two decimal, is under 0, which is the same for the whole horizontal column.

Hence, the logarithm of 3725 is 3.571126
of 37250 is 4.571126
of 3.725 is 0.571126
&c., &c.

Find the logarithm of 1176. We find 117 at the side, and 6 at the top, and opposite the former and under the latter we find .407

3

The point here demands a cipher, and is put in to arrest attention to make the operator look to the next horizontal line below for the first two decimals. Thus, we find .070407 for the decimal part of the logarithm required.

Hence, the log. of 1176 is 3.070407

1. What is the log. of .001176 ? Ans. —3.070407
2. What is the log. of 13.81 ? Ans. 1.140194
3. What is the log. of 72.55 ? Ans. 1.860637
4. What is the log. of .6762 ? Ans. —1.830075
5. What is the logarithm of the number 834785 ?

This number is so large that we cannot find it in the table, but we can find the numbers 8347 and 8348. The logarithms of these numbers are the same as the logarithms of the numbers 834700 and 834800, except the indices.

	834700	log.	5.921530
	834800	log.	5.921582
Difference,	100		52

Now, our proposed number, 834785, is between the two preceding numbers; and, of course, its logarithm lies between the two preceding logarithms; and, without further comment, we may proportion to it thus, 100 : 85=52 : 44.2

Or, 1. : .85=52 : 44.2

To the logarithm	5.921530
Add	44
Hence, the logarithm of 834785 is	5.921574
the logarithm of 83.4785 is	1.921574

From this we draw the following rule to find the logarithm of any number consisting of more than four places of figures.

Rule.—*Take out the logarithm of the four superior places directly from the table, and take the difference between this logarithm and the next greater logarithm in the table. Multiply this difference by the inferior places in the number as a decimal, and add the result to the logarithm corresponding to the superior places, the sum will be the logarithm required.*

Example. Find the log. of 357.32514.

The four superior digits are 3573; the logarithm of these

corresponds to the decimal, .553033, for its decimal part. The inferior digits, taken as a decimal, are

```
   .2514
    122
   -----
    5028
   5028
  2514
  -------
  30.6708
```

This result shows that 30, or more nearly, 31, should be added to the logarithm already found, thus giving .553064 for the decimal part of the logarithm 357.32514.

Therefore, as three digits of the given number are whole numbers, the index must be 2, and the logarithm

of	357.32514	is	2.553064
of	3573251.4	is	6.553064
of	.035732514	is	—2.553064

The change between the place of the decimal point in a number, and the corresponding change of the index to its logarithm, should be strongly impressed on the mind of a learner.

Example 2. What is the log. of 366.25636 ? Ans. 2.563785

3. What is the log. of 39.37079 ? Ans. 1.595174

4. What is the log. of 2.37581 ? Ans. 0.375812

Art. 7. We now give the converse of the last article; that is, we give the decimal part of a logarithm to find its corresponding number.

Taking the decimal in Example 1, (Art. 6,) .553064, we demand its corresponding number.*

The next less logarithm in the table, is .553033, corresponding to the figure 3573. The difference between this given logarithm and the one next less in the table, is 31 ; and the difference between two consecutive logarithms in this part of the table, is 122. Now divide 31 by 122, and write the quotient after the number 3573.

* To take out a number from its logarithm, never enter the first part of the table between 1 and 100. Go to the main table, as it contains many more logarithms.

That is,

```
122)31.(254
    244
    ---
     660
     610
     ---
      500
      488
      ---
```

The figures, then, are 3573254, which corresponds to the decimal logarithm .553064; and the value of these figures will, of course, depend on the index to the logarithm.

If this given logarithm contained an index, such index would point out how many of these figures must be taken for whole numbers, the others will be decimals; thus, if the index had been 4, the number would be 35732.54

If the given decimal had been .553063.67, which is the exact converse of example 1, then we should have found that number, 35732514; but we did not give that decimal logarithm, because the table contains only six decimal places. From this obvious operation we derive the following rule to find the number corresponding to a given logarithm.

RULE.—*If the given logarithm is not in the table, find the one next less, and take out the four figures corresponding; and if more than four figures are required, take the difference between the given logarithm and the next less in the table, and divide that difference by the difference of the two consecutive logarithms in the table, the one less, the other greater than the given logarithm; and the figures arising in the quotient, as many as may be required, must be annexed to the former figures taken from the table.*

EXAMPLES.

1. Given, the logarithm 3.743210, to find its corresponding number true to three places of decimals. Ans. 5536.182

2. Given, the logarithm 2.633356, to find its corresponding number true to two places of decimals. Ans. 429.89

3. Given, the logarithm —3.291742, to find its corresponding number. Ans. .0019577

MULTIPLICATION BY LOGARITHMS.

ART. 8. If the principle first laid down in (ART. 1) is true, the sum of the exponents will be the exponent of the product of any number of factors. In other words,

The sum of the logarithms of any number of factors will be the logarithm of the product of those factors.

N. B. The logarithmic table corresponds to this principle, and we may see by the following

EXAMPLES.

The log. of 3 (taken from the table,) is	0.477121
The log. of 4 " " " " is	0.602060
Therefore the log. of 12 must be	1.079181

Given, the log. of 7 and the log. of 9, to find the logarithm of 63. Because $7 \times 9 = 63$, therefore,

To log. 7=	0.845098
Add log. 9=	0.954243
Sum	1.799341

By inspecting the table, we shall find this logarithm stands opposite 63, and by this process the logarithms of all the composite numbers have been found. In this we may consider that the logarithm pointed out the product 63.

Hence we have the following rule for obtaining the product of any number of factors.

RULE.—*Find the logarithm of each factor, add those logarithms together and the sum will be the logarithm of the product. The number corresponding to this last logarithm taken from the table, will be the product itself.*

EXAMPLES.

1. To multiply 23.14 by 5.062.

	Numbers.	Logs.
	23.14	1.364363
	5.062	0.704322
Product	117.1347	2.068685

2. To multiply 2.581926 by 3.457291.

	Numbers.	Logs.
	2.581926	0.411944
	3.457291	0.538736
Product.	8.92648	0.950680

3. To mult. 3.902 and 597.16 and .0314728 all together.

Numbers.	Logs.
3.902	0.591287
597.16	2.776091
.0314728	—2.497935
Prod. 73.3333	1.865313

Here the—2 cancels the 2, and the one to carry from the decimals is set down.

4. To mult. 3.586, and 2.1046, and 0.8372, and 0.0294 all together.

Numbers.	Logs.
3.586	0.554610
2.1046	0.323170
0.8372	—1.922829
0.0294	—2.468347
Prod. 0.1057618	—1.268956

Here the 2 to carry cancels the —2, and there remains the —1 to set down.

DIVISION BY LOGARITHMS.

ART. 9. As division is the converse of multiplication we draw the following rule for division by use of logarithms.

N. B. Addition and subtraction is to be understood in the algebraic sense.

RULE.—*From the logarithm of the dividend subtract the logarithm of the divisor, and the number corresponding to the remainder is the quotient required.*

EXAMPLES.

1. Divide 327.5 by 2207

log.	327.5	2.515211
log.	2207	3.342028
Quotient	.14839	—1.173183

2. Divide .054 by 1.75

log.	.054	—2.732394
log.	1.75	0.243038
Quotient	.030857	—2.489356

ART. 10. The preceding examples in multiplication and division were adduced only to show the nature of logarithms: had our object been results, the common arithmetical operations would have been more convenient for some of them; but there are cases that demand the use of logarithms, and such cases mostly occur in Involution and Evolution.

Rule for Involution.—*Take out the logarithm of the given number, and multiply it by the index of the proposed power. Find the number corresponding to the product, and it will be the power required.*

EXAMPLES.

1. What is the 2d power of 351?

	log. 351	2.545307
		2
Ans.	123201	5.090614

2. What is the cube of 1.72?

	log. 1.72	0.235528
		3
Ans.	5.0884	0.706584

3. What is the 4th power of .0916?

	log. .0916	—2.961895
		4
Ans.	.000070401	—5.847580

Here 4 times the negative index is —8, adding the 3 to carry gives —5.

4. What is the 17th power of 1.04?

	log. 1.04	0.017033
		17
		0.119231
		0.17033
Ans.	1.9476	0.289561

N. B. This last example begins to disclose the utility of logarithms.

5. What is the 6th power of 1.037?

	log. 1.037	0.015779
		6
Ans.	1.243+	0.094674

6. What is the 21st power of 2.02?

	log. 2.02	0.305351
		21
		.305351
		6.10702
Ans.	2584454.6	6.412371

EVOLUTION.

Art. **11.** Evolution is the converse of Involution; hence we have the following rule for the extraction of roots:

Take the logarithm of the given number out of the table. Divide the logarithm, thus found, by the index of the required root; then the number corresponding is the root sought.*

EXAMPLES.

1. What is the cube root of 125?

	log. 125	3)2.096910
Ans.	5	0.698970

2. What is the cube root of 200?

	log. 200	3)2.301030
Ans.	5.848+	0.767010

3. What is the 4th root of 751?

	log. 751	4)2.875640
Ans.	5.235+	0.718910

4. What is the 20th root of 1.035?

	log. 1.035	20)0.014940
Ans.	1.001718	0.000747

5. What is the cube root of the decimal .00048

log. .00048 —4.681241

To the inexperienced here would be a difficulty, as the index is *negative*, and the decimal part *positive*. How then shall we divide by 3? Add —2 and +2 to the index; and this is, in effect, adding *nothing;* it merely changes the form of the index, thus, —6+2.681241 Now, we can divide by 3, and the quotient is —2.893747. The corresponding number, or root, sought, is .07829+ Ans.

REMARK.—In the preceding articles we have taught all the preliminary rules for the use of logarithms; "*but there is a wisdom beyond rules*," and he who does not arrive at it, attains only the burdens of knowledge without its benefits. Rules are necessary through the first rudiments of any science; but he who can instantly fall back on to first principles, and do the most advantageous thing at the most advantageous point of time, has a practical tact of the highest value.

To awaken this faculty in the mind of the learner, we give what follows on the subject of logarithms. It may not be necessary for some, but to many it will be interesting and new

ART. **12.** Persons who possess both theoretical knowledge and practical skill, rarely, if ever, use the first part of the logarithmic table of numbers, *except for exact numbers*, especially if they pretend to any thing like accuracy. Such persons take some artifice to *throw* their logarithm into the last part of the table, *where the variation of the logarithm is slower* than in the first part of the table.

To illustrate these remarks let us take into consideration the resulting logarithm to Example, 4, Art. 11.

It would be troublesome, and, indeed, quite impossible, to take out the number corresponding to this logarithm 0.000747 if we simply apply the usual rules, and go directly to the table with the logarithm.

From the given log.		0.000747
Subtract the log. of	1.01	0.004321
This log. corresponds to	.9918	—1.996426

Subtracting the logarithm of 1.01 was equivalent to dividing the number by 1.01. We must, therefore, multiply .9918 by 1.01 to produce the number corresponding to the log. 0.000747.

Thus,

```
0.9918
  9918
--------
1.001718
```

Subtracting the logarithm of 1.01 produced a logarithm having a large decimal part, and this was the object. We can, then, take it into the table, and find its number, to great accuracy, by mere inspection. We might have added the logarithm of 9, and thus produced a *large decimal*, and then have divided the corresponding number by 9, for the required result.

Again: Suppose the logarithm of the number 101248, was required to as great exactness as our tables will allow. The operator who exercises no original thought, and depends only upon rules, will go directly to the table for the logarithm required; but he cannot find it there without some trouble to proportion to it; *and even then his result will not be accurate*, because the logarithms vary by no exact numerical ratio.

Take the number and divide by some number that will give a *large integer* quotient. Thus,

```
102)101248(992.6¼
    918
    ---
     944
     918
     ---
      268
      204
      ---
       640
       612
       ---
```

Now, we can find the logarithm of 992.6¼ very accurately, by inspection. Not regarding the fraction $\frac{1}{4}$ it would be very accurate; but the practical man always makes a little correction for such fractions, *without taking any proportion to do so,* or using any formality about the matter. In this case, we perceive that 10 or 11, added to the last two decimal figures, will correct it; hence,

	log. $992.6\frac{1}{4}$		2.996785
	log. 102		2.008600
Therefore,	log. 101248	is	5.005385

We give one more example.

What number corresponds to the log.		2.111497
Add log. of 7		0.845098
Num. 904.88	log.	2.956595

Dividing this by 7, gives 129.27 for the number required.

REMARK.—The foregoing comments and illustrations are sufficient for all practical problems, that can come before the surveyor, navigator, or engineer. The common table of logarithms, to six decimal places, extending from page two to twenty, of tables, is sufficiently accurate for all common problems; but there are cases in Astronomy, and in very delicate scientific investigations, where it is important to have the logarithms extend to a greater number of decimal places. Accordingly I have computed a table to 12 decimal places, corresponding to the consecutive whole numbers up to 110, and the prime numbers from thence to 1129. These logarithms, together with the auxiliary logarithms on page 71, are sufficient to find the logarithm of any number that can be proposed, and to find the number corresponding to any given logarithm containing ten decimal places.

It is a general impression that a table of logarithms must be practically useless, unless it is voluminous and complete; and for constant practical use it should be so; but for occasional service, the tables here given are sufficient, and for *educational purposes* they are better than they would be if they were more full, for now they demand proper theoretical knowledge to use them with success. On the contrary, whoever cannot use them with success, must be deficient in theoretical knowledge, or wanting in practical tact to bring such knowledge into immediate use.

To show the importance and practical utility of these tables, is the object of the following illustrations and examples.

ART. **13.** To make a table of logarithms anew, to contain any particular number of decimal places, the following formula taken from algebra, appears to be the most practical and convenient.

The investigation of the formula belongs to the science of algebra, and not to a work like this.

$$\log(z+1)-\log z=$$
$$0.8685889638\left(\frac{1}{2z+1}+\frac{1}{3(2z+1)^3}+\frac{1}{5(z+1)^5}\ \&c.\right)$$

By this formula we perceive that the log. of $(z+1)$ becomes known when that of z is known; but the log. of z is known when $z=1$, 10, 100, 1000, &c. Then the formula will give the logarithms of 2, 11, 101, 1001, &c.

After a commencement has been made and the logarithms of a few numbers obtained, the logarithms of others can be deduced from them — hence the formula is used for the prime numbers only.

When z is large, over 100, the series converges very rapidly and only two terms need be used. When z is over 2000 only one term is required *even for twelve decimal places.*

The auxiliary logarithms marked *A*, *B*, *C* page 71, were computed by this formula. For example, the log. of 1000 is 3,000000: make $z=1000$, then $(z+1)=1001$, and $(2z+1)=2001$. The formula now readily gives the log. of 1001; and the log. of 1.001 is the same, if we suppose the index or rather make the index 0. Having the log. of 1001, we find that of 1002, and thus we run through *A*. In the same manner we run through *B* and *C*.

The greater the number the more readily can its log. be computed.

That the learner may fully comprehend the application of the auxiliary logarithms *A*, *B*, and *C*, he must call to mind the following principle.

ART. **14.** *The product of any number of factors consisting of one and a small fraction, is very nearly equal to one and the sum of those fractions.*

Thus the product of (1.0001) (1.00002) (1.000003) is very nearly equal to 1.000123.

If this be true, we can immediately separate 1.000123 or any other similar quantity in the following factors.

(1.0001) (1.00002) (1.000003)

The number 1.00008 may be taken as the product of (1.00002) (1.00005) (1.00001) without any material error.

This principle may be proved algebraically, thus: Let a, b, and c represent very small fractions, then the product of

$$(1+a)\,(1+b)=1+a+b+ab.$$

But a and b being very small fractions their product ab is still much less, and the material part of the whole product is $1+a+b$. Multiply this by $(1+c)$ making the same consideration, and we shall have $1+a+b+c$ for the essential value of $(1+a)(1+b)(1+c)$.

Try it by numbers, thus: multiply 1.0001 by 1.00004.

$$\begin{array}{r} 1.00004 \\ 1.0001 \\ \hline 100004 \\ 1.00004 \\ \hline 1.000140004 \end{array}$$

But the value of this is extremely near 1.00014, *the sum of unity and the fractional parts of the factors.*

ART. **15.** When the difference of two quantities of the same kind, is very small in relation to the quantities themselves, such a difference is called a differential.

Thus, the difference between 1.000140004 and 1.00014 is .000000004, and it may be called the differential of 1.00014, and in reference to it may be omitted.

The difference between 8 and 9 is 1, but in this case 1 cannot be considered the differential of 8, it is too large.

But the difference between 8000 and 8001 is 1, and here 1 is sufficiently small to be considered as the differential of 8000.

There is no exact line of demarkation where a difference may be taken for a differential, that depends on the nature of the case; hence the prejudice in a certain class of minds against the calculus.

Now, if we take the logarithmic formula from Art. 13, and conceive z to be very large, then the difference between $(z+1)$ and z which is 1, may be considered as the *differential* between the two numbers; and in that case log. $(z+1)$—log. z, is the same as the differential of the logarithm of z.

Making this supposition, the formula in Art. 13 becomes

$$(\text{dif.})\ \log.z = 0.8685889638 \times \frac{(\text{dif.})\ z}{(2z+1)}$$

We take only one term of the series, because the other terms are of no essential value, compared with the first; moreover, as z is very large, $2z+1$ is comparatively so little greater than $2z$, that *for all practical purposes* it may be taken as $2z$; this being admitted, the preceding equation reduces to

$$(\text{dif.})\ \log. z = \frac{0.4342944819\ (\text{dif.})\ z}{z}$$

The symbol (dif.) stands for the *differential* of the quantity. Observe that the decimal 0.4342944819 is the *modulus* of our system of logarithms.

Now this equation put in words, is the following:

The differential of a logarithm is equal to the modulus, into the differential of the number divided by the number.

This equation also gives

$$(\text{dif.})\ z = \frac{z\ (\text{dif.})\ \log. z}{0.4342944819}$$

Or in words,

The differential of a number is equal to the number into the differential of the logarithm of the number divided by the modulus.

The practical use of these principles will be shown in the following articles.

Art. **16.** When the diameter of a circle is 1, the circumference is 3.14159265359. Find the log. of this number, true to at least ten decimal places.

When the logarithm is found its index will be 0. Now, consider the digits as composing one whole number, and then pay no attention to the index.

Take the three superior digits 314. Its factors are 2.157. (Table commencing on page 67.)

	Log. 157		2.195899653409
	Log. 2		0.301029995644
	Log. 314		2.496929649053
Table *B.*	1.0005	log.	0.000217099966
Prod.	3141570	log.	.497146749019
Table *C.*	1.000007	log.	3040058 (7*b*)
	2199099		
	314157		
Prod.	314159199099—	log.	.497149789077

REMARK.—Let the learner take hold of the preceding problem with great deliberation, understand the reason of every part of the process — and make every necessary consideration, and then he will understand how to manage every problem of the like kind.

Here then, we have the exact decimal part of the logarithm to these digits. If the value of that superior digit 3 is simply three, then the index to the log. is 0; if it is 30, 1, if 300, 2, &c. Knowing the value to be 3, we shall in the end, put the index at 0.

We have obtained the exact logarithm of a certain number, but it is not the number required; it is a number however, very near the number required. Take their difference

From	314159265359
Take	314159199099
	66260

This difference, great as it may appear by itself, is so small in relation to 314159265359 that it may be taken as its *differential.*

But corresponding to this differential of the number, there is a differential for the logarithm, which is given by the equation in Art. 15.

That is, $(\text{dif.})\ \log.\ z = \frac{(0.4342944819)(66260)}{314159265359}$

It would be a tedious operation to draw out the result of this expression arithmetically. We will, therefore, use the common table of logarithms, which will answer every purpose.

We have the logarithm of the *modulus* as a constant quantity on page 71 of tables; and having the logarithm of the denominator, we have only to look for the logarithm of 66260, the operation stands thus,

	log. 66260	4.821251
	log. *m*.	—1.637784
		4.459035
	log. *z*.	11.497150
Num. 0.000000091596	log.	—8.961885

To	0. 497 149 789 077
Add	0. 000 000 091 596
Log. of 3.14159265359 =	0. 497 149 880 673

The factor 1.0005 was obvious enough from inspection; but the other factor 1.000007 is not obvious. The question then arises, how did we find it?

Wanting a factor, which with the other factor 314157, would produce the given number, we represented it by x, then

$$314157x=314159265359$$

By division $x=1000007+$

The sidereal year consists of 365.2563744 mean solar days. What is the logarithm of this number?

We know that the index must be 2, therefore pay no attention to the index during the operation.

Take the three superior digits 365: its factors are 73 and 5.

	73	log.	1. 863 322 860 120	
	5	log.	698 970 004 336	
Prod.	365	log.	2. 562 292 864 456	
Table B,	10007	log.	0. 000 303 836 798	
Prod.	3652555	log.	2. 562 596 701 254	
C	1.000002	log.	868 588	(2b)
	1.0000003	log.	130 287	(3c)
	1.00000009	log.	39 087	(9d)
Prod. = given number nearly,		log.	=2. 562 597 739 216	

We found these factors by taking the value of x out of the following equation:

$$3652555x=3652563744$$

Whence, $x=1.00000239.$

And by Art. 14, $x=(1.000002)(1.0000003)(1.00000009)$

In place of these last factors we might have taken the differential equation (Art. 15), and that would have been more direct and to the point. The factor method but approximates to the given number: the differential method comes directly to it, but it will not be so generally understood as the factor method.

The differential equation applied in place of the last three factors, gives 2.562 597 740 854 for the required logarithm. But, if *very great* accuracy was required, the factor (1.000002) should be employed and then the differential equation.

When the radius of a circle is 1, the natural sine of 7° 30′ is expressed by the decimal 0.1305261921 what is the logarithm of this number, true to nine places of decimals, consider the decimal a whole number?

Take its four superior digits 1305: the factors of these are 15 and 87; therefore,

	.87	log.	—1. 939 519 252 619
	.15	log.	—1. 176 091 259 059
Prod.	.1305	log.	—1. 115 610 511 678
	10002	log.	0. 000 086 850 213
	13052610	log.	—1. 115 697 361 891
		Correction,	306 450
Num.	.1305261921	log.	—1. 115 697 668 341

For the radius of our common tables add, 10

Hence, log. sine of 7° 30′= 9. 115 697 668 341

In this manner the *logarithmic sine* of any other arc can be found when we have its *natural sine*.

The correction was found by the differential equation, Art. 15, thus,

$$(\text{dif.})\ \log.\ z = \frac{(0.4342944819)\ (921)}{1305261000}$$

Having the logarithm of the modulus, and of the denominator, we can readily deduce the result by logarithms, using the common table to find the logarithm of 921.

	921	log.	2.964260
	m.	log.	—1.637784
	Numerator,	log.	2.602044
	Denom.	log.	9.115697
Num. .000000306450		log.	—7.486347

We give one more porblem of this kind.

The mean distance between the Sun and Earth is 1; the greatest distance in 1800 was 1.01685317: what is the logarithm of this number?

Take the whole as the whole number, 101685317, and pay no attention to the index during the process.

The four superior digits are 1016; the factors of this number are 8 and 127: hence,

	8	log.	.903 089 986 992
	127	log.	.103 803 720 956
Prod.	1016	log.	.006 893 707 948
Table *B*.	10008	log.	.000 347 233 698
Prod.	10168128	log.	.007 240 941 646
Table *C*.	100004	log.	17 371 430
	10168128		
	40672512		
Prod.	1016853472512	log.	.007 258313 076

From	1016853472512
Take	101685317
	30.25

Taking the given quantity as a whole number, our last product exceeds it by $30\frac{1}{4}$, which is the differential of the number. The differential of the logarithm will therefore be $\frac{m.\ 30.\ 25}{z}$

By log.	log. m	—1. 637 784
	log. 30. 25	1. 480 725
		1. 118 509
	log. z*	8. 007 257
log. 0.0000001292		—7. 111 252

From	.007 258 313 076
Take	.000 000 129 200
Log. sought	0. 007 258 183 876

From the foregoing problems, the reader will perceive that the logarithm of any number whatever, can be found by these tables — and to any degree of accuracy within ten decimal places. We are now prepared to take the converse problem ; that is,

Given a logarithm, to find its corresponding number.

What number corresponds to the log. 4. 636 747 519 487 ? Carrying this log. to the table, we find it corresponds to a number a little greater than 43, but the index being 4, the number must be a little over 43000. Let this be *one factor* of the number sought, and the reason for the following operation must be obvious.

* z=101685317, a number now considered as a whole number, its log. is taken approximately, and its index is 8.

		Given log.	4.636747519487
1st Factor, 43000		log.	4.633468455579
			3279063908
2nd Factor, 1.007	table *A*.	log.	3030465635
			248598273
3rd Factor, 1.0005	table *B*.	log.	217099966
			31498307
4th Factor, 1.00007	table *C*.	log.	30399546
			1098761
5th Factor, 1.000002		log.	868588 (2*b*)
			230173
6th Factor, 1.0000005		log.	217145 (5*c*)
			13028
7th Factor, 1.00000003		log.	13029 (3*d*)

The product of these factors is the number required, and that product can be obtained with great facility.

Prod. of 1st and 2nd factors=	43301
This, by the 3rd	10005
	43301
	21.6505
	43322.65050000
Prod. of 4th, 5th, 6th, 7th	1.00007253
	43322.6505
	3.032585535
	866453010
	2166132525
	1299679515
Num. required	43325.792691840765

2. What number corresponds to log. 2. 563 785 181 020 ?

Taking this decimal log. to the table, we find that its place is between 36. and 37. and as the index is 2., it must be between 360. and 370., nearer the latter than the former. Suppose it near 366. It is not 367., for the table gives a log. of 367., greater than our given log.

366=6×61		
	From given log.	2. 563 785 181 020
Sub. log. 61		1. 785 329 835 011
		0. 778 455 346 009
log. 6		0. 778 151 250 384
		304 095 625
3rd Factor 1.0007		303 836 798
		258 827

The product of these three factors is 366.2562, which is very near the number required.

When a logarithm is reduced below its sixth decimal place, what is left may be taken as a differential of the given logarithm.

This differential will give a corresponding differential, to be applied to the number by using the equation in Art 15.

$$\text{That is (dif.) N} = \frac{\text{N (dif.) log.}}{\text{modulus}}$$

By the common table of log.		Given log.	2. 563 785
	.0000002588	log.	—7. 412 964
			—5. 976 749
	log. m		—1. 637 784
	0.0002182	log.	—4. 338 965
Add	366.2562		
Number sought	366.2564182		

In this manner, the number to any logarithm may be found.

EXAMPLES.

1. What number corresponds to the log. 2.204923118054?
Ans. 160.29616

2. What number corresponds to the log. 4.133409102?
Ans. 13595.93

3. What number corresponds to the log. 3.278902074620?
Ans. 1900.64967

CHAPTER III.

ELEMENTARY PRINCIPLES OF PLANE TRIGONOMETRY.

Trigonometry in its literal and restricted sense, has for its object, the measure of triangles. When the triangles are on planes, it is plane trigonometry, and when the triangles are on, or conceived to be portions of a sphere, it is spherical trigonometry. In a more enlarged sense, however, this science is the application of the principles of geometry, and numerically connects one part of a magnitude with another, or numerically compares different magnitudes.

As the *sides* and *angles* of triangles are quantities of different kinds, they cannot be *compared* with each other; but the *relation* may be discovered by means of other complete triangles, to which the triangle under investigation can be compared.

Such other triangles are numerically expressed in Table II, and all of them are conceived to have one common point, the center of a circle, and as all possible angles can be formed by two straight lines drawn from the center of a circle, no angle of a triangle can exist whose measure cannot be found in the table of trigonometrical lines.

The measure of an angle is the arc of a circle, intercepted between the two lines which form the angle—the center of the arc always being at the point where the two lines meet.

The arc is measured by *degrees, minutes,* and *seconds,* there being 360 degrees to the whole circle, 60 minutes in one degree, and 60 seconds in one minute. Degrees, minutes, and seconds, are designated by °, ′, ″. Thus 27° 14′ 21″, is read 27 degrees, 14 minutes, and 21 seconds.

All circles contain the same number of degrees, but the greater the radii the greater is the absolute length of a degree; the circumference of a carriage wheel, the circumference of the earth, or the still greater and indefinite circumference of the heavens, have the same number of degrees; yet the same number of degrees in each and every circle is precisely the same angle in amount or measure.

As triangles do not contain circles, we can not measure triangles by circular arcs; we must measure them by *other triangles*, that is, by *straight lines*, drawn in and about a circle.

Such straight lines are called trigonometrical lines, and take particular names, as described by the following

DEFINITIONS.

1. The *sine* of an angle, or an arc, is a line drawn from one end of an arc, perpendicular to a diameter drawn through the other end. Thus, *BF* is the sine of the arc *AB, and also* of the arc *BDE. BK* is the sine of the arc *BD,* it is *also* the cosine of the arc *AB,* and *BF,* is the cosine of the arc *BD.*

N. B. The *complement* of an arc is what it wants of 90°; the *supplement* of an arc is what it what it wants of 180°.

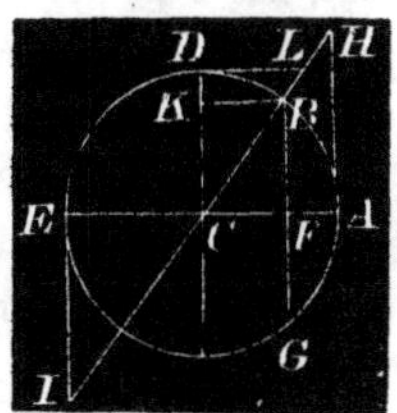

2. The *cosine* of an arc is the perpendicular distance from the center of the circle to the sine of the arc, or it is the same in magnitude as the sine of the complement of the arc. Thus, *CF,* is the cosine of the arc *AB;* but *CF=KB,* the sine of *BD.*

3. The *tangent* of an arc is a line touching the circle in one extremity of the arc, continued from thence, to meet a line drawn through the center and the other extremity.

Thus, *AH* is the tangent to the arc *AB,* and *DL* is the tangent of the arc *DB,* or the cotangent of the arc *AB.*

N. B. *The* co, *is but a contraction of the word complement.*

4. The *secant* of an arc, is a line drawn from the center of the circle to the extremity of its tangent. Thus, *CH* is the secant of the arc *AB,* or of its supplement *BDE.*

5. The *cosecant* of an arc, is the secant of the complement. Thus, *CL,* the secant of *BD,* is the cosecant of *AB.*

6. The versed sine of an arc is the difference between the cosine and the radius; that is, *AF* is the versed sine of the arc *AB,* and *DK* is the versed sine of the arc *BD.*

For the sake of brevity these technical terms are contracted thus: for sine *AB,* we write *sin.AB,* for cosine *AB,* we write *cos.AB,* for tangent *AB,* we write *tan.AB,* &c.

From the preceding definitions we deduce the following obvious consequences :

1st, That when the arc AB, becomes so small as to call it nothing, its sine tangent and versed sine are also nothing, and its secant and cosine are each equal to radius.

2d, The sine and versed sine of a quadrant are each equal to the radius ; its cosine is zero, and its secant and tangent are infinite.

3d, The chord of an arc is twice the sine of half the arc. Thus, the chord BG, is double of the sine BF.

4th, The sine and cosine of any arc form the two sides of a right angled triangle, which has a radius for its hypotenuse. Thus, CF, and FB, are the two sides of the right angled triangle CFB.

Also, the radius and the tangent always form the two sides of a right angled triangle which has the secant of the arc for its hypotenuse. This we observe from the right angled triangle CAH.

To express these relations analytically, we write

$$\text{sin.}^2+\text{cos.}^2=R^2 \qquad (1)$$

$$R^2+\text{tan.}^2=\text{sec.}^2 \qquad (2)$$

From the two equiangular triangles CFB, CAH, we have

$$CF : FB = CA : AH$$

That is, $\text{cos.} : \text{sin} = R : \text{tan.} \qquad \text{tan.}=\frac{R\,\text{sin.}}{\text{cos.}} \qquad (3)$

Also, . $CF : CB = CA : CH$

That is, . $\text{cos} : R = R : \text{sec.} \qquad \text{cos. sec.}=R^2 \qquad (4)$

The two equiangular triangles CAH, CDL. give

$$CA : AH = DL : DC$$

That is, . $R : \text{tan.} = \text{cot} : R \qquad \text{tan. cot.}=R^2 \qquad (5)$

Also, . $CF : FB = DL : DC$

That is, . $\text{cos.} : \text{sin.} = \text{cot} : R \qquad \text{cos. } R=\text{sin. cot.} \qquad (6)$

By observing (4) and (5), we find that

$$\text{cos. sec.}=\text{tan. cot.} \qquad (7)$$

Or, $\text{cos.} : \text{tan.} = \text{cot.} : \text{sec.}$

The *ratios* between the various trigonometrical lines are always the same for the same arc, whatever be the length of the radius ; and therefore, we may assume radius of any length to suit our convenience ; and the preceding equations will be more concise, and more readily applied, by making radius equal unity. This supposition being made, the preceding become

$$\sin.^2+\cos.^2=1 \qquad (1)$$

$$1+\tan.^2=\sec.^2 \qquad (2)$$

$$\tan.=\frac{\sin.}{\cos.} \quad (3) \qquad\qquad \cos.=\frac{1}{\sec.} \quad (4)$$

$$\tan.=\frac{1}{\cot.} \quad (5) \qquad\qquad \cos.=\sin.\ \cot. \quad (6)$$

The center of the circle is considered the absolute *zero* point, and the different directions from this point are designated by the different signs + and —. On the right of C, toward A, is commonly marked plus (+), then the other direction, toward E, is necessarily minus (—). Above AE is called (+), below that line (—).

If we conceive an arc to commence at A, and increase continuously around the whole circle in the direction of ABD, then the following table will show the mutations of the signs.

	sin.	cos.	tan.	cot.	sec.	cosec.	vers.
1st quadrant.	+	+	+	+	+	+	+
2d "	+	—	—	—	—	+	+
3d "	—	—	+	+	—	—	+
4th "	—	+	—	—	+	—	+

PROPOSITION 1.

The chord of 60° *and the tangent* 45° *are each equal to radius; the sine of* 30° *the versed sine of* 60° *and the cosine of* 60° *are each equal to half the radius.*

(The first truth is proved in problem 15, book 1).

On C=, as radius, describe a quadrant; take AD=45°, AB =60°, and AE=90°, then BE=30°.

Join AB, CB, and draw Bn, perpendicular to CA. Draw Bm, parallel to AC. Make the angle CAH=90°, and draw CDH.

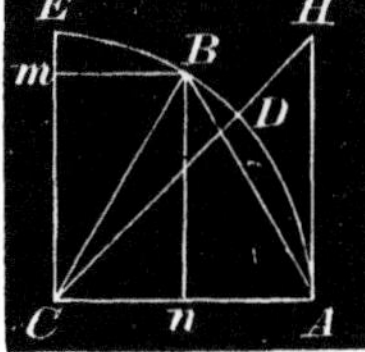

In the △ ABC, the angle ACB=60° by hypothesis; therefore, the sum of the other two angles is (180—60)=120°. But CB=CA, hence the angle CBA= the angle CAB, (th. 15 b. 1), and as the sum of the two is 120°, each one must be 60°; therefore, each of the angles of triangle ABC, is 60°

and the sides opposite to equal angles are equal; that is, AB, the chord of 60°, is equal to CA, the radius.

In the △ CAH, the angle CAH is a right angle; and by hypothesis, ACH, is half a right angle; therefore, AHC, is also half a right angle; consequently, $AH=AC$, the tangent of 45°= the radius.

By th. 15, book 1, cor. $Cn=nA$; that is, the cosine and versed sine of 60° are each equal to the half of the radius. As Bn and EC are perpendicular to AC, they are parallel, and Bm is made parallel to Cn; therefore, $Bm=Cn$, or the sine 30°, is the half of radius.

PROPOSITION 2.

Given the sine and cosine of two arcs to find the sine and cosine of the sum, and difference of the same arcs expressed by the sines and cosines of the separate arcs.

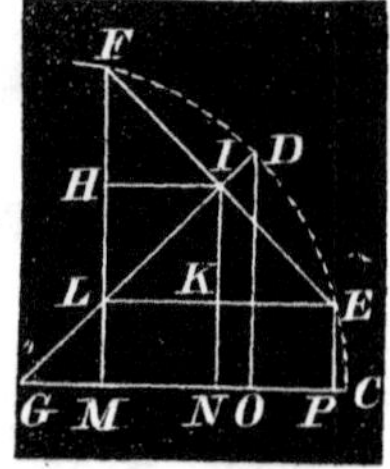

Let G be the center of the circle, CD, the greater arc which we shall designate by a, and DF, a less arc, that we designate by b.

Then by the definitions of sines and cosines, $DO=\sin.a$; $GO=\cos.a$; $FI=\sin.b$; $GI=\cos.b$. We are to find FM, which is

$$=\sin.(a+b);\ GM=\cos.(a+b);$$
$$EP=\sin.(a-b);\ GP=\cos.(a-b).$$

Because IN is parallel to DO, the two △s GDO, GIN, are equiangular and similar. Also, the △ FHI, is similar to GIN; for the angle FIG, is a right angle; so is HIN; and, from these two equals take away the common angle HIL, leaving the angle $FIH=GIN$. The angles at H and N, are right angles; therefore, the △ FHI, is equiangular, and similar to the △ GIN, and, of course, to the △ GDO; and the side HI, is homologous to IN, and DO.

Again, as $FI=IE$, and IK, parallel to FM,

$$FH=IK, \text{ and } HI=KE.$$

By similar triangles we have

$$GD:DO=GI:IN.$$

That is, $R:\sin.a=\cos.b:IN$, or $IN=\dfrac{\sin.a\ \cos.b}{R}$

Also, $GD:GO=FI:FH$

That is, $R:\cos.a=\sin.b:FH$, or $FH=\frac{\cos.a\ \sin.b}{R}$

Also, $GD:GO=GI:GN$

That is, $R:\cos.a=\cos.b:GN$, or $GN=\frac{\cos.a\ \cos.b}{R}$

Also, $GD:DO=FI:IH$

That is, $R:\sin.a=\sin.b:\ IH$, or $IH=\frac{\sin.a\ \sin.b}{R}$

By adding the first and second of these equations, we have

$$IN+FH=FM=\sin.(a+b)$$

That is, . $\sin.\ (a+b)=\frac{\sin.a\ \cos.b+\cos.a\ \sin.b}{R}$

By subtracting the second from the first, we have

$$\sin.\ (a-b)=\frac{\sin.a\ \cos.b-\cos.a\ \sin.b}{R}$$

By subtracting the fourth from the third, we have

$GN-IH=GM=\cos.(a+b)$ for the first member.

Hence, . $\cos.(a+b)=\frac{\cos.a\ \cos.b-\sin.a\ \sin.b}{R}$

By adding the third and fourth, we have

$$GN+IH=GN+NP=GP=\cos.(a-b)$$

Hence, . $\cos.\ (a-b)=\frac{\cos.a\ \cos.\ b+\sin.a\ \sin.b}{R}$

Collecting these four expressions, and considering the radius unity, we have

$$(A)\quad\left\{\begin{array}{ll}\sin.(a+b)=\sin.a\ \cos.b+\cos.a\ \sin.b & (7)\\ \sin.(a-b)=\sin.a\ \cos.b-\cos.a\ \sin.b & (8)\\ \cos.(a+b)=\cos.a\ \cos.b-\sin.a\ \sin.b & (9)\\ \cos.(a-b)=\cos.a\ \cos.b+\sin.a\ \sin.b & (10)\end{array}\right.$$

Formula (A), accomplishes the objects of the proposition, and from these equations many useful and important deductions can be made. The following, are the most essential:

By adding (7) to (8), we have (11); subtracting (8) from (7), gives (12). Also, (9)+(10) gives (13); (9) taken from (10) gives (14).

$$(B)\quad\left\{\begin{array}{ll}\sin.(a+b)+\sin.(a-b)=2\sin.a\ \cos\ b & (11)\\ \sin.(a+b)-\sin.(a-b)=2\cos.a\ \sin.\ b & (12)\\ \cos.(a+b)+\cos.(a-b)=2\cos.a\ \cos.b & (13)\\ \cos.(a-b)-\cos.(a+b)=2\sin.\ a\ \sin.b & (14)\end{array}\right.$$

If we put $a+b=A$, and $a-b=B$, then (11) becomes (15), (12) becomes (16), 13 becomes (17), and (14) becomes (18).

$$
(C)\left\{\begin{array}{ll}
\sin.A+\sin.B=2\sin.\left(\frac{A+B}{2}\right)\cos.\left(\frac{A-B}{2}\right) & (15)\\
\sin.A-\sin.B=2\cos.\left(\frac{A+B}{2}\right)\sin.\left(\frac{A-B}{2}\right) & (16)\\
\cos.A+\cos.B=2\cos.\left(\frac{A+B}{2}\right)\cos.\left(\frac{A-B}{2}\right) & (17)\\
\cos.B-\cos.A=2\sin.\left(\frac{A+B}{2}\right)\sin.\left(\frac{A-B}{2}\right) & (18)
\end{array}\right.
$$

If we divide (15) by (16), (observing that $\frac{\sin.}{\cos.}=\tan.$ and $\frac{\cos.}{\sin.}=\cot.=\frac{1}{\tan.}$ as we learn by equations (6) and (5) trigonometry), we shall have

$$
\frac{\sin.A+\sin.B}{\sin.A-\sin.B}=\frac{\sin.\left(\frac{A+B}{2}\right)}{\cos.\left(\frac{A+B}{2}\right)}\times\frac{\cos.\left(\frac{A-B}{2}\right)}{\sin.\left(\frac{A-B}{2}\right)}=\frac{\tan.\left(\frac{A+B}{2}\right)}{\tan.\left(\frac{A-B}{2}\right)} \quad (19)
$$

Whence,

$$
\overline{\sin.A+\sin.B}:\overline{\sin.A-\sin.B}=\tan.\left(\frac{A+B}{2}\right):\tan.\left(\frac{A-B}{2}\right)
$$

or in words. *The sum of the sines of any two arcs is to the difference of the same sines, as the tangent of the half sum of the same arcs is to the tangent of half their difference.*

By operating in the same way with the different equations in formula (C), we find,

$$
(D)\left\{\begin{array}{ll}
\frac{\sin.A+\sin.B}{\cos.A+\cos.B}=\tan.\left(\frac{A+B}{2}\right) & (20)\\
\frac{\sin.A+\sin.B}{\cos.B-\cos.A}=\cot.\left(\frac{A-B}{2}\right) & (21)\\
\frac{\sin.A-\sin.B}{\cos.A+\cos.B}=\tan.\left(\frac{A-B}{2}\right) & (22)\\
\frac{\sin.A-\sin.B}{\cos.B-\cos.A}=\cot.\left(\frac{A+B}{2}\right) & (23)\\
\frac{\cos.A+\cos.B}{\cos.B-\cos.A}=\frac{\cot.\left(\frac{A+B}{2}\right)}{\tan.\left(\frac{A-B}{2}\right)} & (24)
\end{array}\right.
$$

These equations are all true, whatever be the value of the arcs designated by A and B; we may therefore, assign any possible value to either of them, and if in equations (20), (21) and (24), we make $B=O$, we shall have,

$$\frac{\sin.A}{1+\cos.A}=\tan.\frac{A}{2}=\frac{1}{\cot.\frac{1}{2}A} \qquad (25)$$

$$\frac{\sin.A}{1-\cos.A}=\cot.\frac{A}{2}=\frac{1}{\tan.\frac{1}{2}A} \qquad (26)$$

$$\frac{1+\cos.A}{1-\cos.A}=\frac{\cot.\frac{1}{2}A}{\tan.\frac{1}{2}A}=\frac{1}{\tan.^2\frac{1}{2}A} \qquad (27)$$

If we now turn back to formula (A), and divide equation (7) by (9), and (8) by (10), observing at the same time, that $\frac{\sin.}{\cos.}=\tan.$ we shall have,

$$\tan.(a+b)=\frac{\sin a\cos.b+\cos.a\sin.b}{\cos.a\cos.b-\sin.a\sin.b}$$

$$\tan.(a-b)=\frac{\sin.a\cos.b-\cos.a\sin.b}{\cos.a\cos.b+\sin.a\sin.b}$$

By dividing the numerators and denominators of the second members of these equations by (cos.a cos.b), we find,

$$\tan.(a+b)=\frac{\dfrac{\sin.a\cos.b}{\cos.a\cos.b}+\dfrac{\cos.a\sin.b}{\cos.a\cos.b}}{\dfrac{\cos.a\cos.b}{\cos.a\cos.b}-\dfrac{\sin.a\sin.b}{\cos.a\cos.b}}=\frac{\tan.a+\tan.b}{1-\tan.a\tan.b} \qquad (28)$$

$$\tan.(a-b)=\frac{\dfrac{\sin.a\cos.b}{\cos.a\cos.b}-\dfrac{\cos.a\sin.b}{\cos.a\cos.b}}{\dfrac{\cos.a\cos.b}{\cos.a\cos.b}+\dfrac{\sin.a\sin.b}{\cos.a\cos.b}}=\frac{\tan.a-\tan.b}{1+\tan.a\tan.b} \qquad (29)$$

If in equation (11), formula (B), we make $a=b$, we shall have,

$$\sin.2a=2\sin.a\cos.a \qquad (30)$$

Making the same hypothesis in equation (13), gives,

$$\cos.2a+1=2\cos^2.a \qquad (31)$$

The same hypothesis reduces equation (14), to

$$1-\cos.2a=2\sin^2.a \qquad (32)$$

The same hypothesis reduces equation (28), to

$$\tan.2a=\frac{2\tan.a}{1-\tan^2.a} \qquad (33)$$

The secants and cosecants of arcs are not given in our table, because they are very little used in practice; and if any particular secant is required, it can be determined by subtracting the cosine from 20; and the cosecant can be found by subtracting the sine from 20.

PROPOSITION 3.

In any right angled plane triangle, we may have the following proportions:

1st. *As the hypotenuse is to either side, so is the radius to the sine of the angle opposite to that side.*

2d. *As one side is to the other side, so is the radius to the tangent of the angle adjacent to the first-mentioned side.*

3d. *As one side is to the hypotenuse, so is radius to the secant of the angle adjacent to that side.*

Let CAB represent any right angled triangle, right angled at A. AB and AC are called the sides of the △, and CB is called the hypotenuse.

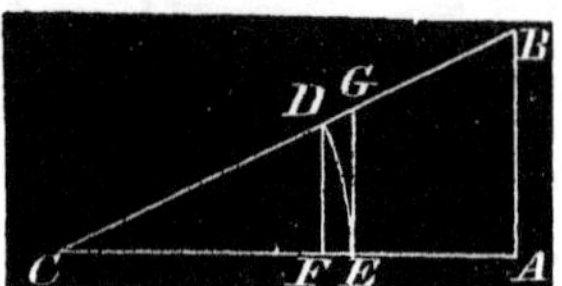

(Here, and in all cases hereafter, we shall represent the angles of a triangle by the large letters A, B, C, and the sides opposite to them, by the small letters a, b, c.)

From either acute angle, as C, take any distance, as CD, *greater* or *less* than CB, and describe the arc DE. This arc measures the angle C. From D, draw DF parallel to BA; and from E, draw EG, also parallel to BA or DF.

By the definitions of sines, tangents, and secants, DF is the sine of the angle C; EG is the tangent, CG the secant, and CF the cosine.

Now, by proportional triangles we have,

$$\left.\begin{array}{lll} CB : BA = CD : DF & \text{or, } a : c = R : \sin. C \\ CA : AB = CE : EG & \text{or, } b : c = R : \tan. C \\ CA : CB = CE : CG & \text{or, } b : a = R : \sec. C \end{array}\right\} Q.\ E.\ D.$$

Scholium. If the hypotenuse of a triangle is made radius, one side is the sine of the angle opposite to it, and the other side is the cosine of the same angle. This is obvious from the triangle CDF.

PROPOSITION 4.

In any triangle, the sines of the angles are to one another as the sides opposite to them.

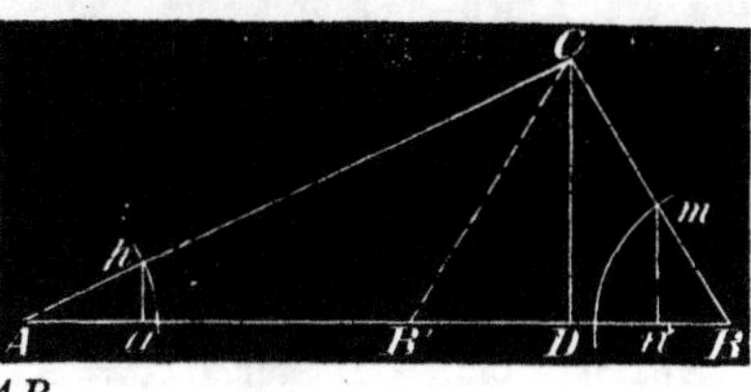

Let ABC be any triangle. From the points A and B, as centers, with any radius, describe the arcs measuring these angles, and draw pa, CD, and mn, perpendicular to AB.

Then, . . $pa=\sin.A$, $mn=\sin.B$

By the similar $\triangle$s, Apa and ACD, we have,

$$R : \sin.A=b : CD; \text{ or, } R(CD)=b\sin.A \quad (1)$$

By the similar $\triangle$s Bmn and BCD, we have,

$$R : \sin.B=a : CD; \text{ or, } R(CD)=a\sin.B \quad (2)$$

By equating the second members of equations (1) and (2).

$$b\sin.A=a\sin.B.$$

Hence, . $\sin.A : \sin.B=a : b$ \
Or, . . $a : b=\sin A : \sin. B$ } *Q. E. D.*

Scholium 1. When either angle is 90°, its sine is radius.

Scholium 2. When CB is less than AC, and the angle B, acute, the triangle is represented by ACB. When the angle B becomes B', it is obtuse, and the triangle is ACB'; but the proportion is equally true with either triangle; for the angle $CB'D=CBA$, and the sine of $CB'D$ is the same as the sine of $AB'C$. In practice we can determine which of these triangles is proposed by the side AB, being greater or less than AC; or, by the angle at the vertex C, being large as ACB, or small as ACB'.

In the solitary case in which AC, CB, and the angle A, are given, and CB less than AC, we can determine both of the $\triangle$s ACB and ACB'; and then we surely have the right one.

PROPOSITION 5.

If from any angle of a triangle, a perpendicular be let fall on the opposite side, or base, the tangents of the segments of the angle are to one another as the segments of the base.

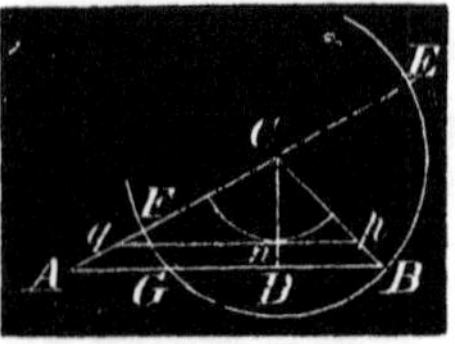

Let ABC be the triangle. Let fall the perpendicular CD, on the side AB.

Take any radius, as Cn, and describe the arc which measures the angle C. From n, draw qnp parallel to AB. Then it is obvious that np is the tangent of the angle DCB, and nq is the tangent of the angle ACD.

Now, by reason of the parallels AB and qp, we have,

$$qn : np = AD : DB$$

That is, $\tan.ACD : \tan.DCB = AD : DB$ *Q. E. D.*

PROPOSITION 6.

If a perpendicular be let fall from any angle of a triangle to its opposite side or base, this base is to the sum of the other two sides, as the difference of the sides is to the difference of the segments of the base.

(See figure to proposition 5.)

Let AB be the base, and from C, as a center, with the shorter side as radius, describe the circle, cutting AB in G, AC in F, and produce AC to E.

It is obvious that AE is the sum of the sides AC and CB, and AF is their difference.

Also, AD is one segment of the base made by the perpendicular, and $BD=DG$ is the other; therefore, the difference of the segments is AG.

As A is a point without a circle, by theorem 18, book 3, we have,

$$AE \times AF = AB \times AG$$

Hence, . . $AB : AE = AF : AG$ *Q. E. D.*

PROPOSITION 7.

The sum of any two sides of a triangle, is to their difference, as the tangent of the half sum of the angles opposite to these sides, to the tangent of half their difference.

Let ABC be any plane triangle. Then, by proposition 4, trigonometry, we have,

$$CB : AC = \sin.A : \sin.B$$

Hence,

$$CB + AC : CB - AC = \sin.A + \sin.B : \sin.A - \sin.B \text{ (th. 9 b. 2)}$$

But, $\tan.\left(\frac{A+B}{2}\right) : \tan.\left(\frac{A-B}{2}\right) = \sin.A+\sin.B : \sin.A-\sin.B$ (eq. (1), trig.)

Comparing the two latter proportions (th. 6, b. 2), we have,

$$CB+AC : CB-AC = \tan.\left(\frac{A+B}{2}\right) : \tan.\left(\frac{A-B}{2}\right) \quad Q.\ E.\ D.$$

PROPOSITION 8.

Given the three sides of any plane triangle, to find some relation which they must bear to the sines and cosines of the respective angles.

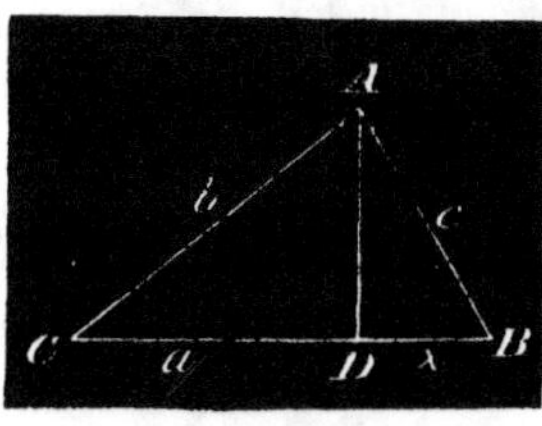

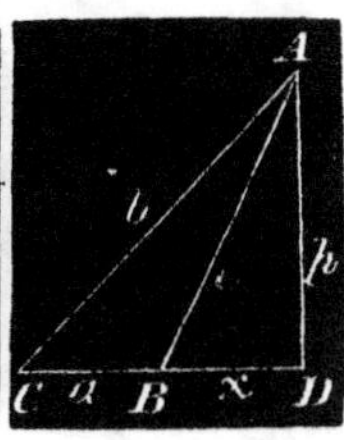

Let ABC be the triangle, and let the perpendicular fall either upon, or without the base, as shown in the figures; and by recurring to theorem 38, book 1, we shall find

$$CD = \frac{a^2+b^2-c^2}{2a} \qquad (1)$$

Now, by proposition 3, trigonometry, we have,

$$R : \cos.C = b : CD$$

Therefore, . $$CD = \frac{b\cos.C}{R} \qquad (2)$$

Equating these two values of CD, and reducing, we have,

$$\cos.C = \frac{R(a^2+b^2-c^2)}{2ab} \qquad (m)$$

In this expression we observe that the part of the numerator which has the minus sign, is the side opposite to the angle; and that the denominator is twice the rectangle of the sides adjacent to the angle. From these observations we at once draw the following expressions for the cosine A, and cosine B.

Thus, . . $$\cos.A = \frac{R(b^2+c^2-a^2)}{2bc} \qquad (n)$$

$$\cos.B = \frac{R(a^2+c^2-b^2)}{2ac} \qquad (p)$$

As these expressions are not convenient for logarithmic computation, we modify them as follows:

If we put $2a = A$, in equation (31), we have,

$$\cos. A + 1 = 2 \cos.^2 \tfrac{1}{2}A$$

In the preceding expression (n), if we consider radius, unity, and add 1 to both members, we shall have,

$$\cos. A + 1 = 1 + \frac{b^2 + c^2 - a^2}{2bc}$$

Therefore, $$2 \cos.^2 \tfrac{1}{2}A = \frac{2bc + b^2 + c^2 - a^2}{2bc}$$

$$= \frac{(b+c)^2 - a^2}{2bc}$$

Considering $(b+c)$ as one quantity, and observing that we have the difference of *two squares*, therefore

$(b+c)^2 - a^2 = (b+c+a)(b+c-a)$; but $(b+c-a) = b+c+a-2a$

Hence, . $$2 \cos.^2 \tfrac{1}{2}A = \frac{(b+c+a)(b+c+a-2a)}{2bc}$$

Or, . . $$\cos.^2 \tfrac{1}{2}A = \frac{\left(\frac{b+c+a}{2}\right)\left(\frac{b+c+a}{2} - a\right)}{bc}$$

By putting $\frac{a+b+c}{2} = s$, and extracting square root, the final result for radius unity, is

$$\cos. \tfrac{1}{2}A = \sqrt{\frac{s(s-a)}{bc}}$$

For any other radius we must write,

$$\cos. \tfrac{1}{2}A = \sqrt{\frac{R^2 s(s-a)}{bc}}$$

By inference, $$\cos. \tfrac{1}{2}B = \sqrt{\frac{R^2 s(s-b)}{ac}}$$

Also, . . $$\cos. \tfrac{1}{2}C = \sqrt{\frac{R^2 s(s-c)}{ab}}$$

In every triangle, the sum of the three angles must equal 180°; and if one of the angles is small, the other two must be comparatively large; if two of them are small, the third one must be large. The greater angle is always opposite the greater side; hence, by merely inspecting the given sides, any person can decide at once which is the greater angle; and of the three preceding equations, *that one* should be taken which applies to the greater angle, whether that be the particular angle required or not; because the equations bring out the *cosines* to the angles; and the cosines, to very small arcs vary so slowly, that it may be impossible to decide, with sufficient numerical accuracy to what particular arc the cosine belongs. For instance, the cosine 9.999999, carried to the table, applies to several arcs; and, of course, we should not know which one to take; but this difficulty does not exist when the angle is large; therefore, compute the largest angle first, and then compute the other angles by proposition 4.

But we can deduce an expression for the sine of any of the angles, as well as the cosine. It is done as follows:

EQUATIONS FOR THE SINES OF THE ANGLES.

Resuming equation (m), and considering radius, unity, we have,

$$\cos. C=\frac{a^2+b^2-c^2}{2ab}$$

Subtracting each member of this equation from 1, gives

$$1-\cos. C=1-\left(\frac{a^2+b^2-c^2}{2ab}\right) \qquad (1)$$

Making $2a=C$, in equation (32), then $a=\frac{1}{2}C$,

And . . $1-\cos. C=2\sin.^2\frac{1}{2}C \qquad (2)$

Equating the right hand members of (1) and (2),

$$2\sin.^2\frac{1}{2}C=\frac{2ab-a^2-b^2+c^2}{2ab}$$

$$=\frac{c^2-(a-b)^2}{2ab}$$

$$=\frac{(c+b-a)(c+a-b)}{2ab}$$

Or, . . . $\sin.^2\frac{1}{2}C=\dfrac{\left(\frac{c+b-a}{2}\right)\left(\frac{c+a-b}{2}\right)}{ab}$

But, . $\dfrac{c+b-a}{2}=\dfrac{c+b+a}{2}-a$ and $\dfrac{c+a-b}{2}=\dfrac{c+a+b}{2}-b$

Put . $\dfrac{a+b+c}{2}=s$, as before; then,

$$\sin.\frac{1}{2}C=\sqrt{\frac{(s-a)(s-b)}{ab}}$$

By taking equation (p), and operating in the same manner, we have . . . $\sin.\frac{1}{2}B=\sqrt{\dfrac{(s-a)(s-c)}{ac}}$

From (n) . . $\sin.\frac{1}{2}A=\sqrt{\dfrac{(s-b)(s-c)}{cb}}$

The preceding results are for radius unity; for any other radius, we must multiply by the number of units in such radius. For the radius of the tables, we write R; and if we put it under the radical sign, we must write R^2; hence, for the sines corresponding with our logarithmic table, we must write the equations

thus, . . . $\sin.\frac{1}{2}A=\sqrt{\frac{R^2(s-b)(s-c)}{bc}}$

$$\sin.\tfrac{1}{2}B=\sqrt{\frac{R^2(s-a)(s-c)}{ac}}$$

$$\sin.\tfrac{1}{2}C=\sqrt{\frac{R^2(s-a)(s-b)}{ab}}$$

A large angle should not be determined by these equations, for the same reason that a small angle should not be determined from an equation expressing the cosine.

In practice, the equations for cosine are more generally used, because more easily applied.

In the preceding pages we have gone over the whole ground of theoretical plane trigonometry, although several particulars might have been enlarged upon, and more equations in relation to the combinations of the trigonometrical lines, might have been given; but enough has been given to solve every possible case that can arise in the practical application of the science.

By the application of equations (1), (31), and (32), the table of natural sines and cosines has been computed.

The operation is as follows. *The sine of* 30° *is half radius;* making the radius unity, equation (1) gives

$\frac{1}{4}+\cos.^2 30°=1$: whence $\cos.^2 30°=\frac{3}{4}$ or $\cos. 30°=\frac{1}{2}\sqrt{3}$.

From (32) we have, $\sin.a=\sqrt{\frac{1-\cos. 2a}{2}}$

Making $2a=30°$, then $\sin. 15°=(\frac{1}{2}-\frac{1}{4}\sqrt{3})^{\frac{1}{2}}=0.25881904$

From (31) we have, $\cos.a=\sqrt{\frac{1+\cos. 2a}{2}}$

Making $2a=30°$ as before, $\cos.a=(\frac{1}{2}+\frac{1}{4}\sqrt{3})^{\frac{1}{2}}=0.96592582$

Having sine and cosine of 15° the second application of these equations will give the sine and cosine of the half of 15°, and so on through as many bisections as we please.

Being desirous of giving a full exposition of the formation of table II, we give the following geometrical demonstration of equation 30, by the help of the figure in the margin.

Let the arc $AD=2a$

Then $DG=\sin. 2a$, $CG=\cos. 2a$,

$DI=\sin. a$, $AD=2\sin. a$,

$CI=\cos. a$, $DB=2DO=2\cos. a$.

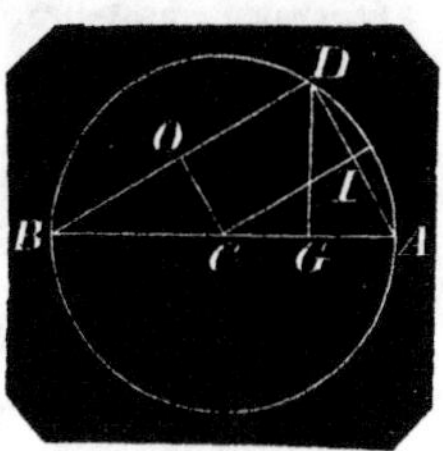

The angle DBA being at the circumference, is measured by half the arc AD, or by a.

Now, by applying proposition 4 to the triangle DBG, we have

$$\sin. DBG : DG=\sin. 90° : BD.$$

The sin. $DBG=\sin. a$, and sin. $90°=1$, the radius being unity; therefore, the preceding proportion becomes,

$$\sin. a : \sin. 2a=1 : 2\cos. a.$$

Whence $\quad 2\cos. a \sin. a=\sin. 2a.$ (Same as eq. 30.)

PROBLEM.

Given the sine and cosine of an arc, to find the sine and cosine of one half that arc.

Designate the given arc by $2a$, the radius by unity, and whatever be the value of a, equation (1) gives

$$\cos.^2 a+\sin.^2 a=1 \qquad (m)$$

It is proved in proposition 1, that the sine of 30° is half the radius: therefore, let $2a=30°$, then $\sin. 2a=0.5$: and equation 30, just demonstrated, gives

$$2\cos. a \sin. a=0.5. \qquad (n)$$

Add (m) and (n), and extract the square root of both members.

Then $\quad \cos. a+\sin. a=1.22474486 \qquad (o)$

Subtracting (n) from (m), and extracting square root, gives

$$\cos. a-\sin. a=0.70710678 \qquad (p)$$

By subtracting, and adding (p) and (o), and dividing by 2, we find

$$\sin. a=\sin. 15°=0.25881904$$
$$\cos. a=\cos. 15°=0.96592582$$

Now let $2a=15^p$. Then

$$\cos.^2 a+\sin.^2 a=1.$$

and $$2\cos. a\sin. a=0.25881904$$

Operating as before, we find

$$\sin. a=\sin. 7^\circ 30'=0.1305261921$$
$$\cos. a=\cos. 7^\circ 30'=0.9914447879$$

Again, put $2a=7^\circ 30'$ then as before,

$$\cos.^2 a+\sin.^2 a=1$$
$$2\cos. a\sin. a=0.1305261921$$

These equations give

$$\sin. a=\sin. 3^\circ 45'=0.0654031291$$
$$\cos. a=\cos. 3^\circ 45'=0.9978589222$$

Thus we can bisect the arc as many times as we please. After *five* more bisections, we have

$$\sin. a=\sin. 7' 1'' 52\tfrac{1}{2}'''=0.0020453077$$
$$\cos. a=\cos. 7' 1'' 52\tfrac{1}{2}'''=0.99999799$$

As the sines of all arcs *under* 10′, may be considered as coinciding with the arc, and varying with it, we can now find the sine of 1′ by proportion.

Thus, $7' 1'' 52\tfrac{1}{2}''' : 1' :: 0.0020453077 : \sin. 1$

Or, $25312.5''' : 3600 ::$

Or, $10125 : 1440 :: 0.0020453077 : \sin. 1'$

Whence $$\sin. 1'=0.0002908882$$
$$\sin. 2'=0.0005817764$$
$$\sin. 3'=0.0008726646$$

In formula (B) equation (11), we find

$$\sin. (a+b)+\sin. (a-b)=2\sin. a\cos. b$$

Now, if $a=3'$ and $b=1'$

$$\sin. 4'+\sin. 2'=2\sin. 3'\cos. 1'$$

We have already sin. 2′ and sin. 3′, and cos. 1′ does not sensibly differ from *unity*, therefore

$$\sin. 4'=2\sin. 3'-\sin. 2'=0.0011635528$$
$$\sin. 5'=2\sin. 4'\cos. 1-\sin. 3' \text{ \&c. \&c. to } 15'$$

When the sine of any arc is known, its cosine can be found by the following formula, which is, in substance, equation (1) trigonometry $$\cos. a=\sqrt{(1+\sin. a)(1-\sin. a)}$$

In formula (A) equation (7) we find that

$$\sin.(a+b)=\sin.a\cos.b+\cos.a\sin.b$$

Now, if we make $a=30°$ and $b=4'$ Then

$$\sin.a=0.5 \qquad \cos.a=\tfrac{1}{2}\sqrt{3}=0.8660254$$
$$\sin.b=0.00116355 \quad \cos.b=0.999999323$$

Whence

$$\sin.(30°\ 4')=(0.5)\ (0.999999323)+(0.8660254)\ (0.00116355)$$
$$=0.499999661 \qquad + \ 0.0010007620$$
$$=0.501007281$$

Equation (8) gives

$$\sin.(29°\ 56')=0.498992041$$

When the sine and cosine of any arc are both known, the sine and cosine of the half or double of the arc can be determined by equation 30;—and thus, from equations (30), (7), (8), (11), and (1), the sines and cosines of all arcs can be determined.

But these sines and cosines are expressed in natural numbers, to radius unity, hence they are called *natural sines* and *natural cosines*, and they are all decimals, except the sine of 90° and the cosine of 0°, each of which is unity.

To form table II, we require logarithmic sines, and cosines, which are found by taking the logarithms of the natural sines and cosines, and increasing the indices by 10, to correspond to the radius of 10000000000. The radius of this table might have been greater or less, but custom has settled on this value.

To find the logarithmic sine of 1′, we proceed thus,

Nat. sin. 1′=0.0002908882	log.	—4. 463 726
	To which add	10.
The log. sine of 1′, therefore is		6. 463 726
Nat. sin. 3′=0.0008726646	log.	—4. 950 847
	Add	10.
Log. sin. 3′ therefore is		6. 940 847

Thus the logarithmic sine and cosine of all arcs are found. After the logarithmic sine and cosine of any arc have been found, the tangent and cotangent of the same arc can be found by equations (3) and (5), and the secants by (4); that is,

$$\tan.a=\frac{R\sin.a}{\cos a} \qquad \cot.a=\frac{R\cos.a}{\sin.a} \qquad \sec.a=\frac{R^2}{\cos.a}$$

For example, the logarithmic sine of 6° is 9.019235, and its cosine 9.997614. From these, find tan., cot., and secant.

R sin. - - - - -		19.019235
Cos. - - - -	subtract	9.997614
Tan. is - - - - -		9.021621
R cos. - - - - -		19.997614
Sin. - - -	subtract	9.019235
Cotan. is - - - -		10.978379
R^2 is - - - - -		20.000000
Cos. - - - -	subtract	9.997674
Secant is - - - - -		10.002326

Thus we find all the materials for

TABLE II.

This table contains logarithmic sines and tangents, and natural sines and cosines. We shall confine our explanations to the logarithmic sines and cosines.

The sine of every degree and minute of the quadrant is given, directly, in the table, commencing at 0° and extending to 45°, at the head of the table; and from 45° to 90°, at the foot of the table, increasing backward.

The same column that is marked sine at the top, is marked cosine at the bottom; and the reason for this is apparent to any one who has examined the definitions of sines.

The difference of two consecutive logarithms is given, corresponding to *ten* seconds. Removing the decimal point one figure will give the difference for *one* second; and if we multiply this difference by any proposed number of seconds, we shall have a difference corresponding to that number of seconds, above the logarithm, corresponding to the preceding degree and minute.

For example, find the sine of 19° 17′ 22″.

The sine of 19° 17′, taken directly from the table, is	9.518829
The difference for 10″ is 60.2 ; for 1″, is 6.02×22	133
Hence, 19° 17′ 22″ sine is	9.518952

From this it will be perceived that there is no difficulty in obtaining the sin. or tan., cos. or cot., of any angle greater than 30′.

Conversely. Given the logarithmic sine 9.982412, to find its corresponding arc. The sine next less in the table, is 9.982404, and gives the arc 73° 48′. The difference between this and the given sine, is 8, and the difference for 1″, is .61; therefore, the number of seconds corresponding to 8, must be discovered by dividing 8 by the decimal .61, which gives 13. Hence, the arc sought is 73° 48′ 13″.

These operations are too obvious to require a rule. When the arc is very small, such arcs as are sometimes required in astronomy, it is necessary to be very accurate; and for that reason we omitted the difference for seconds for all arcs under 30′. Assuming that the sines and tangents of arcs under 30′ vary in the same proportion as the arcs themselves, we can find the sine or tangent of any very small arc to great accuracy, as follows:

The sine of 1′, as expressed in the table, is . .	6.463726
Divide this by 60; that is, subtract logarithm . .	1.778151
The logarithmic sine of 1″, therefore, is . . .	4.685575
Now, for the sine of 17″, add the logarithm of 17 .	1.230449
Logarithmic sine of 17″, is	5.916024

In the same manner we may find the sine of any other small arc.

For example, find the sine of 14′ 21½″; that is, 861″5

To logarithmic sine of 1″, is, ...	4.685575
Add logarithm of 861.5	2.935255
Logarithmic sine of 14′ 21½″	7.620830

Without further preliminaries, we may now preceed to practical

EXAMPLES.

2. In a right angled triangle, ABC, given the base, AB, 1214, and the angle A, 51° 40′ 30″, to find the other parts.

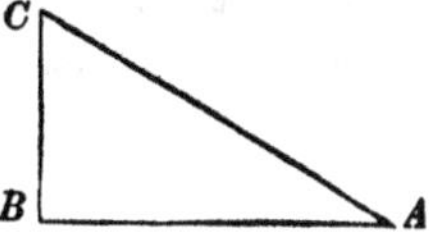

To find BC.

As radius	. .	10.000000
: tan.A	51° 40′ 30″	10.102119
:: AB	1214 .	3.084219
: BC	1535.8 .	3.186338

N. B. When the first term of a logarithmic proportion is radius, the resulting logarithm is found by adding the second and third logarithms, rejecting 10 in the index, which is dividing by the first term.

In all cases we add the second and third logarithms together; which, in logarithms, is multiplying these terms together; and from that sum we subtract the first logarithm, whatever it may be, which is dividing by the first term.

To find AC.

As sin. C, or cos. A 51° 30′ 40″	- - -	9.792477
: AB 1214	- - -	3.084219
: : Radius	- - -	10.000000
: AC 1957.7	- - -	3.291742

To find this resulting logarithm, we subtracted the first logarithm from the second, conceiving its index to be 13.

Let ABC represent any plane triangle, right angled at B.

1. Given AC 73.26, and the angle A 49° 12′ 20″; required the other parts.

Ans. The angle C 40° 47′ 40″, BC 54.46, and AB 47.87.

2. Given AB 469.34, and the angle A 51° 26′ 17″, to find the other parts.

Ans. The angle C 38° 33′ 43″, BC 588.5, and AC 752.9.

3. Given AB 493, and the angle C 20° 14′; required the remaining parts. Ans. The angle A 69° 46′, BC 1338, and AC 1425.

It is not necessary to give any more examples in right angled plane trigonometry, for *every distance* in the *traverse table* is but the hypotenuse of a right angled triangle, and its corresponding latitude and departure form the sides of the triangle.

If any one should suspect an error in the traverse table, let him test it by computing the triangle anew.

OBLIQUE ANGLED TRIGONOMETRY.

Of the six parts of a triangle, three sides and three angles, three of them must be given and one of the given parts must be a side.

The subject presents four cases.

1. *When two sides are given, and an angle opposite one of them.*
2. *When two sides are given, and the included angle.*
3. *When one side and two angles are given.*
4. *When the three sides are given.*

The principles previously demonstrated are sufficient, and indeed ample, to give all solutions that can come under any one of these cases. The operator must use his own judgment in applying these principles.

We give an example in each case, which, with the incidental examples, will be sufficient to fix the principles in the mind of the operator.

EXAMPLE 1.

In any plane triangle, given one side and the two adjacent angles, to find the other sides and angle.

In the triangle ABC, given $AB=376$, the angle $A=48° \ 3'$, and the angle $B=40° \ 14'$, to find the other parts.

As the sum of the three angles of every triangle is equal to 180°, the third angle C must be $180°-88° \ 17'=91° \ 43'$.

INSTRUMENTALLY.

Take 376 from the scale, by means of the dividers, and place it on paper; making one extremity of the line A, and the other extremity B. From A, by means of the protractor (or otherwise), make the angle $A=48° \ 3'$, and from B, make the angle $B=40° \ 14'$. The intersections of the lines AC, BC, will give the angle C, which being measured will be found to be a little more than a right angle.

Take AC in the dividers, and apply it to the scale, and it will be found to be 243; and BC will be found to be 279.8, if the projection is *accurately made; but no one should expect numerical accuracy from this mechanical method.*

N. B. Our figures in the book do not pretend to accuracy, they should be drawn on paper on a larger scale.

BY LOGARITHMS.

To find AC.

As sin. 91° 43′ - - -	9.999805
: AB 376 - - - -	2.575188
: : sin. AB 40° 14′ - - -	9.810167
	12.385355
: AC 243 - - - -	2.385550

Observe, that the sine of 91° 43′ is the same as the cosine of 1° 43′.

To find BC.

As sin. 91° 43′	- - - -	9.999805
: AB 376	- - - - -	2.575188
: : sin. A 48° 3′	- - -	9.871414
		12.446602
: BC 279.8	- - - -	2.446797

EXAMPLE 2.

In a plane triangle, given two sides, and an angle opposite one of them, to determine the other parts.

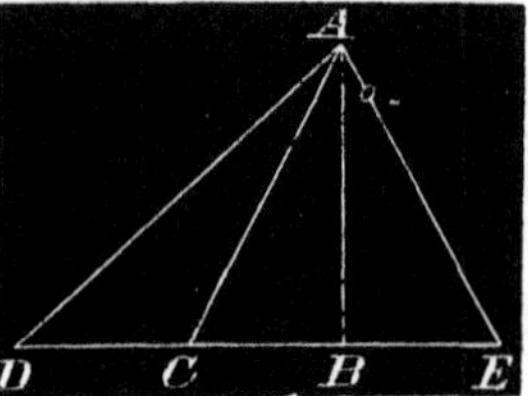

Let AD=1751. feet, one of the given sides. The angle D=31° 17′ 19″, and the side opposite, 1257.5. From these data, we are required to find the other side, and the other two angles.

In this case we do not know whether AC or AE represents 1257.5, because AC=AE. If we take AC for the other given side, then DC is the other required side, and DAC is the vertical angle. If we take AE for the other given side, then DE is the required side, and DAE is the vertical angle; but in such cases we determine both triangles.

INSTRUMENTALLY.

Draw DE indefinitely—from the point D make the angle D=31° 17′. AD=1751., but call it 175.1, which take from the scale. Place one foot of the dividers at D, the other foot will extend to A, thus finding the point A.

Take 125.75 in the dividers, place one foot at A as a center, and with the other strike an arc, cutting DE in C and E. Join AC, AE, and one or the other of the triangles ACD ADE, will be the triangle required. DC and DE applied to the scale, will give *one-tenth* of the required side, and the angle E or DCA, measured, will be one of the required angles.

We can also take one hundredth part of the sides, as well as the tenth; this will make no difference with the angles, the triangles thus formed will be similar.

In that case AD=17.51, and the side sought will be 23.64, which can be changed to 2364.

BY LOGARITHMS.

To find the angle $E=C$.

(Prop. 4.)	As $AC=AE=1257.5$	log.	3.099508
	: D 31° 17′ 19″	sin.	9.715460
	: : AD 1751	log.	3.243286
			12.958746
	$E=C$: 46° 18′	sin.	9.859238

From 180° take 46° 18′, and the remainder is the angle DCA =133° 42′.

The angle $DAC=ACE-D$ (th. 11, b. 1); that is,

DAC=46° 18′—31° 17′ 19″=15° 0′ 41″.

The angles D and E, taken from 180°, give DAE=102° 24′ 41″.

To find DC.

As sin. D 31° 17′ 19″	log.	9.715460
: AC 1257.5	log.	3.099508
: : sin. DAC 15° 0′ 41″	log.	9.413317
		12.512825
: DC 626.86		2.797165

To find DE.

As sin. D 31° 17′ 19	9.715460
: AC 1257.5	3.099508
: : sin. 102° 24′ 41″	9.989730
	13.089238
: DE 2364.5	3.373778

N. B. To make the triangle possible, AC must not be less than AB, the sine of the angle D, when DA is made radius.

EXAMPLE 3.

In any plane triangle, given two sides and the included angle, to find the other parts.

Let AD=1751 (see last figure), DE=2364.5, and the included angle D=41° 17′ 19″. We are required to find DE, the angle DAE, and angle E. Observe that the angle E must be less than the angle DAE, because it is opposite a less side.

INSTRUMENTALLY.

Take DE=236.45 from the scale (as near as possible), and from D draw DA, making the given angle 41° 17′ 19″.

Take 175.1 from the scale, in the dividers, and with it mark off DA. Join AE; and ADE will be the triangle in question, and AE applied to the scale will give the *tenth* part of the side sought; and measuring the angle E with the protractor (or otherwise), will determine its value.

BY LOGARITHMS.

From - - - -	180°
Take D - - - -	31° 17′ 19″
Sum of the other two angles	=148° 42′ 41″ (th. 11, b. 1)
½ sum - - -	= 74° 21′ 20″

By proposition 7,

$DE+DA : DE-DA$=tan. 74° 21′ 20″ : tan. ½ $(DAE-E)$

That is,

4115.5 : 613.5=tan. 74° 21′ 20″ : ½$(DAE-E)$

Tan. 74° 21′ 20″ - - -	10.552778
613.5 - - - - - -	2.787815
	13.340593
4115.5 log. (sub.)	3.614423
½$(DAE-E)$ tan. 28° 1′ 36″	9.726170

But the half sum and half difference of any two quantities are equal to the greater of the two; and the half sum, less the half difference, is equal the less.

Therefore, to	74° 21′ 20″
Add	28 1 36
DAE=	102° 22′ 56″
E=	46 19 44

To find AE.

As sin. E 46° 19′ 44″ - -	9.859323
: DA 1751 - - - -	3.243286
: : sin. D 31° 17′ 19″ - -	9.715460
	12.958746
: AE 1257.2 - - -	3.099423

EXAMPLE 4.

Given the three sides of a plane triangle to find the angles.

Given $AC=1751$, $CB=1257.5$, $AB=2364.5$

If we take the formula for cosines, we will compute the greatest angle, which is C.

INSTRUMENTALLY.

Construct a triangle with the three given sides 236.45, 125.75, and 175.1, according to problem 16, chapter 1. The angles then measured will show their value.

BY LOGARITHMS.

R^2		20.000000
$s=2686.5$		3.429187
$s-c=322$		2.507856
Numerator,	log.	25.937043
a 1257.5	3.099508	
b 1751.	3.243286	
Denominator, log.	6.342794	6.342694
		2)19.594249
$\frac{1}{2}C=$ 51° 11′ 10″ cos.		9.797124
$C=$102 22 20		

The remaining angles may now be found by problem 4.

We give the following examples for practical exercises:

Let ABC represent any oblique angled triangle.

1. Given AB 697, the angle A 81° 30′ 10″, and the angle B 40° 30′ 44″, to find the other parts.

Ans. AC 534, BC 813, and the angle C 57° 59′ 4″.

2. If $AC=720.8$, the angle $A=70°$ 5′ 22″, and the angle $B=$ 59° 35′ 36″, required the other parts.

Ans. AB 643.2, BC 785.8, and the angle C 50° 19′ 6″.

3. Given BC 980.1, the angle A 7° 26′ 26″, and the angle B 106° 2′ 23″, to find the other parts.

Ans. AB 7284, AC 7613.3, and the angle C 66° 51′ 11″.

SURVEYING.

SURVEYING is the art of running definite lines on the surface of the earth, measuring them, and finding the contents of lands; and the subject necessarily includes the measure of surfaces generally. We shall therefore commence with mensuration.

Mensuration is the application of the principles of Geometry, to the measure of surfaces and solids, and when lands are measured it is a part of surveying. We shall be very brief on mensuration proper, because the rules are so simple and obvious. For the demonstration of the rules, we refer to (Legendre and Robinson's Geometry.)

All surfaces are measured by the number of square units which they contain. The unit may be taken at pleasure; it may be an inch, foot, yard, rod, mile, &c., as convenience and propriety may dictate.

The *square unit* is always the square or the *linear* unit.

PROBLEM I.

To find the area of a square, or a parallelogram.

RULE.— *Multiply the length by the perpendicular breadth, and the product will be the area.*

(Leg. b. IV, prop. V. Rob. book I, th. 29).

1. What is the area of a square whose sides are 6 feet 3 inches?
Ans. $39\frac{1}{16}$ square feet. *

2. How many square feet are in a board that is $13\frac{1}{2}$ feet long and 10 inches wide? Ans. $11\frac{1}{4}$ square feet.

3. A lot of land is 80 rods long, and 45 rods wide, how many square rods does it contain, and how many acres?
Ans. 3600 rods, $22\frac{1}{2}$ acres.

---

* NOTE.— Reductions from one measure to another have no reference to the rules here given.

4. A man bought a farm 198 rods long, and 150 rods wide, at \$32 per acre; what did it come to? Ans. \$5940.

PROBLEM II.

To find the area of a triangle, when the base and altitude are given.

Rule.— *Multiply one of these dimensions by half the other, and the product will be the area required.*

(Leg. book IV, p. VI. Rob. book I, th. 30).

1. How many yards in a triangle whose base is 148 feet, and perpendicular 45 feet? Ans. 370 yards.

2. What is the area of a triangle whose base is $18\frac{1}{3}$ feet and altitude $25\frac{1}{4}$ feet? Ans. $231\frac{11}{24}$ feet.

PROBLEM III.

Investigate and give a rule for finding the area of a triangle when two sides and their included angle are given.

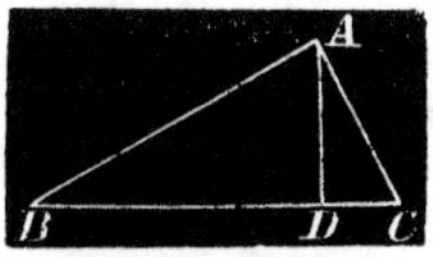

Let ABC be the triangles, AB, BC the given sides, and B the given angle.

Represent the side opposite to the angle A, by a, opposite C, by c, and opposite B, by b.

Now a and c are the given sides, and by problem II, the area is

$$\tfrac{1}{2}a(AD) \qquad (1)$$

The trigometrical value of AD can be found from the right angled triangle ABD.

Thus, $\sin. ADB : c :: \sin. B : AD.$

That is, $1 : c :: \sin. B : AD.$

Whence $AD = c \sin. B.$

This value of AD substituted in (1) gives

$$\tfrac{1}{2}ac \sin. B = \textit{area } \triangle \qquad (2).$$

This expression is the area of the triangle, and from it we draw the following rule.

Rule.— *Take half the product of the two given sides and multiply it by the natural sine of the included angle, and the last product will be the area required.*

1. One side of a triangle is 82 feet, another 90 feet, and their included angle is 27° 31′. What is the area?

Ans. 1749.4 square feet.

27° 31′ Nat. sine. - - -	46201
$\frac{1}{2}ac$ - - - - -	3780
	3996080
	323407
	138603
	1749.39780 Ans.

When we use logarithms we have the following rule:

RULE.—*To the logarithms of the two sides, add the log. sine of the included angle, and the sum rejecting* 10, *in the index, is the logarithm of twice the area of the triangle.*

2. A certain triangle has one side 125.81, another equal 57.65, and their included angle 57° 25′, what is its area? Ans. 3055.7.

125.81	log.	2.099715
57.65	log.	1.760799
57° 25′	sine	9.925626
2 Area, 6111.4	log.	3.786140

3. How many square yards in a triangle, two sides of which are 25 and $21\frac{1}{4}$ feet, and their included angle 45°? Ans. 20.8695.

PROBLEM IV.

Investigate and give a rule for finding the area of a triangle when the three sides are given.

(See figure to problem III). Let A represent the area of any plane triangle, then by problem III

$$A=\tfrac{1}{2}ac \sin. B. \qquad (1)$$

But $\sin. B=2 \sin. \tfrac{1}{2}B \cos. \tfrac{1}{2}B.$ (Eq. 30, trigonometry).

Therefore, $A=ac, \sin. \tfrac{1}{2}B \cos. \tfrac{1}{2}B. \qquad (2)$

Now in proposition 8, trigonometry, we find

$$\text{Sin. } \tfrac{1}{2}B=\sqrt{\frac{(s-a)(s-c)}{ac}}, \qquad (3)$$

and $$\cos. \tfrac{1}{2}B=\sqrt{\frac{s(s-b)}{ac}} \qquad (4)$$

The product of (3) into (4) is

$$\sin.\tfrac{1}{2}B\cos.\tfrac{1}{2}B=\sqrt{\frac{s(s-a)(s-b)(s-c)}{a^2c^2}}, \qquad (5)$$

or $$ac\sin.\tfrac{1}{2}B\cos.\tfrac{1}{2}B=\sqrt{s(s-a)(s-b)(s-c)}. \qquad (6)$$

By comparing (2) and (6) we perceive that

$$A=\sqrt{s(s-a)(s-b)(s-c)}.$$

Here s represents the half sum of a, b, and c, therefore, we have the following rule to find the area when the three sides are given.

RULE.—*Add the three sides together and take half the sum. From the half sum take each side separately, thus obtaining three remainders. Multiply the said half sum and the three remainders together; the square root of this product is the area required.*

1. Find the area of a triangle whose sides are 20, 30, and 40.

Ans. 290.47.

$\frac{1}{2}$ sum $=45$, 1st Rem. $=25$, 2d$=15$, 3d$=5$.

$$\sqrt{45.25.15.5}=\sqrt{225.25.15}=15.5\sqrt{15}=75(3.873)=290.474.$$

2. How many acres in a triangle whose sides are severally 60, 50, and 40 rods? Ans. $6\frac{1}{3}$ nearly.

3. How many square yards are there in a triangle whose sides are 30, 40 and 50 feet? Ans. $66\frac{2}{3}$.

4. There is a triangular lot of land containing 8 acres, two of its sides are 64, and 46 rods respectively; what is the angle between these sides, and what is the length of the remaining side?

Ans. The angle is 60° 27′, or is supplement 119° 33′.

The side is 57.37 rods, or 95.535 rods; the less angle corresponding to the lesser side.

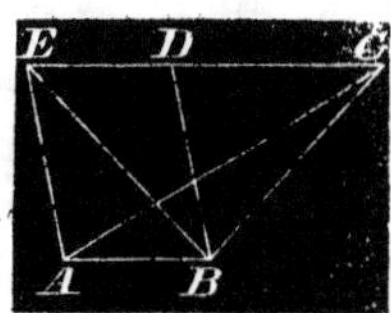

In short, there are two triangles answering to the conditions, the one is ABE, the other ABC. They are equal because they are on the same base and between the same parallels. $AE=57.37$, $AC=95.535$.

7

PROBLEM V.

To find the area of a trapezoid.

RULE.— *Add the two parallel sides together, and take half the sum. Multiply this half sum by the perpendicular distance between the sides.*

Or, The sum of the parallel sides multiplied by their distance asunder will give twice the area.

(Leg. book IV, prop. VII. Rob. b. I, th. 31).

REMARK.—The application of this problem is the most important of any in general surveying, as will appear in the sequel, and if the geometrical theorem is not familiar to the student he should again review it.

Ex. 1. In a trapezoid, the parallel sides are 750 and 1225, and the perpendicular distance between them 1540 links: to find the area.

1225
750

1975×770=152075 square links =15 acr. 33 perches.

Ex. 2. How many square feet are contained in a plank, whose length is 12 feet six inches, the breadth at the greater end 15 inches, and at the less end 11 inches? Ans. $13\frac{13}{24}$ feet.

Ex. 3. In measuring along one side AB of a quadrangular field, that side, and the two perpendiculars let fall on it from the two opposite corners, measured as below, required the content.

$AP =$ 110 links
$AQ =$ 745
$AB =$ 1110
$CP =$ 352
$DQ =$ 595

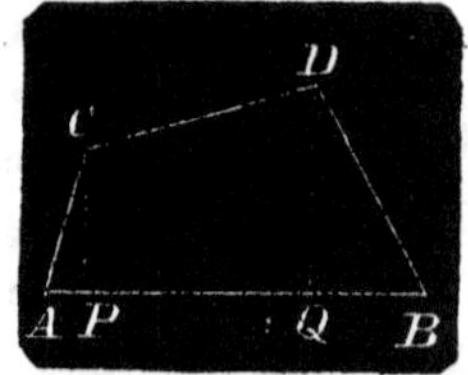

Ans. 4 acres, 1 rood, 5.792 perches.

Here we perceive a trapezoid and two right angled triangles.

N. B. A chain is 4 rods, and contains 100 links; 10 square chains make an acre.

PROBLEM VI.

To find the Area of any Trapezium.

DIVIDE the trapezium into two triangles by a diagonal; then find the areas of these triangles, and add them together.

Or thus, let fall two perpendiculars on the diagonal from the other two opposite angles; then add these two perpendiculars together, and multiply that sum by the diagonal, taking half the product for the area of the trapezium.

Ex. 1. To find the area of the trapezium, whose diagonal is 42, and the two perpendiculars on it 16 and 18.

Here $16+18=34$, its half is 17.
Then $42\times17=714$ the area.

Ex. 2. How many square yards of paving are in the trapezium, whose diagonal is 65 feet; and the two perpendiculars let fall on it 28 and $33\frac{1}{2}$ feet? Ans. $222\frac{1}{12}$ yards.

When the sides of a trapezium, and two of its opposite angles are given, the most convenient rule for finding its area is found in problem III.

Conceive CB joined, then the whole figure consists of two triangles and the whole area is found in the following expression $(AB\times AC\times \sin.A)+(CD\times DB\times \sin.CDB.)$

EXAMPLE.

In the quadrilateral $ACDB$ we have AC 15.7, CD 20.4, DB 14.24, and BA 27.7 rods. The angle A 78° 15′ and the opposite angle CDB 97° 30′. What is the area enclosed?

Ans. 356.65 square rods.

PROBLEM VII.

To find the area of an irregular figure bounded by any number of right lines.

Rule.—*Draw diagonals dividing the figure into triangles. Find the areas of the triangles so formed and add them together for the area of the whole.*

Let it be required to find the area of the adjoining figure of five sides. On the supposition that $AC=36.21$ $EC=39.11$ $Aa=4.18$ $Bb=4$ and $Dd=7.26$

Ans. 296.129.

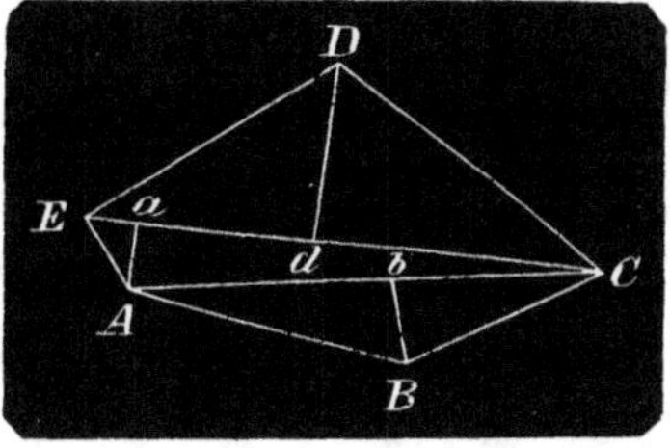

PROBLEM VIII.

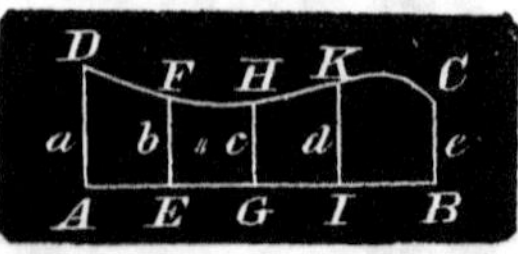

To find the area of a long irregular figure like the one represented in the margin, it is necessary to divide it into *trapezoids*. Then find the area of each one of the trapezoids (by problem V.) and add them together for the whole area.

If however the trapezoids have equal distances between their parallel sides we can take a more summary process, which we discover by the following investigation.

The trapezoid $AEFD=\frac{1}{2}(a+b)\times AE.$
" " $EGHF=\frac{1}{2}(b+c)\times EG.$
" " $GIKH=\frac{1}{2}(c+d)\times GI.$
" " $IBCK=\frac{1}{2}(d+e)\times IB.$

On the supposition that AE, EG, GI, &c. are all equal to each other the sum of these is $\left(\frac{a}{2}+b+c+d+\frac{e}{2}\right)\times AE,$

which represents the area of the whole figure.

From this we draw the following rule to find the area of a long and narrow figure bounded by a right line on one side, and a broken or curve line on the other, to which *off sets* are made at equidistant points along the right line.

RULE.—*Add the intermediate breadths or off sets together, and the half sum of the extreme one: then multiply this sum by one of the equal parts of the right line, the product will be the area required, very nearly.**

1. The breadths of an irregular figure at five equidistant places, being 6.2, 5.4, 9.2, 3.1, 4.2, and the length of the base 60, what is the area?

Mean of the Extremes	5.2
Sum of 5.4, 9.2, 3.1	17.7
Sum	22.9
One of the equal parts	1 5
	114 5
	229
	Area=343.5

* In case DF, FH, &c. are right lines we shall have the area exactly, if they are other than right lines the area will be nearly.

2. The length of an irregular figure being 84, and the breadths at six equidistant places 17.4 20.6 14.2 16.5 20.1 24.4; what is the area? Ans. 1550.64.

PROBLEM IX.

To find the area of a circle, also any sector or segment of a circle.

RULE 1. — *The area of a circle is found by multiplying the radius by half the circumference.*

(Leg. book V, prop. 12. Rob. book V, th. 1.)

RULE 2. *Multiply the square of the diameter by the decimal* .7854.

When the radius of a circle is 1, the length of one degree on the circumference is 0.01745 and the whole circumference is 3.1416.

The radius and the circumference increase and decrease by the same ratio, therefore the length of any arc corresponding to any radius is easily computed.

A sector of a circle is to the whole circle as the number of degrees it contains is to 360.

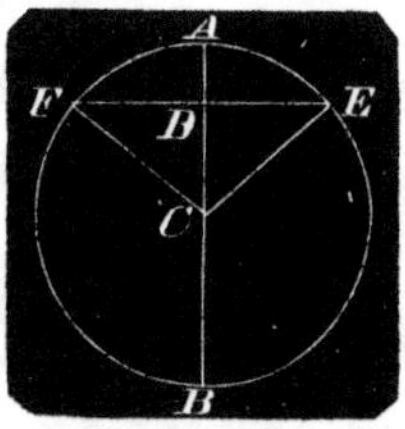

The area of a segment of a circle as *FAE*, may be found by first finding the sector *FCE*, and from it taking the area of the triangle *FCE*.

This same triangle added to the greater sector will give the greater segment.

These principles and rules are sufficient to solve the following examples which are given merely as educational Exercises.

1. What is the area of a circle whose diameter is 10? Ans. 78.54.

2. What is the area of a circle whose diameter is 20? Ans. 4 times 78.54.

3. What is the area of a circle whose circumference is 12? Ans. 11.4595.

4. How many square yards are in a circle whose diameter is $3\frac{1}{2}$ feet? Ans. 1.069.

5. Find the length of an arc of 20°, the radius being 9 feet. Ans. 3.141.

6. Find the length of an arc of 60°, the radius being 18 feet. Ans. 18.846.

7. To find the length of an arc of 30 degrees, the radius being 9 feet. Ans. 4.7115.

8. To find the length of an arc of 12° 10′, or $12°\frac{1}{6}$, the radius being 10 feet. Ans. 2.1231.

9. What is the area of a circular sector whose arc is 18° and the diameter 3 feet? Ans. 0.35343.

10. To find the area of a sector, whose radius is 10, and arc 20° Ans. 100.

11. Required the area of a sector, whose radius is 25, and its arc containing 147° 29′. Ans. 804.3986.

12. What is the area of the segment, whose height is 18, and diameter of the circle 50? Ans. 636.375.

13. Required the area of the segment whose chord is 16, the diameter being 20? Ans. 44.728.

14. What is the length of a chord which cuts off one-third of the area from a circle whose diameter is 289? Ans. 278.6716.

15. The radius of a certain circle is 10; what is the area of a segment whose chord is 12? Ans. 16.35.

16. What is the area of a segment whose height is 2 and chord 20? Ans. 26.88.

17. What is the area of a segment whose height is 5, the diameter of the circle being 8? Ans. 33.0486.

PROBLEM X.

To find the Area of an Ellipse.

RULE.—*Multiply the two semi-axes together and their product by* 3.1416. (See conic sections).

1. Required the area of an Ellipse whose two semi-axes are 25 and 20. Ans. 1570.8.

2. The two semi-axes of an Ellipse are 12 and 9, what is its area? Ans. 339.29.

To find the area of any portion of a parabola we *multiply the base by the perpendicular height, and take two-thirds of the product for the area required.* (See conic sections).

Required the area of a parabola, the base being 20, and the altitude 30. Ans. 400.

The surfaces of prisms, cylinders, pyramids, cones, &c., are found by the application of the preceding rules.

From theorem 16, book VII, Geometry, we learn that

The convex surface of a sphere is equal to the product of its diameter into its circumference.

The surface of a segment is equal to the circumference of the sphere, multiplied into the thickness of the segment.

In the same sphere, or in equal spheres, the surfaces of different segments are to each other as their altitudes.

MENSURATION OF SOLIDS.

By the Mensuration of Solids are determined the spaces included by contiguous surfaces; and the sum of the measures of these including surfaces, is the whole surface or superficies of the body.

The measures of a solid, is called its solidity, capacity, or content.

Solids are measured by cubes, whose sides are inches, or feet, or yards, &c. And hence the solidity of a body is said to be so many cubic inches, feet, yards, &c., as will fill its capacity or space, or another of an equal magnitude.

The least solid measure is the cubic inch, other cubes being taken from it according to the proportion in the following table, which is formed by cubing the linear proportions.

Table of Cubic or Solid Measures.

1728	cubic inches make	1 cubic foot
27	cubic feet	1 cubic yard
$166\frac{3}{8}$	cubic yards	1 cubic pole
64000	cubic poles	1 cubic furlong
512	cubic furlongs	1 cubic mile.

As the mensuration of solids has little to do with surveying or navigation, we shall leave this subject after simply stating the following truths, which are demonstrated in solid geometry.

In fact, these truths may be called *rules* for practical operations.

1. *The solidity of a cube, parallelopiped, prism, or cylinder, is found by multiplying the area of its base by the altitude.*

2. *The solidity of a pyramid or cone is found by multiplying the base by the altitude, and taking one-third of the product.*

3. *The solidity of the frustum of a pyramid or cone is found by calculating the solidity of the pyramid when complete, and subtracting from it the solidity of the part removed; or by multiplying the top and bottom diameters together and that product by the altitude, and then to this last product adding a sum produced by squaring the difference between the top and bottom diameters and multiplying it by one-third of the height.*

4. *Guaging is performed by considering a cask to be made up of two frustums of cones placed base to base, and applying the rules for the measurement of such solids.*

5. *The solidity of a sphere is two-thirds of the solidity of its circumscribing cylinder.*

CHAPTER I.

MENSURATION OF LANDS.

Lands are not only measured to find their areas, but their exact positions must be ascertained, the direction which each line makes with the meridians, or *with the north and south lines on the earth.*

The boundaries of each tract of land are referred to that meridian which runs through or by the side of it.

All meridian lines meet at the poles, therefore they are not parallel, (except at the equator,) but the poles are so far distant that no sensible error can arise from supposing them parallel, and all surveys are made on the supposition that the surface of the earth is a plane and the meridians parallel. When large surveys are made, like a county or a state, the spherical form of the earth should be taken into consideration.

Meridian lines in surveys are usually determined by the *magnetic needle*, but the needle does not settle exactly north and south, generally speaking, and the direction which it does settle is called the *magnetic meridian.*

Surveys are often made by the *magnetic meridian* as the true one, and this would answer every purpose, provided the difference between the magnetic and true meridians were *every where* and at *all times* the same, but this is not so.

The magnetic meridian is variable, and for this reason it is very difficult to trace old lines, unless visible monuments are left, or unless the record refers to the true meridian.

Lines are generally measured by a *chain* of 66 feet or 4 rods in length, containing 100 *links,* each link is therefore 7.92 inches.

The area of land is estimated in acres and hundredths, formerly in acres, roods, and perches, but the modern method is more simple and convenient; we have a clearer conception of 35 hundreths of an acre than we have of 1 rood and 16 perches.

An acre is equal to 10 *square chains or* 100,000 *square links.*

We may note down the length of a line in chains and hundreths, or in links only, for it is nearly one and the same thing: thus, 12 chains and 38 links may be written 12.38, or 1238 links.

The area of a field may be found by measuring with the chain only, and dividing it into rectangles and triangles, and computing each of them separately, according to the rules laid down in mensuration.

The most common method for measuring a field for calculation, is, to take the length of all the sides of the field with the chain, and their bearings with the surveyor's compass. With these notes an accurate plan or plot of the field may be made on paper, and then its contents ascertained by cutting it into triangles and measuring their bases and perpendiculars with a scale and dividers. A very little instruction from a teacher will enable the student to practice this method with success; *yet no instrumental measures pretend to be numerically accurate, they are but approximately so.*

TO MEASURE A LINE.

Provide a chain and 10 small arrows or marking pins to fix one into the ground, as a mark, at the end of every chain; two persons take hold of the chain, one at each end of it; and all the 10 arrows are taken by one of them who goes foremost, and is called the leader; the other being called the follower, for distinction's sake.

A picket, or station-staff being set up in the direction of the line to be measured, if there do not appear some marks naturally in that direction, they measure straight towards it, the leader fixing down an arrow at the end of every chain, which the follower always takes up, as he comes at it, till all the ten arrows are used. They are then all returned to the leader, to use over again. And thus the arrows are changed from the one to the other at every 10 chains' length, till the whole line is finished; then the number of changes of the arrows shows the number of tens, to which the follower adds the arrows he holds in his hand, and the number of links of another chain over to the mark or end of the line. So, if there have been 3 changes of the arrows, and the follower hold 6 arrows, and the end of the line cut off 45 links more, the whole length of the line is set down in links thus, 3645.

In all these measures horizontal distances are required, and they are obtained, at least very nearly, by holding the chain in a *horizontal position*, both on ascending and descending ground. If the declivity is too great to admit of measuring a whole chain at a time, take a part of it, and in all cases the proper position of the elevated extremity should be determined by a plumb line. The reason of these operations is obvious by the adjoining figure; we require the line AB, and not the line along the ground as AC.

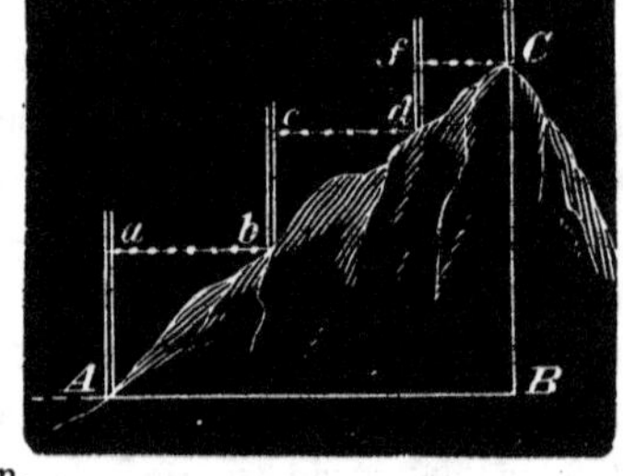

$$AB = ab + cd + fC.$$

It is not only necessary to measure lines but we must also know their direction or the angles which they make with the meridian. This is commonly determined by means of the

SURVEYOR'S COMPASS.

The surveyor's compass consists of a horizontal circle to which are attached sight-vanes and a magnetic needle delicately balanced on its centre.

When the compass *is set*, that is, standing in a free horizontal position and the needle free to move on the center, the needle will keep the magnetic meridian, and the circular plate *may be turned under it* to bring the sight-vanes to any line; — the needle will then point out the degree of inclination which the line makes with the meridian.

It is important that this part of the subject be most clearly understood by the learner, we therefore give the following minute illustration of it.

Let the reader now face the north with the book open before him, his right hand is then toward the east, and his left hand toward the west.

The following figure represents the compass set to the magnetic meridian, that is the sight-vanes *Vv* and the needle lie in the same direction. The degrees on the plate are numbered both ways from *N* and *S* to *E* and *W.*

At the first view of this subject, it has surprised many to find *W* for *west* on the right hand toward the east, and *E* in the direction toward the west. The reason of this is explained by the next figure.

Suppose we wished to find the direction of a line from the center of the compass to the object *B.* We set the compass, that is place it horizontal on its staff or tripod, the needle will take the same direction as in the first figure, parallel to the margin of the paper.

The sight-vanes *Vv* are turned toward the object *B* which turns the whole plate, but the needle retains its position.

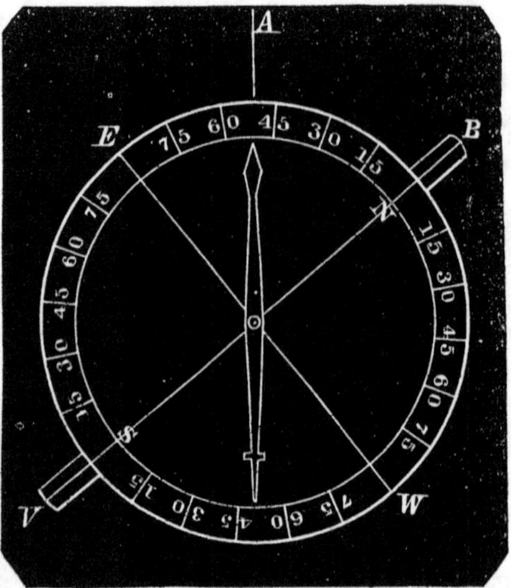

We now *read* the degree pointed out by the north end of the needle, and we find it to be about 50° on the arc between *N* and *E*, *showing that the course or direction from the center of the compass to B* is North about 50° toward the East — a result obviously true.

Turning the sight-vanes toward the north west will bring the arc between *N* and *W* to the north point of the needle. For example, if it were required to run a line North 31° West, from a certain point, all we have to do is to set the compass over that point, level the plate, see that the needle is free to move on its pivot, and so turn the plate that the north end of the needle will settle at 31° between *N* and *W*, *the range* of the sight-vanes will then show the required line. Proceed in the same manner to find any other line.

Care should be taken that no iron or steel comes near the compass while operating with it. To insure a correct position of the needle is the principal difficulty, but if it settles with a free motion, describing nearly equal arcs, slowly decreasing on each side of a given point and finally rests at that point, it operates well, and may be relied upon.

In whatever direction we run, the north point of the needle should always lie on *some part* of the north side of the plate, that is, nearer to *N* than to *S* and this can always be except when we run due east or west per compass.

Lines should be tested by taking *back sights* or reverse bearings, which will be *exactly in the opposite* point of the compass, in case there is no local attraction to disturb the needle. If the line just run over does not correspond to the exact opposite point of the compass, it shows carelessness in running or some local attraction of the needle. We shall show how to overcome this last difficulty further on.

Compasses are usually marked to half degrees, some of them are subdivided to one fourth degrees, but by the aid of a *vernier scale* we can *theoretically* read the arc to *one minute* of a degree.

DESCRIPTION OF THE VERNIER.

The vernier to a compass is on the outer edge of the graduated limb. *It is a slip of metal made to fit the graduated limb of an instrument, and the equal divisions upon it are so made that* n *divisions on the vernier will cover* n±1 *divisions on the limb.*

The vernier of the compass is on the outside of the dial plate and it is firmly attached to the bar that holds the sight-vanes. The dial plate can be moved to and fro along it by means of a screw.

The vernier is used when the needle points between two divisions on the limb : the dial plate is then gently moved by the screw until the needle points exactly to the preceding division on the limb, this being done, that division on the vernier which makes a right-line, that is, coincides with a division on the arc is the number of small divisions (minutes) to be added to the division on the limb now pointed out by the needle.

Practical and experienced men, never use the vernier of the compass, because they can read the compass without it to greater accuracy than they can really run a line.

But as the vernier scale is of the greatest importance attached to several other instruments, which will be referred to in this work, we now make an effort to give the learner a clear comprehension of it.

Let *AB* represent a portion of an arc and *ED* the vernier which is conceived to be attached to an index bar and made to revolve with it.

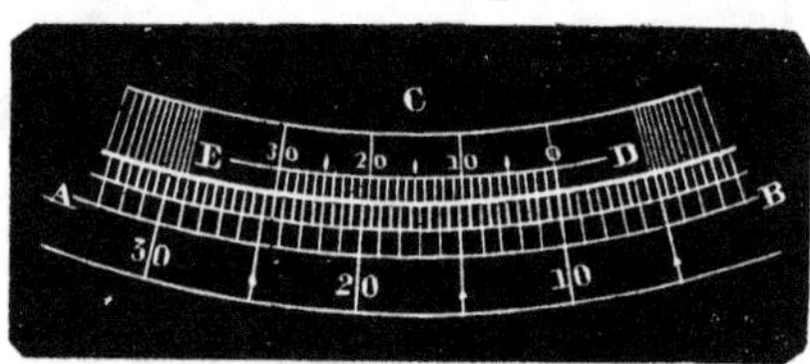

In case 0 on the vernier makes a right line with 10° on the arc as is represented in the figure, then the index marks 10°. But if 0 on the vernier is a little beyond 10° we then look along the vernier scale to see what division of it makes a right line with some division on the limb, suppose the division 8 on the vernier coincided with a division on the limb then the index would mark $10^\circ\ 8'$.

To understand the philosophy of this: Let x represent the value of a division on the vernier and n the number of them which cover $(n-1)$ divisions on the limb, then

$$nx = n - 1$$

$$x = 1 - \frac{1}{n}$$

$$\frac{1}{n} = 1 - x$$

$$\frac{2}{n} = 2 - 2x$$

$$\frac{3}{n} = 3 - 3x$$

&c. &c. to n, number, from which it appears that one division on the vernier beyond a division on the limb corresponds with the nth part of the unit of graduation, two divisions of the vernier above two divisions on the limb, correspond with $2n$th division of the unit of graduation, &c.

In our figure $n=30$, the graduation of the limb is to half degrees or 30 minutes, and this vernier measures minutes. Verniers on many instruments measure as small as *ten seconds* of arc and on some very large instruments as low as *four seconds.*

CHAPTER II.

Having shown in the preceding chapter how to use a compass—to run lines, and to measure them; the next step is to keep a proper record of all the lines run, and compute the areas they enclose.

A line traced on the ground, is called a *course*, the angle that it makes, with the meridian passing through the point of beginning, is called its *bearing*.

A course written *N* 42° *E*, indicates that the line runs between the north and the east, and makes an angle of 42° with the meridian; when between the north and west, we write *N. W.*, putting the number of degrees and minutes between.

Lines from the south point, are also written *S. E.* and *S. W.*; that is, bearings are reckoned from the north and south points, east and west, as the case may be.

Hence, to make a record of a survey, all we have to do is to write the *bearing* and distance of each course, and if the last side runs to the point of beginning, it is a complete survey; otherwise it is not.

Of course, no *area* can be attached to any un-enclosed space. To complete a partial survey, to enclose a space, to find an *area*, or to test the accuracy of a complete survey; the most satisfactory method of investigation, is that known as

LATITUDE AND DEPARTURE.

Latitude is the distance of the end of a line north or south of its beginning, measured on a meridian, and it is called either northing or southing, according as the line runs north or south. Departure is the distance of the end of a line east or west of its beginning, measured perpendicular to a meridian, and it is called easting or westing, according as the line runs east or west.

For example, suppose that we have the following bearings and distances, which enclose a space represented by *ABCD*.

	Bearings.	Distances.
AB	*N.* 23° *E.* - -	17
BC	*N.* 83° *E.* - -	11
CD	*S.* 14° *E.* - -	23
DA	*N.* 77° *W.* - - -	23.66

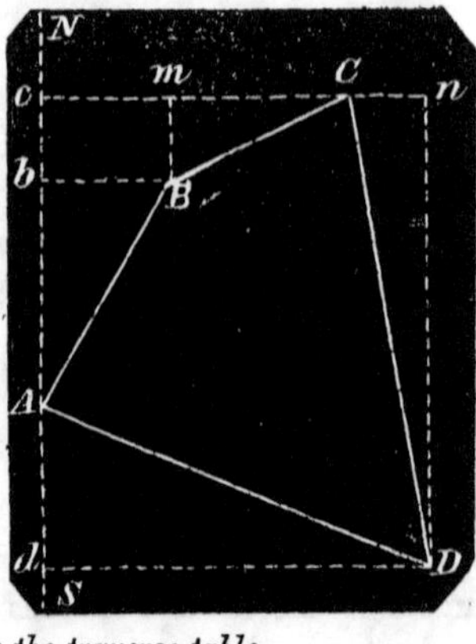

Let NS represent the meridian running through A, the most western point of the field; make the angle $NAB=23°$, and $AB=17$; then Ab is the latitude, and Bb is the departure, corresponding to the course AB. By means of the right angled triangle ABb, having the hypotenuse AB, and the angles, we can compute Ab, and Bb, 15.65, and 6.64, or we can turn to the traverse table, and under 23° and opposite 17, we shall find the value of these lines at once; *and this is the utility of having the traverse table.*

In the same manner we find Bm and mC, the latitude and departure corresponding to the bearing and distance of the line BC. We find $Bm=1.34$, and $mC=10.92$.

Thus we go round the field, taking the latitude and departure of each side, and arrange the whole in a table as follows:

	Bearings.	Dist.	N.	S.	E.	W.
AB	*N.* 23° *E.*	17	15,65		6,64	
BC	*N.* 83° *E.*	11	1,34		10,92	
CD	*S.* 14° *E.*	23		22,32	5,56	
DA	*N.* 77° *W.*	23,66	5,33			23,05
			22,32	22,32	23,12	23,05

When the several operations are performed with perfect accuracy, the sum of the northings will be equal to that of the southings, and the sum of the eastings to that of the westings. This necessarily follows from the circumstance of the surveyor's returning to the place from which he set out; and it affords a means of judging of the correctness of the work. But it is not to be expected that the measurements and calculations in ordinary surveying will strictly bear this test. If there is only a small difference, as in the above example, between the northings and southings, or between the eastings and westings, it may be imputed to slight imperfections in the measurements.

Here the northings and southings agree, but the eastings are a little greater than the westings; we will therefore decrease the eastings by half the error, and increase the westings by the same amount; the sums will then agree.

We do this without any formal statement, but the operation is strictly that of proportion; the greater the line the greater the correction to be applied.

When the errors are considerable, a re-survey should be made, and if the errors are still great, and in the same direction, there is reason to suspect that some local attraction disturbs the free action of the needle; and then, if the importance demands it, a survey can be taken *without* the compass, by methods we shall explain in some following chapter.

We shall make use of this example and this figure to illustrate the

TAKING OF ANGLES BY THE COMPASS.

For this, and for several other operations in practical mathematics, the learner must not expect a written rule: original principles are far more simple and reliable.

We now require the angle ABC; conceive the AB to be produced, then the angle between BC and the produced part, is 83° less 23°, or 60°. Now 60° taken from 180° gives 120° for the angle ABC.

Again the line BC makes an angle with the meridian toward the north of 83°, therefore toward the south it must be 97° on the east side of it. The line BA makes an angle of 23° with the meridian on the west side of it; therefore, the angle $ABC = 97+23 = 120$, the same as before.

To find the angle BCD, we add 83° and 14°. Why?

To find the angle CDA, we subtract 14° from 77°. Why?

To find the angle DAB, we add 14° to 77°.

The sum of these 4 angles must equal 4 right angles.

Suppose now that the surveyor runs the lines AB, BC, CD, and then wishes

TO CLOSE THE SURVEY.

To close a survey is to run the last side so as to strike the first point, when we are not able to see it.

To accomplish this, we sum up the latitude and departure as far as the point D, the result will show Dd and dA. Having then two sides of the right angled triangle, the angle dAD will be the bearing for $dAD = nDA$, because nD and NS are parallel. The side DA can also be computed, but it should be measured also, as a test to the accuracy of the whole survey. If, on running DA, according to computation, we actually strike the point A, or very near to it, and there is little or no difference between actual measure and computation, then we may be sure that all the sides have been run correctly; but if on running DA, we do not strike A, or the distance does not correspond to computation, we may be sure of errors somewhere — either in want of skill or care in the operation, or the action of the compass has not been uniform at all the angular points. In case of material errors a re-survey should be made.

In case that we have no means of making computations in the field, we may take a course as near the true one as our judgment will permit, and run it. This line must bring us near the point of beginning, if it does not strike it, and when we get opposite to that point we must measure to it at right angles from the line run; then we shall have data to correct our course. Running a line thus by guess work, is called *running a random line*, from which the true line can be found as follows:

Suppose that when we arrive at D, we judge the course to the first point to be *N.* 75° *W.*, and run that course, and after measuring 28.65 chains we find that we are passing the first point, which is 83 links in perpendicular distance toward the south; what course should have been taken?

By the following investigation we draw out a rule that will apply to *all such cases.*

Let AB represent a true course, AD a *random line*, and DB its amount of deviation.

Also let DAE equal one degree, and take $Ad=1$, then the deviation at d will be the natural sine of one degree, and may be taken from the table of natural sines (which is .01745).

By proportional triangle we have

$$1 : .01745 : : AD : DE;$$

Whence $$DE = 0.01745(AD).$$

Now DE is contained in DB as often as 1° is contained in the number of degrees in the angle DAB. Let x represent the number of degrees in DAB, then

$$\frac{x}{1}=\frac{(DB)}{.01745(AD)}.$$

That is, $x=\frac{(DB)(57.3)}{(AD)}$, because $\frac{1}{.01745}=57.3$.

Hence, to correct a course, we have the following

RULE.— *Multiply the deviation by* 57.3, *and divide that product by the distance, aud the quotient will be the number of degrees and parts of a degree to add to, or subtract from, the random course.*

This rule, applied to the present example, gives

$$\frac{.83(57.3)}{23.65}=2^\circ.$$

Hence, the true course is $75^\circ+2^\circ=77^\circ$.

Had the deviation of the *random line* been toward the south, we should have subtracted the correction; but for this the operator must rely on his judgment.

We now come to the

COMPUTATION OF AREAS.

By inspecting the figure we perceive that $cCDd$ is a *trapezoid,* from which, if we subtract the triangles ADd, ABb, and the *trapezoid* $bBCc$, the area of the field $ABCD$ will be left.

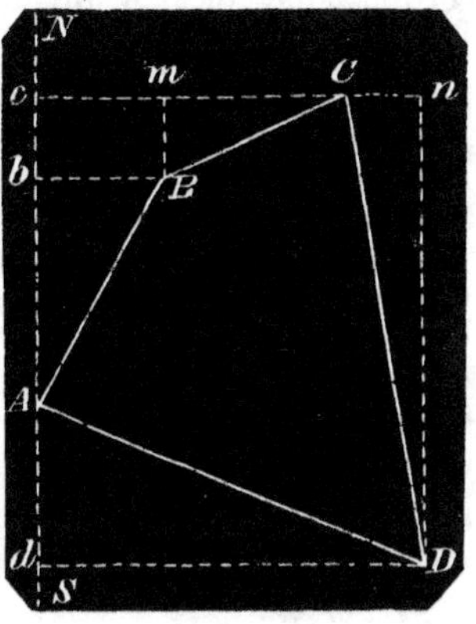

OBSERVATION.— To preserve uniformity of expression, and clearness and brevity in forming a rule, we shall *call triangles, trapezoids,* while discussing this subject. A trapezoid becomes a triangle, when its smallest parallel side is so small as to call it zero — conversely, then, a triangle is a trapezoid, whose smallest parallel side is zero.

We observe that C is the most northern point of the field, and D is the most southern. In traversing from C to D, from the north to the south, we pass along the oblique sides of trapezoids that we shall call *south areas,* and in traversing from D to A, B, and C, from the south to the north, we pass along the oblique sides of trapezoids, which we shall call *north areas*. Now it is obvious *that if we subtract the sum of the north areas from the sum of the south areas, we shall have a remainder equal to the area of the field.*

We now require a systematic method of finding these *areas*, or the area of these several trapezoids. In the first place, we must have latitudes and meridian distances.

Latitude and departure have already been defined and explained.

MERIDIAN DISTANCES.

Meridian distances are the distances of the angular points of the field from the meridian which runs through the most westerly point of the field; thus, bB, cC, dD, are meridian distances.

Double meridian distances are the bases of triangles, or the sum of the parallel sides of the trapezoids.

Thus bB is the *double* meridian distance of the side AB, or it is the double meridian distance of the middle point of the line AB. $bB + cC$ is the *double* meridian distance of the line BC, or double the meridian distance of the middle point of BC. This double meridian distance $(bB+cC)$ multiplied by bC, or Bm, the latitude corresponding to BC, will give the double area of the trapezoid bB, Cc.

We are now prepared to give the following summary or rule for finding the area of any field bounded by any number of right lines:

RULE. — 1. *Prepare a table headed as in the example, namely: Bearings, Distance, North, South, East, West, Meridian distance, Double meridian distance, North areas, South areas.*

2. *Begin at the most western point of the field, and conceive a meridian to pass through that point.*

Find, by the traverse table or by trigonometry, the northings, southings, eastings, and westings of the several sides of the field, and set them in the table opposite their respective stations, under their proper letters N., S., E., or W.

3. *For the first meridian distance take the departure of the first line; for the second, take the first meridian distance and add to it the departure of the second line, if the departure is east, or subtract if west, &c.*

4. *Add each two adjacent meridian distances, and set their sum opposite the last of the two in the column of double meridian distances.*

5. *Multiplg each double meridian distance by the latitude to which it is opposite, and set the product in the column of N. areas, if the latitude is north, and in that of S. areas, if the latitude is south.*

6. *Subtract the sum of the N. areas from that of the S. areas, and take half the remainder, which will be the area of the field in square chains. Dividing this by* 10 *gives the acres; and the roods and rods are found by multiplying the decimal parts by* 4 *and by* 40.

	Bearings.	Dis.	N.	S.	E.	W.	M. D.	D. M. D.	N. areas.	S. areas.
AB	*N* 23° *E*	17	15.65		6.64 (6.63)		6.63	6.63	103.76	
BC	*N* 83° *E*	11	1.34		10.92 (10.90)		17.53	24.16	32.37	
CD	*S* 14° *E*	23		22.32	5.56 (5.55)		23.08	40.61		906.42
DA	*N* 77° *W*	23.66	5.33			23.05 (23.08)	0.00	23.08	123.02	
						23.08			259.15	906.42
										259.15
									Diff.,	647.27
									Half,	323.63
									Dividing by 10,	32.363

Hence the field contains 323 square chains and 63 hundredths, or thirty two acres and a little more than 36 hundredths of an acre.

The numbers in parentheses, as (6.63), and all others in parentheses are the numbers corrected to make the eastings and westings agree, — the numbers above them are taken from the table.

Before we give any more examples, it is proper to give some examples to show the practical utility of the

TRAVERSE TABLE.

This table is computed to every half degree, but if a course is between two courses in the table, the operator can use his judgment and take out the proper intermediate numbers.

Those who are not satisfied with this method, can use the table of natural sines and cosines, as we shall subsequently explain.

The distances are consecutive to 30, then 35, 40, &c., to 100. But a little thought in the operator will enable him to use the table for *any distance* whatever.

The following examples will illustrate.

1. A course is *N.* 22° 30′ *E.* distance 62.43; what is the corresponding latitude and departure, as found in the table?

We shall regard the distance as 6243 links, and separate it into parts.

		Lat.	Dep.
Thus:	6000	5543	2296
	240	221.7	91.8
	3	2.77	1.15
	6243	5767.47	2388.95

If we now return to chains and links, the latitude is 57.67 and the departure 23.89.

We entered 240 in the table, as 24 chains, and took the numbers corresponding.

2. A course is *N.* 48° 30′ *W.*, distance 187.61; what is the corresponding latitude and departure?

Dis.	Lat.	Dep.
180.00	119.30	134.80
7.60	5.036	5.692
1	066	075
187.61	124.3426	139.4995

If we now take the distance as 187 chains and 61 links, the Lat. is 124 ch. 34 links, and the Dep. is 139 ch. and 50 links.

3. A course is *S.* 81° *W.*, distance 76.87; what is the corresponding latitude and departure?

Dis.	Lat.	Dep.
75.00	1173.	7408.
1.80	28.2	177.8
7	1.10	6.91
76.87	1202.3	7592.71

If 76 ch. 87 lin., Lat. 12 ch. 2 lin., Dep. 75 ch. 93 links.

Thus we can find the latitude and departure for any distance corresponding to any degree and half degree.

We can find it to any degree and minute of a degree by the table of natural sines and cosines.

The common tables containing natural cosines and sines are nothing more than latitude and departure corresponding to unity of distance.

Therefore, a double distance will correspond to a double distance in latitude and departure, a treble distance will give a treble amount of latitude and departure, and so on in proportion. Lat. = Nat. cosine. Dep. = Nat. sine.

Hence: *The natural cosine of any course taken as a decimal, multiplied by any distance, will give the latitude corresponding to that course and distance.* Also, *the natural sine taken as a decimal, multiplied by a given distance, will give the departure corresponding to that course and distance.*

N. B. Nat. sines and cosines are found in table II., pages 21–65 of tables. For common purposes, four places of decimals are sufficient.

1. The bearing of a certain line is *N.* 35° 18′ *E.*; distance 12 chains; what is the corresponding latitude and departure?

Angle 35° 18′	N. cos. .81614	N. sin.	.57786
Dis. (multiplier)	12		12
Diff. Lat=	9.79368	Dep.	6.93432

2. A certain line runs *S.* 4° 50′ *E.*; distance 74.40; what is the corresponding latitude and departure?

Angle 4° 50′	N. cos. .9964	N. sin.	.0842
Distance	74.4		74.4
	39856		3368
	39856		3368
	69748		5894
Lat.	74.13216	Dep.	6.26448

3. A line makes an angle with the meridian of 75° 41′, at a distance of 89.75 chains; what is the latitude and departure?

75° 47′	cos.	.2456	sin.	.9694
Distance		89.75		89.75
Prod. Diff. Lat.		22.042	Dep.	87.001

4. A line bearing *N.* 7° 40′ *W.*; distance 31.20 chains; required the difference of latitude and departure.

7° 40′	cos.	.98106	sin.	.13341
Multiplier		31.2		31.2
Diff. Lat.		30.92	Dep.	4.16

5. A line running *S.* 80° 10′ *E.* distance 35.25 chains; what is the difference of latitude and departure?

80° 10′	cos.	.17078	sin.	.9853
Multiplier		35.25		35.25
Diff. Lat.		6.02	Dep.	34.72

In the last three examples, we have given only the results of the multiplications to two places of decimals; that is, to the nearest link, which is a degree of accuracy sufficient for all practical purposes.

We are now prepared to estimate the areas of the following general surveys, given as

EXAMPLES.

1. In May, 1845, the following measures of a field were taken. Beginning at the *western-most* point of the field; thence *N.* 20° 30′ *E.* 5 chains 83 links; thence *S.* 79° 45′ *E.* 10 chains 15 links; thence *S.* 27° 30′ *W.* 9 chains 45 links; thence *N.* 63° 15′ *W.* 8 chains 28 links; thence *N.* 15° 30′ *W.* 1 chain and 4 links, to the place of beginning; required the area.

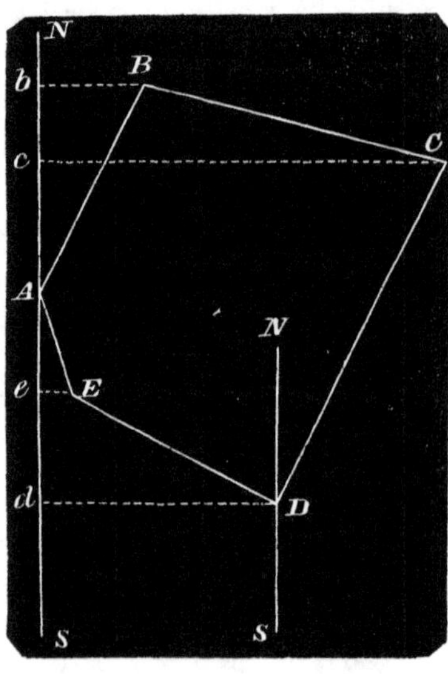

It is not absolutely necessary to make a plot or figure of the field, but for the sake of perspicuity, it is best to do so; yet no reliance is placed on the accuracy of the constructed figure.

We perceive by the figure, that there are two *south areas*, *bBCc*, and *cCDd*; and three *north areas*, *eEdD*, *AeE*, and *ABb*.

Let the reader observe that all the *north areas* are on the outside of the field.

	Bearings.	Dis.	N.	S.	E.	W.	M. D.	D.M.D.	N. areas.	S. areas.
AB	*N.*20° 30′*E.*	5.83	5.46		2.04		2.04	2.04	11.1384	
BC	*S.*79° 45′ *E.*	10.15		1.81	9.99		12.03	14.07		25.4667
CD	*S.*27° 30′W.	9.45		8.38		4.36	7.67	19.70		165.0860
DE	*N.*63°15′W.	8.28	3.73			7.39	0.28	7.95	29.6535	
EA	*N.*15°30′W.	1.04	1.00			0.28	0.00	0.28	0.2300	
			10.19	10.19	12.03	12.03			41.0719	190.5527

41.0719

2)149.4808

10)74.7204

Area in acres, 7.472

The operator can rely on the rule used in the last operation, whatever be the number of sides, or whatever be the shape of the figure, provided that the lines are right lines, from one angular point to another.

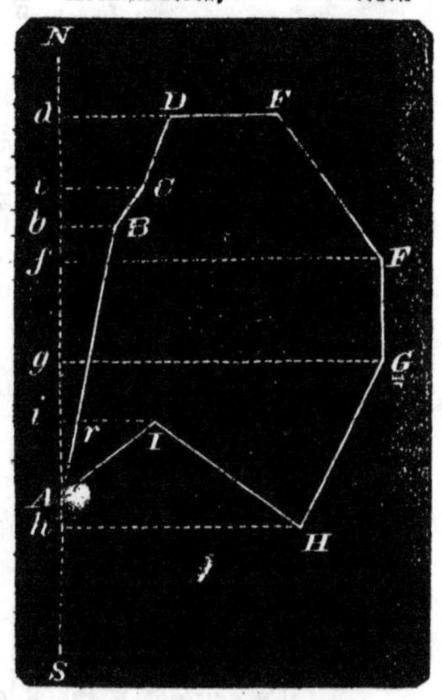

In case of *re-entering* angles, like *HIA*, represented in the adjoining figure, a portion of the figure *AIr*, is reckoned twice. But this is corrected by the subtractive space *HIih*, which includes not only the exterior portion *hHIA*, but also the whole additive triangle *AIi*, belonging to the last side of the figure *IA*.

In the following example are several such re-entering angles.

2. Find the area of a lot of land, of which the following are the field notes.

Beginning at the south west corner, the ancient land mark; thence

1. *N.* 27° 15′ *E.* distance 9.42 chains.
2. *S.* 80° 00′ *E.* " 1.15 "
3. *S.* 69° 00′ *E.* " 12.73 "
4. *S.* 15° 45′ *W.* " 5.00 "
5. *N.* 66° 45′ *W.* " 1.05 "
6. *S.* 31° 00′ *W.* " 2.90 "
7. *N.* 70° 45′ *W.* " 8.92 "
8. *S.* 41° 45′ *W.* " 2.08 "
9. *N.* 63° 00′ *W.* " 4.18 "

We can contract the operation in reference to the space it will occupy, by putting the difference of latitude in one column, and all the departures in another column.

Marking all the northings by the sign +, and all the southings by the sign —. Also, all the eastings by the sign +, and all the westings by the sign —.

This being understood, the work will appear as follows :

	Bearings.	Dis.	Lat.	Dep.	M. D.	D.M.D.	N. areas	S. areas
1	*N.* 27° 15′ *E.*	9.42	+8.37	+ 4.31	4.31	4.31	36.0747	
2	*S.* 80° 00′ *E.*	1.15	—0.20	+ 1.13	5.44	9.75		1.9500
3	*S.* 69° 00′ *E.*	12.73	—4.57	+11.89	17.33	22.77		104.0589
4	*S.* 15° 45′ *W.*	5.00	—4.81	— 1.36	15.97	33.30		160.1730
5	*N.* 66° 45′ *W.*	1.05	+0.41	— 0.96	15.01	30.98	12.7018	
6	*S.* 31° 00′ *W.*	2.90	—2.49	— 1.49	13.52	28.53		71.0397
7	*N.* 70° 45′ *W.*	8.92	+2.94	— 8.42	5.10	18.62	54.7428	
8	*S.* 41° 45′ *W.*	2.08	—1.55	— 1.38	3.72	8.82		13.6710
9	*N.* 63° 00′ *W.*	4.18	+1.90	— 3.72	0.00	3.72	7.0680	
		Sum	0.00	0.00			110.5873	350.8926
								110.5873
								2)240.3053
								10)120.1526
					12 acres, and a small fraction over.			12.01526

3. Having the following field notes, it is required to find the closing side and the area of the field.

	Bearings.	Dis.	N.	S.	E.	W.
1	*S.* 75° *W.*	13.70		3.54		13.24
2	*S.* 20° 30′ *W.*	10.30		9.65		3.60
3	*W.*	16.20				16.20
4	*N.* 33° 30′ *E.*	35.30	29.51		19.49	
5	*N.* 76° *E.*	16.00	3.87		15.52	
6	South	9.00		9.00		
			33.38	22.19	35.01	33.04
			22.19		33.04	
			11.19		1.97	

This result shows, that if we commence at the first station, and traverse round to the sixth, we shall then be 11.19 chains to the north of the place of beginning, and 1.97 chains east of it. This is sufficient data to compute the course and distance.

To compute the area, however, it is not necessary to find either the course or the distance.

We do know, however, by merely inspecting the traverse table, that the course to the place of beginning, is south about 10° 20′ west, and distance near 12 chains.

We are now prepared to compute the area, and as we wish to commence at the western-most point of the field, we shall begin at the 4th station, calling it the first: thus,

	Bearings.	Dis.	Lat.	Dep.	M. D.	D.M.D.	N. area.	S. area.
1	*N.* 33° 30′ *E.*	35.30	+29.51	+19.49	19.49	19.49	575.1499	
2	*N.* 76° *E.*	16.00	+ 3.87	+15.52	35.01	54.50	210.9150	
3	South	9.00	— 9.00	0. 0	35.01	70.02		630.1800
4	*S. W.*	——	—11.19	— 1.97	33.04	68.05		761.4795
5	*S.* 75° *W.*	13.70	— 3.54	—13.24	19.80	42.84		151.6536
6	*S.* 20° 30′ *W.*	10.30	— 9 65	— 3.60	16.20	36.00		347.4000
7	*W.*	16.20	0. 0	—16.20	0. 0	16.20		
—							786.0649	1890.7131
								786.0649
							2)	1104.6482
							10)	552.3241
								55.23241

This result shows that the field contains 55 acres, and a little more than 23 hundredths of an acre.

4. What is the area of a survey, of which the following are the field notes.

Stations.	Bearings.	Distances.
1	*S.* 46° 30′ *E.*	80 rods.
2	*S.* 51° 45′ *W.*	34.16
3	West.	85.00
4	*N.* 56° *W.*	110.40
5	*N.* 33° 15′ *E.*	75.20
6	*S.* 74° 30′ *E.*	123.80

Ans. 104.35 acres.

5. Required the contents and plot of a piece of land, of which the following are the field notes.

Stations.	Bearings.	Distances.
1	*S.* 34° *W.*	3.95 ch.
2	*S.*	4.60
3	*S.* 36½°*E.*	8.14
4	*N.* 59½°*E.*	3.72
5	*N.* 25° *E.*	6.24
6	*N.* 16° *E.*	3.50
7	*N.* 65° *W.*	8.20

Ans. 10*A.* 0*R.* 5*P.*

6. Required the contents and plot of a piece of land, from the following field notes.

Stations.	Bearings.	Distances.
1	*S.* 40° *W.*	70 rods.
2	*N.* 45° *E.*	89
3	*N.* 36° *E.*	125
4	*N.*	54
5	*S.* 81° *E.*	186
6	*S.* 8° *W.*	137
7	*W.*	130

Ans. 207*A.* 3*R.* 33*P.*

7. Given the following bearings and distances of the several sides of a field, namely,

1.	*N.* 58° *E.*	19 ch.
2.	*E.* 6° *S.*	20
3.	*S.* 17° *W.*	20
4.	*W.*	20
5.	*N.* 42° 35′ *W.*	15.10

to find the area.

Ans. 54.9 acres.

8. Given the following bearings and distances, namely,

1.	*N.* 45° *E.*	40 ch.
2.	*S.* 30° *W.*	25
3.	*S.* 5° *E.*	36
4.	*W*	29.60
5.	*N.* 20° *E.*	31

to find the corrected difference of latitude and departure, and the area.

N. B.—In this last example, as in most others, the northings and southings will not exactly balance ; nor will the eastings and westings balance. This arises from inaccuracies in the data. In such cases (if the errors are but trifling) we balance off the errors.

When a course is east or west, as the 4th in this example, some operators have expressed doubts, whether any correction should be applied to latitude in that course.

We reply, that errors do really exist, and, therefore, we cannot say that the course marked west, in the example, was really west, or not; the probability is, that the course was not exactly due west, and it is therefore proper to put a correction in the latitude column, as shown in the following results:

CORRECTED LATITUDES AND DEPARTURES.

	N.	S.	E.	W.
1.	28.30		28.30	
2.		21.63		12.49
3.		35.84	3.16	
4.	0.02			29.59
5.	29.15		10.62	
	57.47	57.47	42.08	42.08

On inspecting these latitudes and departures, we perceive station 5 is the most westerly point of the field, therefore to find the area, we will arrange these results in the following order:

	Lat.	Dep.	M. D.	D. M. D.	N. areas.	S. areas.
5	+29.15	+10.62	10.62	10.62	309.5730	
1	+28.30	+28.30	38.92	49.54	1400.9820	
2	—21.63	—12.49	26.43	65.35		1403.5205
3	—35.84	+ 3.16	29.59	56.02		2007.7568
4	+00.02	—29.59	0. 0	29.59	0.5918	
					1711.1468	3410.2773
						1711.1468
						2)1699.1305
						10)849.5652
					Ans. areas,	84.956

9. What is the area of a survey of which the following are the field notes.

From the place of beginning, *N.* 31° 30′ *W.*, distance 10 chains: thence *N.* 62° 45′ *E.*, 9.25 chains: thence *S.* 36° *E.*, 7.60 chains: thence *S.* 45° 30′ *W.*, 10.60 chains, to the place of beginning.

Ans. $8\frac{54}{100}$ acres.

10. Do the following bearings and distances enclose a space? If not, give an additional bearing and distance that will, then determine the area so enclosed.

Stations.	Bearings.	Distances.
1	*S.* 40° 30′ *E.*	31.80 ch.
2	*N.* 54° 00′ *E.*	2.08
3	*N.* 29° 15′ *E.*	2.21
4	*N.* 28° 45′ *E.*	35.35
5	*N.* 57° 00′ *W.*	21.10

Ans. These bearings and distances *do not* enclose a space. A line run from the further extremity of the 5th to the first station will bear south 46° 43′ *W.*, distance 31.21 chains, and the area thus enclosed will contain 92.9 acres.

11. Do the following bearings and distances enclose a space? If not, determine the additional line that will, and the area of the space so enclosed.

Stations.	Bearings.	Distances.
1	*S.* 85° 00′ *W.*	46.4 rods.
2	*N.* 53° 30′ *W.*	46.4 "
*3	*N.* 36° 30′ *E.*	76.8 "
4	*N.* 22° 00′ *E.*	56.0 "
5	*S.* 76° 30′ *E.*	48.0 "

Ans. These bearings and distances do not enclose a space. A line run from the last station to the first would bear *S.* 30° 25′ *W.*, distance 128.6 rods. Area 56.86 acres nearly.

The operation for the area is as follows:

We commence at station 3, for reasons that have been several times explained. Station 6 is the one we supplied.

Sta.	Lat.	Dep.	M. D.	D. M. D.	N. area.	S. area.
3	+ 61.73	+45.65	45.65	45.65	2817.9745	
4	+ 51.92	+20.98	66.63	112.28	5829.5776	
5	— 11.21	+46.67	113.30	179.93		2017.0153
*6	—125.18	—29.85	83.45	196.75		24629.1650
1	— 4.85	—46.15	37.30	120.75		585.6375
2	+ 27.59	—37.30	0. 0	37.30	1029.1070	
					9676.6590	27231.8178
						9676.6590
					2)	17555.1588

160)8777.5794(54.86.

* The stations marked with a * are those supplied.

12. What is the area of a survey of which the following are the field notes.

Stations.	Bearings.	Distances.
1	*N.* 75° 00′ *E.*	54.8 rods.
2	*N.* 20° 30′ *E.*	41.2 "
3	*E.*	64.8 "
4	*S.* 33° 30′ *W.*	141.2 "
5	*S.* 76° 00′ *W.*	64.0 "
6	*N.*	36.0 "
7	*S.* 84° 00′ *W.*	46.4 "
8	*N.* 53° 15′ *W.*	46.4 "
9	*N.* 36° 45′ *E.*	76.8 "
10	*N.* 22° 30′ *E.*	56.0 "
11	*S.* 76° 45′ *E.*	48.0 "
12	*S.* 15° 00′ *W.*	43.4 "
13	*S.* 16° 45′ *W.*	40.5 "

In this survey 4 is the most easterly and 8 the most westerly station. The area is equal to 110*A.* 2*R.* 23*P.* It may vary a little, on the account of the way in which the balancing is done.

CHAPTER III.

ON THE MERIDIAN LINE AND THE VARIATION OF THE COMPASS.

THE meridian is an astronomical line, having no necessary connection with the magnetic needle. Meridians would be primary lines to which we would refer all surveys, if there were no such thing as magnetism, or a magnetic needle.

It is only a coincidence, that the magnetic needle settles near the meridian, so near, that for a long time it was considered the meridian itself, but accurate observations have shown that the needle does not point rigorously north and south, but has a *variation* which is not the same at all times in the same place, therefore a line run by the compass is still unknown in respect to the meridian, unless we know the variation of the compass, that is, the declination of the needle.

In the year 1657 the needle at London pointed due north, since that time its variation has been west, previous to that time the variation was east.

In the Atlantic ocean, between Europe and the United States, the variation is from 12° to 18° west.

The needle seems to point to the region of greatest cold, which is in the northern part of America, and not the north pole, and if this be true, if the point of *minimum* heat changes its position, there will be a corresponding change in the direction of the magnetic needle.

The needle has a small annual and also a diurnal variation, corresponding to the temperature of the different seasons of the year, and of the different times of the day, but these variations are too small to trouble the common operations of surveying.

Some of the variations of the compass are regular, others irregular; some amount to many degrees and require a long period of time, others are small in amount and require but short intervals to pass through all their changes.

The daily variation consists of an oscillation eastward and westward of the mean position, and is different in different places. Generally the greatest oscillation eastward is between six and nine in the morning, and westward about one in the afternoon, gradually returning toward the east until eight P. M. At night it is stationary.

On the subject of magnetism we know nothing, beyond facts drawn from observation, but there is no doubt that the earth is a great magnet, made so by the action of the sun, and the poles of this great magnet are near the poles of the equator. Indeed, all observations made correspond to this hypothesis, for changes of the weather, clouds, and storms, all have an influence on the needle.

These facts are sufficient to convince any reader that to survey correctly we must know the

VARIATION OF THE COMPASS.

As the true meridian is an astronomical line, we must find it by astronomical observations, and then by comparing the meridian of the compass with it, we shall have the variation of the compass.

When the sun is on the equator, it rises due east, and sets directly in the west. Should we then observe the direction of its center, just as it was rising or setting, at the time it had no declination, and trace that line a short distance on the ground, we should then have a due east and west line.

If from any point in that line we draw another line at right angles, we should then have the true meridian.

If we now put the compass on this meridian, and make the sight-vanes range with it, the needle will also range with it, if there is no variation, but if the north point of the needle is to the west of the sight-vane, the variation is westerly, if to the east, easterly, and the number of degrees and parts of a degree that the needle deviates from the direction of the sight-vanes shows the amount of the variation.

But it is not to be supposed that any particular observer can be at the points and places, where the sun is either rising or setting just at the time the sun is on the equator. We must have a broader basis, and in fact by means of the latitude of the observer and the declination of the sun, any observer has the means of knowing the precise direction in which the sun will rise or set, any day in any year.

Let us suppose that the sun on a certain day, observed from a certain place, must have arisen *S.* 81° *E.*, but by the compass it was observed to rise *S.* 79° *E.*, the variation of the compass was therefore 2° west.

These observations are called taking an azimuth. Azimuths are often taken at sea to determine the variation of the compass.

On land, however, the horizon is rarely visible, and very few observations on sun rise or sun set can be made, besides there are other objections arising from atmospherical refraction; it is therefore best, most convenient, and more conducive to accuracy, to take the sun when up 10, 15 or 25° above the horizon, and observe its direction per compass, and compare the result to the computed bearing for the same moment, and if the two results agree the compass has no variation; if they disagree the amount of such disagreement is the amount of the variation of the compass.

By means of spherical trigonometry the true bearing of the sun can be determined at any time, on the supposition that the observer knows his latitude, the declination of the sun, and its altitude ; these three conditions furnish a triangle like PZS.

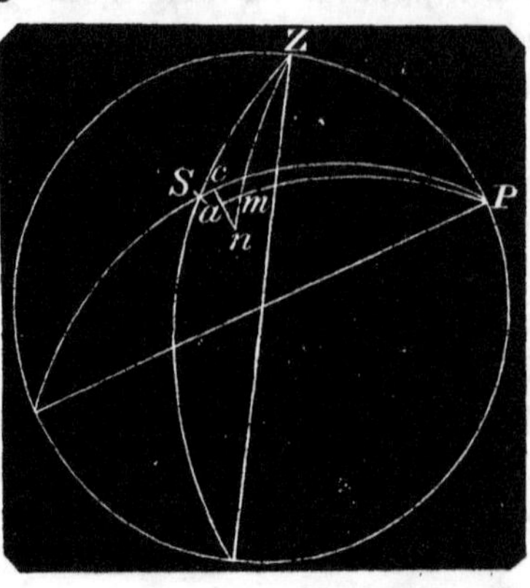

The altitude subtracted from 90°, gives ZS, the latitude from 90°, gives ZP, and PS is found by adding or subtracting the sun's declination to 90°, according as it is north or south.

ZS is co-altitude, ZP is co-latitude, and PS is the sun's polar distance ; the angle PZS is required, and it can be found by the following rule.

1. *Add the three sides of the triangle together and take the half sum. From the half sum, subtract the sun's polar distance, thus finding the remainder.*

2. *Add the sin-complement of the co-altitude, the sin-complement of the co-latitude, the sine of the half sum, and the sine of the remainder. The sum of these four logarithms divided by 2, will be the cosine of half the azimuth angle.*

N. B. This rule is the application of equations on page 204 Robinson's Geometry. The sin-complement is the logarithmic sine of an arc, subtracted from 10.

EXAMPLES.

In latitude 39° 6′ 20″ north, when the sun's declination was 12° 3′ 10″ north, the true altitude of the sun's center was observed to be 30° 10′ 40″, *rising*. What was the true bearing of the sun, or its azimuth ?

	90°		90		90
Lat.	39 6 20	Alt.	30. 10. 40	Dec.	12° 3′ 10
co-Lat.	50 53 40	co-Alt.	59. 49. 20	*PD*	77. 56 50

P. D.	77° 56. 50		
co-Lat.	50. 53. 40	sin. com.	0.110146
co-Alt.	59. 49. 20	sin. com.	0.063295
	2)188 39 50		
½*S.*	94 19 55	sin.	9.997758
	77 56 50		
Rem.	16 23 5	sin.	9.450376
			2)19.622575
	49° 38′ 30″	cosine	9.811287

Bearing, 99° 17′ 0 from the north, or 80° 43′ from the south.

If at the time of taking the altitude of the sun, another observer had taken its bearing by the compass, and found it to be *S.* 80° 43′ *E.*, then the compass would have no variation, and whatever it differed from that would be the amount of variation.

If a line were run along the ground, direct toward the center of the sun, at the time the altitude was taken, and sufficiently marked, that would be a standing line of known direction; and if from any point in that line, we could draw another line, making an angle with it of 99° 17′ on the north, or 80° 43′ on the south, such a line definitely marked, would be a permanent meridian line, for all time to come; on which we could at any time place a compass, and observe its variation.

Let *AS* be the line toward the sun, along the ground, *AE* a line due east, and *Mm* a true meridian line. The angle *SAE* must equal 9° 17′.

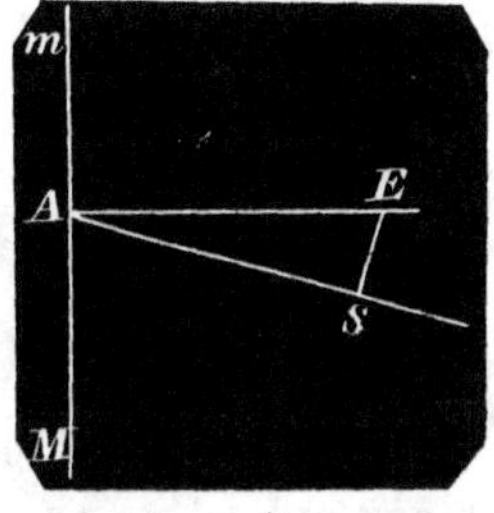

To make that angle, take *AS*, one chain or 100 links; from *S*, draw the line *SE* at right angles to *AS*, by means of a surveyor's cross.*

From *S* take *SE*, of such a value as will make *SAE* 9° 17′, which is determined by trigonometry; as follows,

* Surveyor's cross is nothing more than a pair of sight vanes, set at right angles with each other, for the purpose of making right angles.

As 100. : SE=Rad. : tan. 9° 17′

Whence $SE=\frac{100. \ (\text{tan. } 9^\circ \ 17')}{R.}=$ 9.213405
2

SE=16.347 links. 1.213405

That is, from S measure off 16 and a little more than $\frac{1}{3}$ of a link, and there is the point E. A line drawn from A to E, is a due east and west line.

If we put the surveyor's cross on this line, at any point as A, and range one branch of it along the line AE, the other branch will mark out the line Mm, a true meridian, if everything has been done to accuracy.

In the afternoon, or some other day, another meridian may in like manner be drawn near this one, and if they are both true meridians, they will be parallel. If not parallel, other observations should be made until some two or three are obtained, that are parallel or very nearly so ; and the mean direction then, may be regarded as the true meridian.

A true meridian will always be a test line for a compass ; and by placing any compass upon it, the declination* of the needle can be determined.

Again. In the triangle PZS, if we compute the angle ZPS (as is done on page 211 Robinson's Geometry), we shall have the sun's distance from the meridian or the apparent time ; then if we have a time piece that can be relied upon, for three or four hours, we can determine the time within a few seconds, when the sun will be on the meridian. A line at that time, run direct toward the center of the sun, will define the meridian.

The objections to these methods are,

1. The sun is a large body, and its center cannot be exactly defined.

2. The sun changes position so rapidly that, unless we are in an observatory, where every thing is prepared and in order, it is difficult to get observation upon it.

---

* Declination of the needle, in common language, is called the variation of the compass ; and, as a general thing, we adhere to common language.

3. The sun is so bright an object that it cannot be viewed without prepared glasses.

4. The majority of persons that have been, and probably will be practical surveyors, have not the instruments to take altitudes of the sun, and they are not and cannot be at home in astronomical observations and computations.

Some of these objections are deserving of little respect, and others can be partially removed.

For instance, if the sun is too large, and too brilliant to be accurately and deliberately observed, we can take the planet Venus, Jupiter, or Saturn, and, by proper observations, determine their directions, during the twilight of evening, when we can see the planet distinctly, and at the same time that other objects are sufficiently distinct to run lines.*

But the method most known and most in favor among practical men, is that of taking the direction of the north star.

The north star is a star of the second magnitude (Polaris), whose right ascension, Jan. 1, 1851, was 1h 5m 18s (at present increasing at the rate of 17s 71 per annum), and declination was then 88° 30′ 55″, with an annual increase of 19″8, it is therefore, but 1° 29′ 5″ from the pole, and it is called the pole star or north star because it is so near the pole.

If the star were situated directly at the polar point, a line toward it would be the true meridian line, but being 1° 29′ 5″ distant, the star apparently makes a circle round the pole in a siderial day, making two transits across the meridian, one above and the other below the pole, — a direction to it, at these times, would be a true meridian line.

To find these times, *subtract the right ascension of the sun from the right ascension of the star;* increasing the latter by 24h, to render the subtraction possible, when necessary.

---

* For example, in the year 1853, from the 25th of July to the 5th of August, the planet Jupiter will pass the meridian in the evening twilight. On the first of August, Jupiter will pass the meridian of New York, at 8h 11m 53s, and it will pass the meridian of Cincinnati, at 8h 11 39s, mean local time ; and, of course, whoever is able to designate that time within a few seconds, and is also prepared to mark the direction of the planet, will have a true meridian line.

The moon is not a good object for this purpose; it changes its place too rapidly.

The difference will be the time of the upper transit, and 11h and 59 minutes from that time will be the time of the lower transit. The right ascension of the sun is to be found in the Nautical Almanacs, for every day in the year; and it is nearly the same, for the same day, in every year.

For example. At what times will the north star make its transits over the meridian on the first day of July, 1853.

	H.	M.	S.
✶ *R. A.*+24h - - - -	25	6	0
⊙ *R. A.* - - - - - - -	6	41	16
	18	24	44

This result shows that the upper transit will occur about 6h 24m, in the morning of the 2d of July. I say *about*, because I took the sun's right ascension for the morning of July 1, and from that time to 6, next morning, is 18 hours: and during this time the right ascension of the sun will increase full 3 minutes,—therefore the upper transit will take place 6h 21m in the morning, and the previous lower transit 11h 59m previous, or at 6h 22m, evening.

But neither of these transits will be visible, as they both occur in broad day light, from any place where the north star is ever distinctly visible.

In summer, then, when most surveying is done, the meridian transits of the north star are not visible, nor is this important: for the transits are seldom used, by reason of two objections:

1. The star changes its direction most rapidly while passing the meridian.

2. Observers, generally, have not the means of knowing the time to sufficient accuracy.*

To obviate these objections, observations may be taken on the star at its greatest elongations; for, about those points and for full

* Note.— Very few persons consider that their clocks and watches, however good and valuable, do not give the exact time, but only approximations to the time.

For any astronomical purpose, like the one under investigation, the character of the time piece should be well tested — its rate of motion known — and its errors established by astronomical observations.

15 minutes before and after, the star does not scarcely change its direction; hence the observer has a sufficient interval to be deliberate, and he can be sufficiently exact as to time without any extra trouble.

The following tables show the times of the greatest eastern and western elongations, which occur in the night season. These tables are not perpetual, but they will serve without correction for 20 years or more to come.

EASTERN ELONGATIONS.

Days.	April.	May.	June.	July.	August.	Sept.
	H. M.	H. M.	H. M.	H. M.	H. M.	H. M.
1	18 18	16 26	14 24	12 20	10 16	8 20
7	17 56	16 03	14 00	11 55	9 53	7 58
13	17 34	15 40	13 35	11 31	9 30	7 36
19	17 12	15 17	13 10	11 07	9 08	7 15
25	16 49	14 53	12 45	10 43	8 45	6 53

WESTERN ELONGATIONS.

Days.	Oct.	Nov.	Dec.	Jan.	Feb.	March.
	H. M.	H. M.	H. M.	H. M.	H. M.	H. M.
1	18 18	16 22	14 19	12 02	9 50	8 01
7	17 56	15 59	13 53	11 36	9 26	7 38
13	17 34	15 35	13 27	11 10	9 02	7 16
19	17 12	15 10	13 00	10 44	8 39	6 54
25	16 49	14 45	12 34	10 18	8 16	6 33

It will be observed that these times are astronomical; the day commencing at noon, and 12h 40 means 40m after midnight, etc.

Now, admitting that the direction of the star can be observed, the next step is to find how much that direction deviates from the meridian — and this is a problem in spherical trigonometry.

A great circle passing through the zenith of the observer to the star, when the star is at one of its greatest elongations will touch the apparent small circle made by the apparent revolution of the star about the pole, and will therefore, with the star's polar distance, form a right angle — and we shall have a right angled spherical triangle, of which the observer's co-latitude is the hypotenuse, the star's polar distance one side, and the angle opposite to this side is the angle required.

EXAMPLE.

What will be the bearing of the north star observed from latitude 42° *N.* in the year 1860, when the star's polar distance will be 1° 26′ 12″? Ans. 1° 56′.

As cos. Lat. 42° - - - -	9.871073
is to radius - - - - - -	10.000000
So is sin. 1° 26′ 12″ - - - -	8.399183
To sin. 1° 56′ - - - - -	8.628110

In this manner the following table was computed. The mean angle only is put down, being computed for the first of July in each year.

AZIMUTH TABLE.

Years.	Lat. 30° Azimuth.	Lat. 35° Azimuth.	Lat. 40° Azimuth.	Lat. 45° Azimuth.	Lat. 50° Azimuth.
1852	1° 42′ 30″	1° 48′ 21″	1° 55′ 52″	2° 5′ 32″	2° 18′ 5″
1854	1° 41′ 45″	1° 47′ 39″	1° 55′ 2″	2° 4′ 30″	2° 17′ 6″
1856	1° 41′ 2″	1° 46′ 49″	1° 54′ 12″	2° 3′ 44″	2° 16′ 9″
1858	1° 40′ 27″	1° 46′ 11″	1° 53′ 30″	2° 3′ 2″	2° 15′ 12″
1860	1° 39′ 43″	1° 45′ 24″	1° 52′ 32″	2° 2′ 4′	2° 14′ 16″
1862	1° 38′ 50″	1° 44′ 29″	1° 51′ 44″	1° 1′ 2″	2° 13′ 18″

This table is given for those who may wish to use it, but we would recommend each observer to follow the example which precedes the table, and compute the azimuth corresponding to his latitude and time.

THE PRACTICAL DIFFICULTY.

The north star is not brilliant, it cannot be seen until it is so dark that all minute terrestrial objects are totally invisible, it is therefore difficult to draw a line and accurately mark it. All these night operations are, at best, perplexing and inaccurate, yet, by the means of lights and artificers, lines can be drawn.

If the observer have a theodolite and an assistant, there will be no difficulty. Let them be at the place from which they wish to take the observation in time, adjust the instrument and direct the telescope to the north star. Now sufficient light must be reflected into the telescope to enable the observer to see the cross hairs, and this may be done by the assistant holding a light before a stiff sheet of white paper, so as to throw the reflected light from the paper into the telescope, or this may be done by means of a stand to hold both the light and the paper.

When the vertical spider's line becomes visible, let the star be brought directly upon it, and if it is near the time of greatest elongation it will appear to remain so, for some time. But if the star has not reached it greatest elongation, it will move from the line more to the east, if the elongation is easterly, and more to the west, if westerly.

The telescope must be continually directed to the star, by means of the tangent screw of the horizontal plate, but for some time the spider line and star will coincide without moving the screw, and then the star will depart from the line in the contrary direction to its former motion, *but the telescope must no longer follow the star*, its position will now show the direction to the star, when the star had its greatest elongation, *and thus it should be left until morning.*

In the morning, carefully range and mark a line through the telescope.

If we now make an angle with this line equal to the azimuth, by means of the theodolite, or by means of measuring a triangle as explained in the former part of this chapter, and mark this new line either to the right or left, as the case may require, we shall then have a permanent meridian line for all future use.

By placing a compass on any well defined and true meridian we can determine its variation by simple observation.

If we have not a theodolite, we can obtain a tolerably accurate direction to the north star by means of illuminated plumb lines suspended in vessels of water, so placed as to range to it.

CHAPTER IV.

TO SURVEY WITHOUT A COMPASS.

The inquiry is sometimes made, whether lands could be surveyed without a compass; we reply in the affirmative. The compass is only a convenience, and if it had never been discovered, it is probable that surveys would have been more accurately made. Too much reliance has been placed on the accuracy of the compass, and in consequence little attention has been paid to defining any astronomical lines.

Were it not for the compass, it is probable, that every country-town, and even every large land holder, would have meridian lines well defined about his premises.

Having a meridian line to start upon, we can find angles and define the position of lines very accurately by means of a

CIRCUMFERENTOR.

The circumferentor consists of a horizontal circular plate divided into 360 degrees, over which an index bar, or another circular plate, is made to revolve. This index bar carries sight-vanes or a telescope. The index bar or the revolving circular plate also carries a vernier scale, which will enable the operator to make an angle to one minute of a degree. The whole instrument is placed on a tripod, and by the aid of spirit levels attached to the lower plate, the horizontal position is attained with a sufficient degree of accuracy.

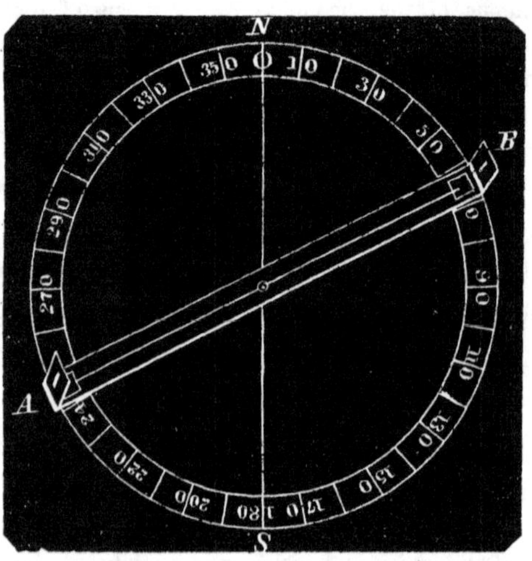

The figure before us, represents the essential parts of a circumferentor. *NS* is considered as the primative or meridian line, and *AB* is the index bar, which turns horizontally on the common center.

Vernier scales are fitted into the index bar, and revolve over the graduated arc. At *A* and *B* are openings, to receive cross hairs or a telescope.

When a vertical semicircle is made to revolve vertically through the plane AB, and the diameter of that circle a telescope, then we have all the essentials of a *theodolite*.

To most theodolites, a magnetic needle is attached, but the magnetic needle is, properly speaking, no part of the instrument.

To show the manner of finding the direction between two given points, by means of the circumferentor, we propose the following problem.

Mr. T. H. Jones wishes me to run the east line of his lot, in the town of A, and give the true bearing, the corners being known. In the public square of the town, about one and a quarter miles distant, a meridian line has been established.

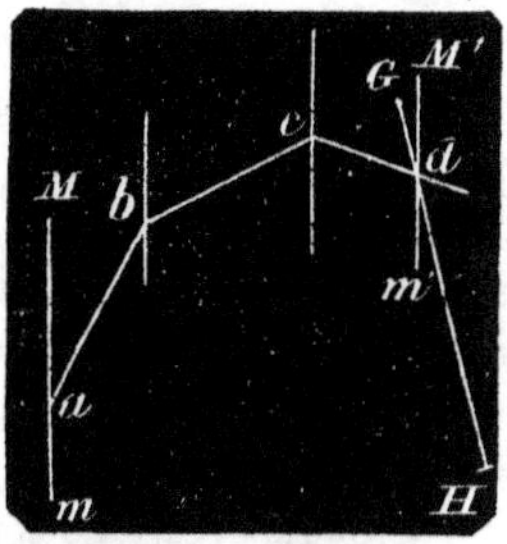

Let Mm be the established meridian in the public square, and GH the direction of the line required.

Place the circumferentor on the meridian line Mm, so that NS of the instrument will coincide with it, the center of the instrument being at a in a road.

The general direction of the road is $a\,b$, and the index bar AB is made to revolve over the plate, which is firmly fixed, until the index bar or sight vanes point out the line ab. The line is run by means of ranging objects; such as flag staffs, if the line is long; or if short, by sending on a flag, and stationing it at b.

Now clasp the index bar on the plate, by the clamp screw under it (made for the purpose). Leave a flag at a; take the instrument to b, and there place it, so that the sight vanes will range back to a; then the position of NS on the instrument will show a meridian line through that point.

Here the road bends a little, unclamp the index (being careful that NS rigidly retains its position), and direct it to the general direction of the road bc. Mark the point c, by an object as before, and mark some other point, so as to secure the line (the other point may or may not be b). Now clamp the index again, and remove the instrument to c. Place the instrument firmly as before, and make the index range along the line bc; the line NS of the instrument, will mark out a meridian line at the point *c*, and thus we can transfer the meridian line *Mm* to any other point whatever.

Thus we may go to any point *d*, in the given line; no matter, theoretically speaking, how many angles we have made during the traverse. Placing the instrument at *d*, with its index to range along the last line, the line *NS* of the instrument gives the meridian *M'm'*.

Now unclamp the instrument, and direct its index along the required line *GH*; the position of the index, on the graduated plate, will give the angle from the north, which, by means of the vernier, can be determined with great exactness.

In this manner we may go to any point, and place a meridian there, and then run any required line whatever; therefore, we can survey any field, farm, or tract of land, without a compass, if we have a circumferentor, and a meridian line.

It would not be safe to transfer meridians, as we have just done, over any very great extent of country, for at every angle, small errors might be made, and the accumulation of many small errors may produce too great inaccuracies to be tolerated or overlooked. When using the magnetic needle, no errors accumulate, for every setting of the compass is primary, and independent of every other.

Therefore, in case no compasses were in existence, primary meridians, astronomically established, would be necessary in every town; and it would be better to have several of them in the same town.

From the foregoing illustrations, we perceive that surveying can be done, and well done, without a compass, yet the compass is an inestimable blessing to mankind; for it is the only index to direction over the wild waste of waters, when the heavens are obscured, and no mariner would dare brave the ocean without it.

CHAPTER V.

ORIGINAL AND SUBSEQUENT SURVEYS.— DIFFICULTIES AND DUTIES OF A SURVEYOR.

In this country, lands were ceded to States, or sold to companies in large tracts, without any definite surveys; the boundaries described, were mountains, rivers, or a certain number of miles along the shores of a lake, and then a certain number of miles back.

The land companies hired surveyors from time to time, to survey off their lands, into lots of 100, 200, and 500 acres; and wherever these surveyors left monuments for the corner of lots, established the corners for all time to come, whether *correctly placed or not.*

These surveys were very loose and inaccurate; it could not be otherwise, for a company of surveyors would frequently run 15 miles in a day; when to run a line accurately, and measure it, four miles is a good day's work.

But, notwithstanding inaccuracies, these surveys are legal and cannot be changed; "thou shalt not move thy neighbor's ancient land mark," and it is right it should be so; for any attempt at correction, would create more trouble, confusion, and injustice, than it could remedy.

Lots originally sold for 100 acres in the state of New York, generally contain from 101 to 106 acres, in consequence of the original surveyors having directions to *have their lots hold out.* Where the lots thus overrun in one portion of the tract, they fall short on another, for the surveyors were probably desirous to show to the company, that their grant actually contained as much land as was anticipated.

The author surveyed one of these lots, that originally sold for 100 acres, and found that it contained but a little over 76 acres.

Some of the companies had their grants laid off into townships 6 miles square, or 6 by 8 miles; then each township into four sections, each section divided off into lots, and the lots numbered, generally beginning at the south-west corner.

The description of the lots in the deeds given, were very loose and indefinite, stating the township, section, and number of the lot, containing 100 acres, *"be the same more or less,"* and in some lots it was more, and in other lots it was less.

As we before remarked, any land mark to the corner of a lot laid down by these original surveyors, must remain; subsequent surveyors can straighten lines between point and point, and decide what the true courses are, and how many acres the lot contains.

When a surveyor is called to survey any farm or estate that has been previously surveyed, he must find some corner as a place of commencing, and from thence run a *random* line, as near the true line as his judgment permits ; and if he strikes another corner he has run the true course, if not, he corrects his course, as taught in chapter II. Thus, he must go round the field from corner to corner. He has a right to *establish corners* only where no corners are to be found, and no evidence can be obtained as to the existence and locality of a former land mark.

It may be the case, that a surveyor is called to survey a lot where no corners are to be found. If a fence or line exists, which has been the undisputed boundary for a long time, that boundary cannot be changed, and the surveyor must establish a corner by ranging some other line to meet the first. Sometimes corners may be found to some neighboring lot, from which lines can be run, to establish a corner to the lot we wish to survey.

Lines of lots in the same town, are generally parallel, and a surveyor who offers his services to the public, must make himself acquainted with the general directions of the lines of lots, over that section of country where his services are required.

When a surveyor is called to divide a piece of land, he is then an original surveyor, and not liable to be embarrassed by old lines and old traditions, he has then only his mathematical problem before him.

Owing to the inaccuracies of original surveys, and the impossibility of leaving proper land marks, in consequence of the great haste in which lands were originally surveyed ; great confusion has followed, in some sections of our country, in respect to lines, and it has been no uncommon thing to have whole neighborhoods at variance, if not in law, in reference to the boundaries of their lands.

In cases of this kind, one, and then another of the disaffected, have successively employed surveyors, and surveyors thus employed, are apt to act the part of advocates, rather than arbitrators, and survey too much according to the direction of their employer ; but all such efforts to settle difficulties, but aggravate them more and more.

On the contrary, however, if the surveyor clearly understands his duties, and can rise above being a special advocate for any one of the parties concerned, he can do more than judges or juries to restore harmony and peace. To illustrate these views, and possibly to give some valuable instruction to some readers, we give a history of a case of this kind, which occured in the year 1837, in the county of Ontario, in the State of New York. A tract of land consisting of about 670 acres, of an irregular shape, was divided *on paper* into five equal parts, and sold to five different individuals.

The whole 670 acres was bounded by four lines, no two of them were equal, and neither of the angles was a right angle. The largest boundary line could not be directly measured on account of an impassable ravine; and the banks of this ravine was so thickly set with hemlocks, that it was impossible even to sight across.

In consequence of the irregular shape of the whole, and the impossibility of directly measuring the principal boundary, they had never been able to agree on their division lines.

Each one imagined that his neighbor was inclined to crowd upon him, and although permanent fences were desirable, none could be made until lines were agreed upon. They had employed several surveyors, but they had not been able to agree on their divisions. In this state of things a surveyor was called upon to go and make a division of this land, but the difficulties of so doing were carefully concealed from him.

When he arrived on the ground, ready for operations, the whole neighborhood was present, and by unmistakable signs he soon learned that an unusual degree of interest was taken in the survey.

He also found that the chief difficulty arose from not being able to measure the line *CD*. All the corners, *A*, *B*, *C*, and *D*, were established. The surveyor commenced at *C* to run a *random* line as near *CD* as possible. After going a few chains, he came to the bank of the ravine at *F*, where it was impossible to pass or sight across. Driving a stake at *F*, he took a direction *FH* along the bank of the ravine, carefully noting the angle, and measuring the line to *H*, a point where objects were clearly to be seen on the other side of the ravine. The surveyor then sent a man over with a flag, stationing his staff, first at *K*,

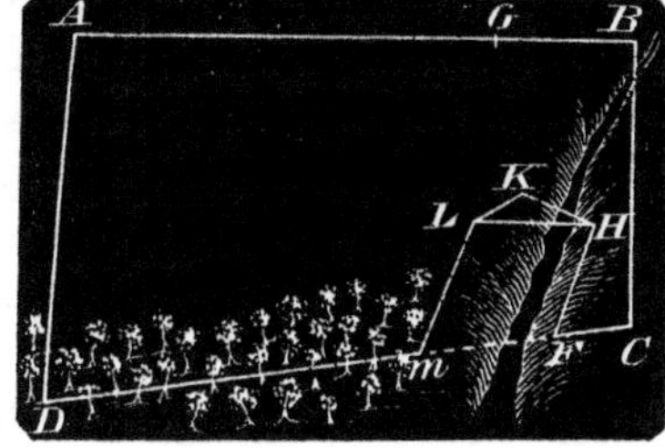

then at *L*, carefully noting the direction of each, and being careful to have the angle *KHL* greater than 30°. He then passed over and set the compass at *K*, took the direction of *KL* and measured it. Having now *KL* one side, and all the angles of the triangle *HKL*, he computed *HL*. He now set the compass at *L* and took a definite direction *Lm*; this definite direction gave him the angle *HLm*, and he now had all the angles of the quadrilateral *LmFH*, and two of its sides. Whence he computed the exact distance to *m*, to strike the line *CF* produced.

He measured that distance and drove a stake at *m*, and computed *mF*. The company now ran through the random line, driving stakes at the end of every third chain, and the random line came out within a few feet of the established corner at *D*. The surveyor measured the perpendicular distance to *D*, and corrected the course by the rule in chapter II. He also computed how far each stake that had been placed on the random line must be moved to transfer it to the true line; this, the reader will perceive, was done by proportional triangles.

At *D* he set the compass, and carefully noted the course and distance to *A*. He then returned to *C*, *taking care not to pass along the line AB*.

At *C* he set the compass, and carefully noted the course and distance to *B*. He now computed the course and distance from *B* to *A*. The line lay in the open fields, over tolerably smooth ground, and it could be directly and accurately measured.

The surveyor now called all the parties interested, including the sour and the belligerent, and told them that the distance from *B* to *A* was a certain number of chains and links, and that they would now measure it, and if they found it to correspond without any material error, they must then be convinced that he had obtained the true length of *CD*, and that he could then divide the land into five equal parts, as required.

To this test they all cheerfully assented; the line was measured and corresponded to the computation within three links; all parties were satisfied, and thus ended a neighborhood quarrel of six years' standing.

Previous surveyors commenced at the point *D* and run *DA*, *AB*, and BC, and then computed CD. This was more direct, simple, and proper, than the method just described, but it left no test behind it, and it is vain to expect that the mass of men will receive theoretical computation as actual measurement.

Here, and in most other cases that involve contention, the surveyor must not only convince himself that the survey is correctly made, but he must, if possible, show others that his conclusions are not only right, but cannot be wrong; hence judicious surveyors must often measure lines, where there is no *mathematical necessity* for so doing.

The next duty of this surveyor was to divide the land into five equal parts. Each one had previously purchased his part, and he knew its locality, but not his exact boundary line on the division.

As CD was not parallel to AB, and AD not exactly parallel to CB, to divide this *mathematically exact* was a problem of considerable difficulty, and this will be explained in the next chapter; but practically we need not apply all mathematical rigor, the surveyor can divide this more strictly conformable to justice without, than with the mathematical rigor.

The persons who had the two most eastern lots, had the worthless part of the land in the ravine, and of course if any one had an excess of *area* it should be those.

To find where or nearly where the divisions come, divide the line AB into five parts and suppose G one of those parts. Now BG is not quite long enough, because the field is a little narrower at this end than at the other; the surveyor took a distance BG a few links greater than one fifth of AB, and from that point run a line in a medium direction between BC and AD, and then computed its area, the result would show whether the area was too great or too small, and if it were within a very small fraction of the area required, the line is left as the true one, otherwise it is moved as the case requires. In the same manner the other division lines were run.

UNITED STATES' LANDS.

Soon after the organization of the present government, several of the States ceded to the United States large tracts of unoccupied land, and these, with other lands, since acquired by treaty and purchase, constitute what is called the public lands.

Previous to 1802, there was no general plan for surveying the public lands, or in fact, no surveys were made, and when grants were made the titles often conflicted with each other, and in some cases different grants covered the same premises.

In the year 1802, Colonel I. Mansfield, then Surveyor General of the north-western territory, adopted the following method:

Through the middle, or about the middle of the tract to be surveyed, a meridian is to be run, called the *principal meridian.* At right angles to this, and near the middle of it, an east and west line is to be run, and called the *principal parallel.*

Other meridians are to be run, six miles distant from the principal meridian, both east and west.

Also, parallels of latitude are to be run, six miles from the principal parallel, both north and south.

When this was done (and it has been on all the public lands east of the Mississippi river), the whole country is divided into squares, six miles on a side, called *townships.*

Each township contains 36 square miles. Each square mile is called a section, and it contains 640 acres. Sections are divided into half sections, quarter sections, and eighths. But these divisions are only made on paper.

When a person makes a purchase of a half or quarter section, it is supposed that he will find it himself, or employ a surveyor to mark it out.

Townships which lie along a meridian, are called *a range,* and numbered to distinguish them from each other.

Sections are regularly numbered in every township, and to designate any particular one, we say, section 13, in township number 4 north, in range 3 east.

This shows that the third range of townships east of the principal meridian, in township No. 4 north of the principal parallel, is the township, and the thirteenth section of this township is the one sought.

Not more than ten townships north or south of a principal parallel should be drawn, before a new principal parallel should be designated, and new measures made between meridians: *because* meridians tend toward the pole, and the north lines of townships will be theoretically shorter than south lines, if the meridians are run by the compass.

Where the public lands extend to rivers and lakes, there will be fractional townships along the shores.

Where the locality of a particular number is found to be occupied by a lake or pond, the sale is void.

CHAPTER VI.

METHODS OF SURVEYING IRREGULAR FIGURES AND OF DIVIDING LANDS.

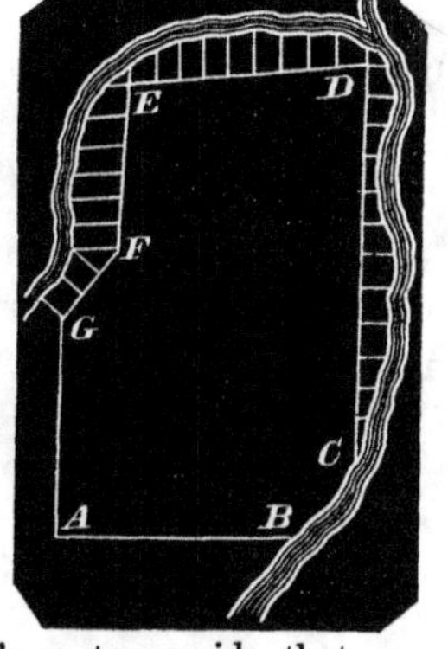

Farms and tracts of lands, wholly or partially bounded by water, as represented in the figure before us, are surveyed and there areas determined by drawing right lines within the tract as near the real boundaries as possible, and from these right lines, at *equal intervals*, measuring the *off-sets* to the real boundary. These off-sets form the parallel sides of trapezoids, and as they are all equally distant from each other, the computation of the areas they occupy will be very easy. A summary rule for finding the united area of all these trapezoids that are bounded by one line, is to be found in Prob. VIII, Mensuration. The area of the right lined figure *ABCDEFG*, is found as directed in Chapter IV, to which add the area of all the trapezoids, and we shall have the area of the whole.

We have now investigated every possible case of computing areas, and we are now prepared to divide them. Commencing with the most simple case of the most simple figure, the triangle, or rather the figure that has the least number of sides.

PROBLEM I.

To divide a triangle into two parts, having a given ratio of m to n.

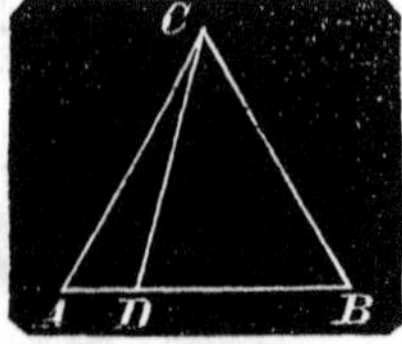

CASE 1. By a line drawn from one angle to its opposite side.

Let ABC represent the triangle; divide its base into two parts, corresponding to the given ratio, and let AD be one of the parts; then we shall have the following proportion.

$$AD : AB :: m : m+n$$

Whence, $AD=\frac{m}{n+m}(AB)$ and $BD=\frac{n}{m+n}(AB)$

Now the two parts are numerically known, and are to each other as m to n. Triangles, having the the same altitudes, are to one another as their bases. Therefore, $ADC : CDB :: m : n$ as required.

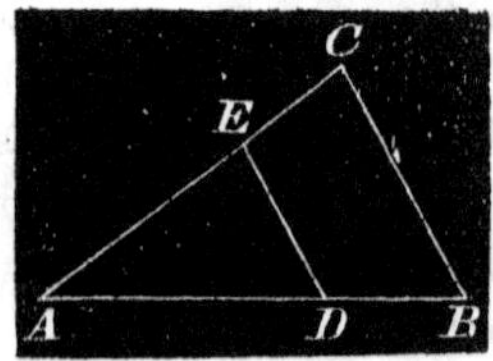

CASE 2. By a line parallel to one of its sides.

Let DE divide the triangle as required, and as similar triangles are to one another as the squares of their homologous sides, therefore:

$$(AB)^2 : (AD)^2 :: m+n : n$$

Whence, $AD=AB\sqrt{\frac{m}{m+n}}$

Which shows that if we have the numerical value of AB, and of n and m, we can find that of AD, and from D draw DE parallel to BC, and the triangle is divided as required.

CASE 3. By a line parallel to a given line, or by a line running in a given direction.

To make this case clear, we commence by giving a definite example:

There is a triangular piece of land, from one of the angular points, A, one line runs N. 25° W., distance 12 chains; another from the same point runs N. 42° E., distance 15 chains. It is required to divide this triangle into two parts in the ratio 2 to 3, by a line running due east and west.

Let ABC be the given triangle, and $B'C'$ the required division line. It is required to find the numerical value of AC' or AB', to make the area $AB'C'$ $\frac{2}{5}$ of the area ABC.

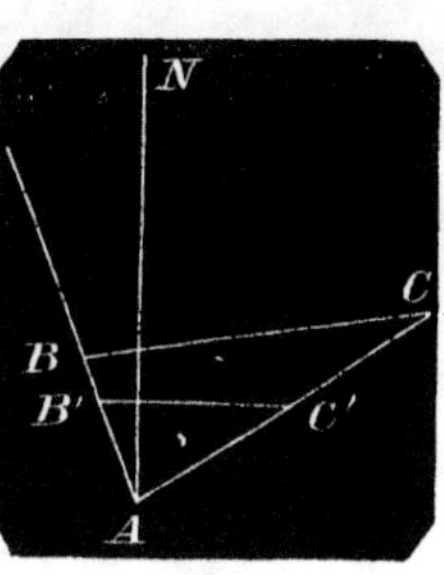

Let b represent the side of the triangle opposite B, and c the side opposite C.

Let $AC'=x$. As AC' and, $C'B'$ have definite directions, the angle $AC'B'$ is given, also $AB'C'$ is given. $AC'B'=48°$, $AB'C'=65°$, $BAC=67°$.

In the triangle $AB'C'$ we have

$$\sin. 65° : x :: \sin. 48° : AB$$

$$\text{Whence } AB'=\frac{\sin. 48°}{\sin. 65°}x \qquad (1)$$

Now, by Prob. III, Mens., area $ABC=\frac{1}{2}bc \sin. A$.

Also, " " " area $AB'C'=\frac{1}{2}\left(\frac{\sin. 48°}{\sin. 65°}\right)x^2 \sin. A$

By the conditions of our problem, we have the following proportion.

$$\tfrac{1}{2}bc \sin. A : \tfrac{1}{2}\left(\frac{\sin. 48°}{\sin. 65°}\right)x^2 \sin. A :: 5 : 2$$

$$\text{Or,} \qquad bc : \frac{\sin. 48°}{\sin. 65°}x^2 :: 5 : 2 \qquad (2)$$

We may here stop, and make the problem general.

If $B'C'$ is given in direction, the angles B' and C' will be given.

We now require the division of the triangle ABC into two parts, in the ratio of m to n by a line opposite to the angle A, running in a given direction.

Represent the sides of the given triangle adjacent the angle A by b and c, b extending from A to B, and c extending from A to C.

Put $x=$ the distance from A to the division line, on the side CA.

Then, by the preceding proportion we have,

$$bc : \frac{\sin. C'}{\sin. B'}x^2 :: m+n : m$$

Whence,
$$x=\left(\frac{m}{m+n}\right)^{\frac{1}{2}}\left(\frac{bc \sin. B'}{\sin. C'}\right)^{\frac{1}{2}}$$

Observe that x is opposite the angle B', the sine of which stands in the numerator of the second fraction. Had x represented AB', sin C' would have been the numerator.

Drawing out the result for (2), we find that

$$\frac{5 \sin 48^\circ}{\sin 65^\circ}x=2bc=2.15.12$$

Or,
$$x=\sqrt{\frac{72 \sin 65^\circ}{\sin 48^\circ}}=9.446 \text{ chains.}$$

CASE 4. By a line that shall pass through a given point within the triangle.

A point in a triangle cannot be given, unless the perpendicular distances from that point to the sides are given, and if these perpendicular distances are given, then we can readily find the three distances from the angular points of the triangle, and the angles which these lines make with the sides of the triangle are known. For instance, if the point P, in the triangle ABC, is known, PR and PT are known, and all the angles of the quadrilateral $ATPR$ are known. These are sufficient data to compute the line AP, and the angles RAP and TAP.

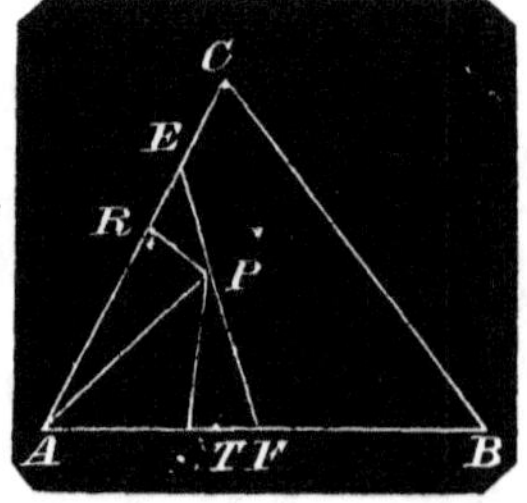

REMARK.—We may now require a triangle to be cut off, by a line running through P, which shall contain any definite portion of the triangle ABC, *not involving an impossibility.*

For instance, if the point P is near the center of the triangle, it would not do to require us to cut off a tenth part of the triangle, or any smaller portion, for it would be impossible to do so. When it is required to cut off a very small portion of the whole triangle, the point P must be near one of the sides, or near one of the angular points. Sometimes the required quantity can be cut off from one angular point, sometimes from another, and sometimes from all three.

Let us now require one-third of the triangle cut off, by a line passing through

***P*, taking the angular point *A*, and let *EF* be that line.** *We are to determine the value of AF.*

In the triangle ABC, the angles A, B, and C, are known, and the sides opposite to them, a, b and c, are also known. AP is known, and call it h. Put the angle $EAP=p$, $PAF=q$. Then, $A=p+q$. Put $AF=x$, $AE=y$.

By Prob. III. Mens. area $ABC=\frac{1}{2}bc \sin. A$
Also " " area $AEF=\frac{1}{2}xy \sin. A$
By the conditions of the problem,

$$\tfrac{3}{2}xy \sin. A=\tfrac{1}{2}bc \sin. A$$

Whence,

$$3xy=bc \qquad (1)$$

The triangle AFE consists of two parts, AFP, APE; therefore,

$$\tfrac{1}{2}hx \sin. q+\tfrac{1}{2}hy \sin. p=\tfrac{1}{2}xy \sin. A$$

Or

$$x \sin. q+y \sin. p=\frac{xy \sin. A}{h} \qquad (2)$$

If we had required the nth part of the triangle ABC, in place of the 3rd part, equation (1) would have been $nxy=bc$.

Making this supposition, to make the problem more general, we have $xy=\frac{bc}{n}$ and $y=\frac{bc}{nx}$. By the aid of these last two equations, (2) becomes

$$x \sin. q+\frac{bc \sin. p}{nx}=\frac{bc \sin. A}{np}$$

Or,

$$x^2-\frac{bc \sin. A}{nh \sin. q}x=-\frac{bc \sin. p}{n \sin. q} \qquad (3)$$

Whence,

$$x=\frac{bc \sin. A}{2nh \sin. q}\pm\left(\frac{b^2c^2 \sin.^2 A}{4n^2h^2 \sin.^2 q}-\frac{bc \sin. p}{n \sin. q}\right)^{\frac{1}{2}}$$

In case n is *large*, that is the part to be cut off *small*, the value of x may be imaginary, corresponding to the preceding remark.

Case 5. When the given point is on one side of the triangle.

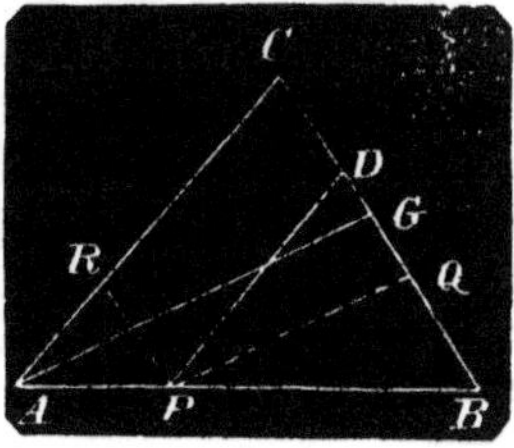

The two parts must be equal, or one of them will be less than half of the whole.

We always compute the less part. Let P be the point in the line AB, PQ and PR perpendiculars to the other sides, are known; BC and AG are both known. Now, through the given point P, it is required to draw PD, so that the triangle BPD, shall be the nth part of ABC. That is

$$PQ.\,BD=\frac{BC.\,AG}{n}$$

Whence, $$BD=\frac{BC.\,AG}{n\,PQ}$$

CASE. 6. When the given point is without the triangle.

Let ABC be the given triangle, and P any given point without it. It is required to run a line from P, to cut off a given portion of the triangle ABC, or (which is the same thing) to divide the triangle into two parts having the ratio of m to n. Let PG be the line required.

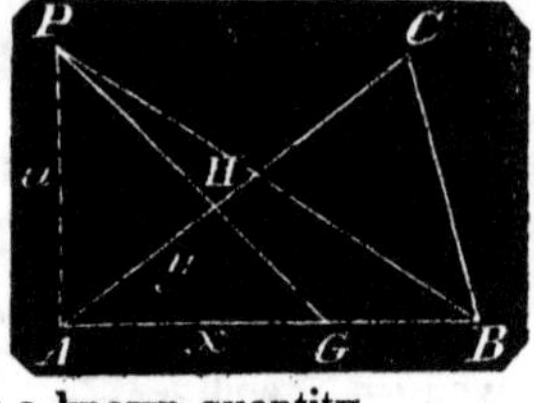

As P is a given point, AP is a line given in distance and position; therefore, the angle PAH is known.

Solution.—Put the angle $PAH=u$, $CAB=v$; then $PAG=(u+v)$. Also put $AG=x$, $AH=y$, $AP=a$, and the area of the triangle $AHG=mc$, mc being a known quantity.

Now, (by Prob. III. Mens.) $\frac{1}{2}xy\ \sin.\ v=mc$ (1)

Also " " $\frac{1}{2}ax\ \sin.\ (u+v)=\text{area}\ APG$

And " " $\frac{1}{2}ay\ \sin.\ u=\text{area}\ PAH$

Therefore, $\frac{1}{2}ax\ \sin.\ (u+v)-\frac{1}{2}ay\ \sin.\ u=mc$ (2)

Or, $$x\ \sin.\ (u+v)-y\ \sin.\ u=\frac{2\,mc}{a} \quad (3)$$

From (1), we find $y=\frac{2\,mc}{x\ \sin.\ v}$ which value substituted in (3) gives

$$x\ \sin.\ (u+v)-\frac{2\,mc\ \sin.\ u}{x\ \sin.\ v}=\frac{2\,mc}{a}$$

Whence, $$x^2\ \sin.\ (u+v)-\frac{2\,mc}{a}x=\frac{2\,mc\ \sin.\ u}{\sin.\ v}$$

Or, $$x^2-\frac{2\,mc}{a\ \sin.\ (u+v)}\ x=\frac{2\,mc\ \sin.\ u}{\sin.\ v\ \sin.\ (u+v)}$$

Therefore, $$x=\frac{mc}{a\ \sin.\ (u+v)}\pm\sqrt{\frac{m^2c^2}{a^2\sin.^2(u+v)}+\frac{2\,mc\ \sin.\ u}{\sin.\ v\ \sin.(u+v)}}$$

EXAMPLES.

1. *In the triangle ABC, the side* $AB=23.645$ *chains,* $AC=17.51$ *chains, and* $BC=12.575$ *chains.*

The given point P from the angle A, is distant 10 *chains, at an angle of* 40° *from the line AC.*

It is required to draw a line from this given point P, through the triangle, so as to divide it into two equal parts.

Whereabouts on AB will PG intersect?

The angle $BAC=31°\ 17'\ 19''=v$. $PAH=40°=u$. Therefore $PAG=71°\ 17'\ 10''=(u+v.)$

The area of the triangle ABC is 107.52 square chains. The part to be cut off by the triangle AHG is therefore $=53.76=m$.

We must use the natural sines, or the logarithmic sines if we omit them in the index.

mc	log. - - -	1.730464
$a\sin.(u+v)$ log. - - -		0.976406
$\dfrac{mc}{a\sin.(u+v)}=5.676$	log.	0.754058
		2
$\dfrac{m^2c^2}{a^2\sin.^2(u+v)}=32.22$	log.	1.508116
$2mc$	log. - - -	2.031494
$\sin. u$	log. - -	—1.808067
$2mc\sin. u$ - - - -		1.839561
$\sin. v\sin.(u+v)$ - -		—1.691866
$\dfrac{2mc\sin. u}{\sin. v\sin.(u+v)}=702.56$	log.	2.147695

The part of the formula under the radical is therefore (32.22+702.56) or 734.78

Whence $x=5.676\pm\sqrt{734.78}=32.7829$ or —21.43.

REMARK. — In geometry plus and minus generally indicate opposite directions, here the sign $\pm$ means, sum and difference of the two numbers, the sum is 32.7829, the difference is 21.43 ; in fact there can be no such thing as minus a line

The value of AG is 21.43 ; the other value is not admissible.

2. *We have a right angled triangle whose base is* 48.87 *chains, and perpendicular* 54.46 *chains. From a given point without it, we are required to run the center of a straight road, to leave one third of the triangle on one side and two thirds on the other.*

From the acute angle at the base, the distance to the given point is 20 *chains, and the line to it makes an angle with the hypotenuse of* 30°.

Remark. — If P is a given point, its distance and direction from one of the angular points must be given; and if the distance and direction from one of the angular points is given, the distances and directions from all of them are virtually given; thus, if we have AP, AC, and the angle PAC, we have CP, and the angle ACP, and we may theorize on the triangles PCH, CHG' as well as on APH and AHG.

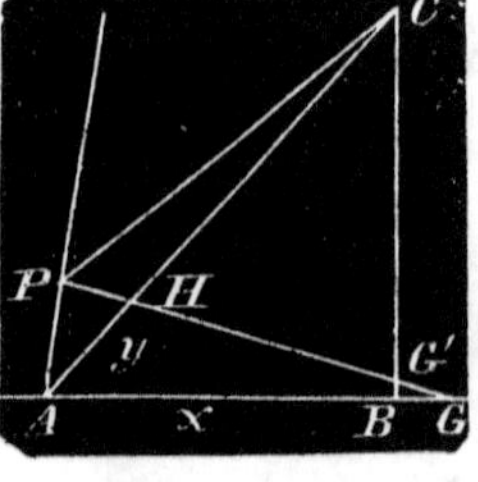

The area of the triangle ABC=1304.94 square chains.

One third of this is mc= 434.97.

$BAC=v=49° 12' 20''$. $PAC=u=30°$. $(u+v)=79° 12' 20''$.
$AP=a=20$. $AH=y$. $AG=x$.

mc log. - - - -			2.638459
$a=20$ log. -	1.301030		1.293268
sin. $(u+v)$ log.	—1.992238		
$\frac{mc}{a \sin. (u+v)}$	=22.142	log.	1.345191
			2
$\frac{m^2 c^2}{a^2 \sin.^2 (u+v)}$	=490.34	log.	2.690382
$2mc$ log. - - - -			2.939489
sin. u - - - - -			—1.698970
$2mc$ sin. u - - - - -			2.638459
sin. v sin.$(u+v)$ - - - -			—1.871339
$\frac{2mc \sin. u}{\sin. v \sin. (u+v)}$	=584.98	log.	2.767120

Whence, $x=22.142 \pm \sqrt{1075.32}=54.932$ or -10.648.

Here $AG=54.932$. $AB=47.87$. Hence $BG=7.062$. Having AP, AG, and the angle PAG, we can compute the angle APG.

PROBLEM II.

To divide a triangle into THREE PARTS *having the ratio of the three numbers m, n, p.*

CASE 1. By lines drawn from one angle of the triangle to the opposite side.

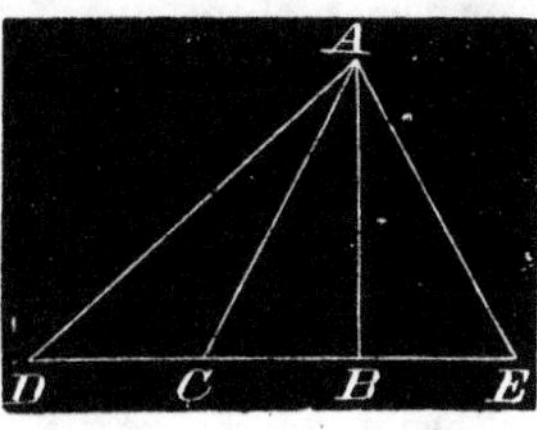

Let ADE be the triangle and A the angle from which the lines are to be drawn.

Divide DE the opposite side into parts in the ratio of m, n, and p, and from the points of division C and B, draw AC, AB, and the triangle is divided as required.

Demonstration. — The areas of triangles are their bases multiplied into their altitudes, but here all the triangles have the *same altitude;* therefore multiplying the bases into that altitude gives the same proportional product, and the areas of the triangles are as m, n, p.

CASE 2. By lines parallel to one of the sides.

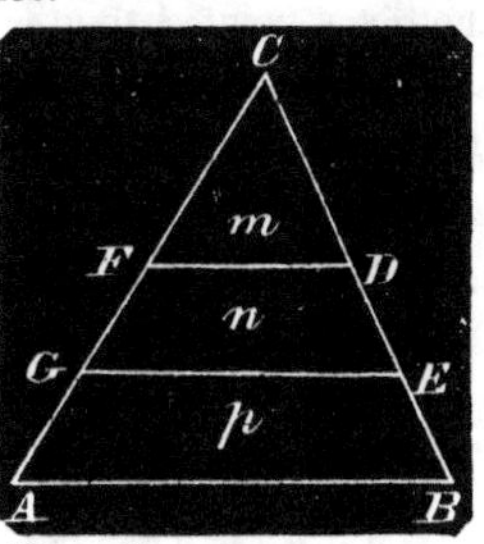

Let ABC be the triangle. Divide its numerical area into three parts in the ratio of m, n, p. Conceive the problem solved and DF, EG, the division lines parallel to AB. *We are to determine the numerical values of CD, and CE.* CB is known, put it equal to a. Put $CD=x$. $CE=y$.

Now as similar triangles are to one another as the squares of their homologous sides, therefore

$$x^2 : a^2 :: m : m+n+p.$$

$$\text{Whence, } x=a\sqrt{\frac{m}{m+n+p}}$$

In the same manner,

$$y=a\sqrt{\frac{m+n}{m+n+p}}$$

In this manner we might divide the triangle into any proposed number of parts, having given ratios.

CASE 3. By lines drawn from a given point on one of the sides of the triangle.

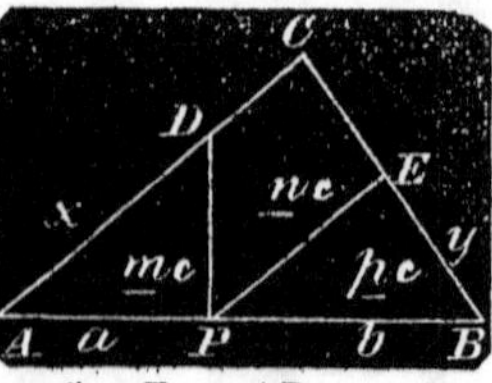

Let ABC be the given triangle, and P the given point on the side AB.

It is required to draw lines from P, as PD and PE, dividing the triangle into three parts mc, nc, pc, that is, assume the given numerical area to be $(mc+nc+pc)$, then the required parts will be mc, nc, and pc.* Put $AD=x$, then (by Prob. III, Mens.)

$$\tfrac{1}{2}ax \sin. A=m \text{ or } x=\frac{2mc}{a \sin. A}$$

By comparison,

$$y=\frac{2pc}{b \sin. B}$$

When mc and pc are cut off, nc is left. Having a and x, the angle ADP is easily determined.

In a similar manner we can divide the triangle into any proposed number of parts, by lines drawn from the given point P.

Case 4. By lines drawn from a given point within the triangle.

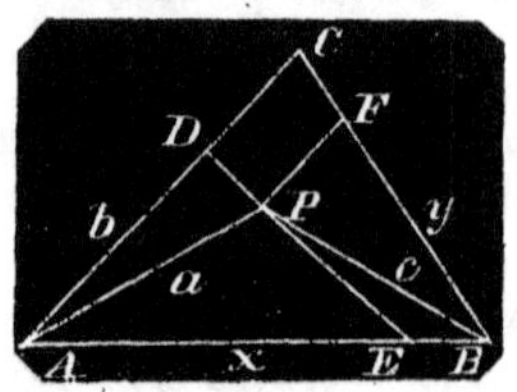

Let ABC be the given triangle, and P the given point within it.

A variety of lines may be drawn from P, to divide the triangle into the parts required. Conceive PD, PF, and PE to make the requisite division.

As P is a given point, AP, PB, and PC are known lines, and the angles DAP, PAE are known angles.

Take $AD=b$, *any convenient assumed value.* Take $AE=x$. Put $AP=a$. Then,

$$\tfrac{1}{2}ab \sin DAP=\text{area } \triangle\ DAP$$

Also,

$$\tfrac{1}{2}ax \sin PAE= \text{area } \triangle\ PAE$$

---

* Suppose we had a triangular piece of ground containing 320 square rods, and we wished to divide it into three parts in the ratio of 2, 3, and 5, what is the area of each of the parts ?

We decide it thus : $m=2$. $n=3$. $p=5$. c is at present unknown, but,

$$mc+nc+pc=320$$

That is,

$$10c=320 \text{ or } c=32.$$

Whence, $mc=64$. $nc=96$. $pc=160$.

The quantity c becomes known on dividing the area, and the parts separately mc, nc, and pc, are *always known.*

Conceive the triangle ABC, divided into three parts, in the ratio of m, n, p; and conceive $ADPE$ to be one of these parts represented by mc. Then

$$ab \sin. DAP + ax \sin. PAE = 2mc$$

Whence,
$$x = \frac{2mc - ab \sin. DAP}{a \sin. PAE}$$

Had we taken b greater than we did, x would have been less, and a variety of lines could be drawn as well as PD and PE, and the same area cut off.

Having x, we have EB as a known quantity, and by the two triangles, PEB and BPF, we determine y in precisely the manner as we found x: thus, we have two parts of the triangle mc and pc, and, consequently, the remainder $DPEC$ corresponds to nc.

CASE 5. By lines drawn from a given point without the triangle.

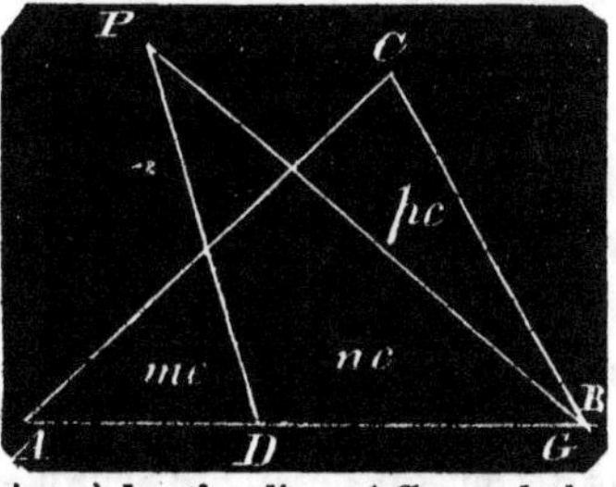

Let ABC be the triangle, and P the given point without it.

Divide the numerical area of the triangle into the three required parts, mc, nc, and pc, as in former cases. Draw PD, cutting off the portion mc, as in Case 6 of the last problem; then cut off the two portions ($mc+nc$) by the line AG: and the portion pc will be left. Or, we may cut off pc, and the portion nc will be left.

PROBLEM III.

To divide a triangle into three parts, having the ratio of m, n, and p, by three lines drawn from the three angular points to some point within.

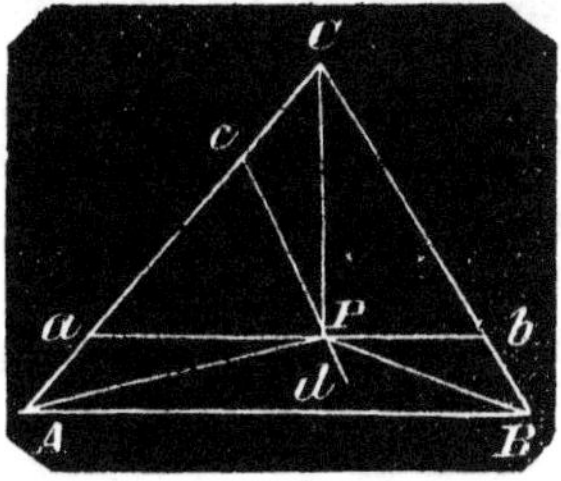

Divide any side, as AC, into three parts in the proportion of m, n, and p.

Let Aa represent the portion corresponding to m, and Cc the part corresponding to p.

Through a, draw ab parallel to AB,

and through c, draw cd parallel to CB. Where these two lines intersect is P, and the triangle ABC is divided into three triangles, APB, CPB, and APC.

Demonstration.— Any triangle having AB for its base, and its vertex in the line, ab will have the same ratio to the triangle ABC, as Aa has to AC, that is, as m to $m+n+p$.

Also, the triangle CPB is to ABC, as Cc is to CA, that is, as p to $m+n+p$, for triangles on the same base are to one another as their altitudes. If these two triangles, APB and CPB, are in due proportion, the third one, APC, is in due proportion, of course.

PROBLEM IV.

To divide a quadrilateral into two parts, having any given ratio m to n.

Case 1. By a line drawn from a given point in the perimeter.

Let $ABCD$ be the given quadrilateral, and P the given point in the side AB.

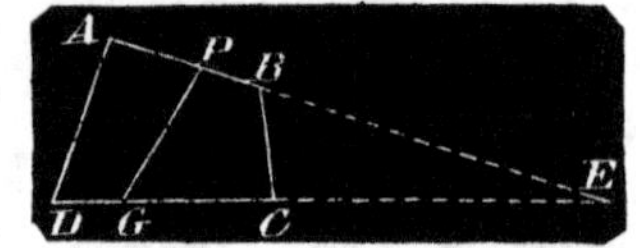

It is required to determine the magnitude, and the direction of the line PG, which divides the figure into parts in the ratio of m to n. All the sides and angles of the quadrilateral are known, and its *area* is known. AB and DC are, or are not parallel; if they are parallel the figure is a *trapezoid,* then the method of finding G, in the opposite side, is easy and obvious. If AB and CD are not parallel, we can produce them and form a triangle in the one direction or the other; by this figure we form the triangle BCE, whose *area* we may represent by t.

As BC, and the angles as B and C are all known, the triangle EBC is determined in all respects.

As PB is known, PE is known, and designate EG by x. Let cm and cn designate the portions of the quadrilateral after it is divided, and let cm represent the part $BPGC$.

Put $PE=a$.

Now (by Problem III, Mensuration), we have

$$\tfrac{1}{2}ax \sin. E=t+cm.$$

Whence, $$x=\frac{2t+2cm}{a \text{ sin. } E}$$

We now have the numerical value of x, from which we subtract EC, and we have CG, which being measured from C will give the point G, through which to draw the line from P, to divide the figure as required.

Case 2. By a line making a given angle with one of the sides.

If the division line makes a given angle with one side, it must also make a known angle with the opposite side.

Taking the last figure, conceiving PG to take a given direction across AB and CD, so as to cut off the area mc.

As in the former case, let t represent the area of the triangle EBC, to this add mc, and we have the area of the triangle PGE. But in this case P is not a given point, and EP is not known.

Put $EG=x$. Let P represent the given angle at P, and G the given angle at G. Now, by trigonometry,

$$\text{sin. } P : x : : \text{sin. } G : EP$$

Or, $$EP=\frac{\text{sin. } G}{\text{sin. } P}x$$

(Prob. III. Mens.) $$\frac{\text{Sin. } E \text{ sin. } G}{2\text{sin.} P}x^2=t+mc$$

Whence, $$x=\sqrt{(2t+2mc)\frac{\text{sin. } P}{\text{sin. } G \text{ Sin. } E}}$$

From x we take EC, measure off the remainder along CD, to the point G, there making the given angle, and the figure will be divided as required.

Case 3. By a line drawn through a given point within the quadrilateral.

Let $ABCD$ be the quadrilateral as before, and P the given point *within* it; and as P is the given point, EP is a known line, and the angles PEH, PEG are known.

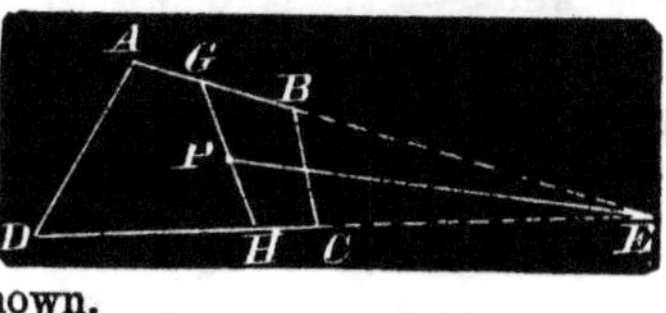

Let t equal the area of the triangle EBC, as before, and mc the area $GHCB$. Put $EH=x$, and $EG=y$.

Now we have a problem *precisely like* Case 4, Problem I, of this chapter; therefore, further explanations would be superfluous.

CASE 4. By a line drawn through a given point without the quadrilateral.

Let $ABCD$ be the quadrilateral as before, and P the given point *without* it.

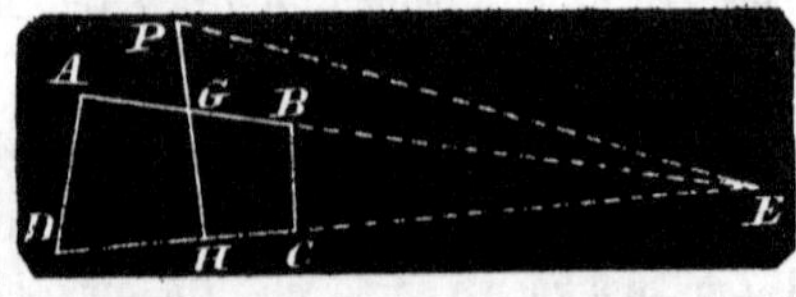

By producing the two sides AB CD, we form the triangle ADE. Let the area of the triangle BCE be represented by t, and the part $GHCB$ by mc, then from a given point P, without a triangle, we are required to draw a line PH, to divide the triangle into two given parts, and this is Case 6, of Problem I of this chapter, which has been fully investigated.

REMARK.—By extending the principles of these several cases we may divide a quadrilateral into three or more parts.

PROBLEM V.

To divide any polygon (regular or irregular) into two parts having a given ratio, m to n, by a line drawn through a given point.

CASE 1. When the given point is on one of the sides of the polygon.

Let $ABCDEF$ be the polygon, and $(mc+nc)$ express its numerical area. Let P be the given point on the side AB. Let the surveyor run a *random* line as near to the line required as his judgment permits, and generally it will be best to run a line from P to one of the opposite angular points.

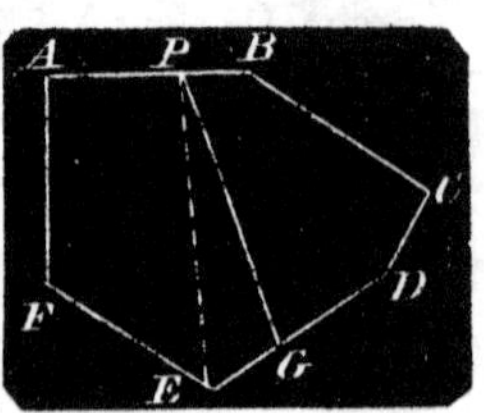

In this figure, let PE represent such a random line, and let the surveyor compute the area of the figure $PAFE$, thus cut off, which area will be *equal to*, or *greater*, or *less* than one of the required parts. We will suppose it *less;* then subtract it from the required portion mc, and let the triangle PEG represent that *known difference*, which we shall designate by t.

PE is known, the angle PEG is known; and put $EG=x$

Then, $\frac{1}{2}PE\,x\,\sin.\,PEG=t$

Whence, $x=\dfrac{2t}{PE\sin.\,PEG}$

This determines the point G, and PG divides the polygon as required.

CASE 2. When the given point is within the polygon.

Let $ABCDEF$ be the polygon as before, and $(mc+nc)$ express its numerical area; also, let P be the given point within. Through P let the surveyor run the *random* line, HPK, measuring from H to P, and from P to K, let him also observe the angles that this line makes with the sides of the polygon AF and CD, and compute the area $HABCK$, and note the *difference* between it and mc, the required portion of the polygon; call this difference d, a known quantity.

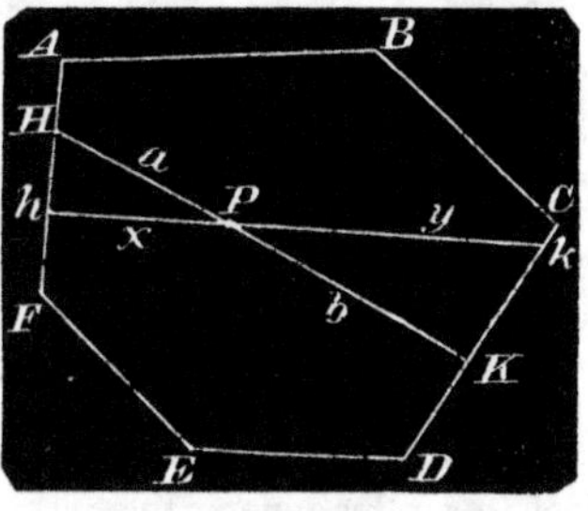

Let hPk represent the true line through P, which divides the polygon as required; but this line diminishes the area $HABCK$, by the triangle PKk, and increases it by the triangle PHh.

Therefore the difference of these triangles must equal d.

The sine of the angle AHP, has the same numerical value as the sine of PHh, and the sine of the angle PKD has the same numerical value as the sine of the angle PKk.

Put the angle $AHP=u$, and the angle $PKD=v$; let the acute verticle angles at P be designated by the letter P.

Let $HP=a$, $PK=b$, $hP=x$, $Pk=y$

In the triangle PhH, we have

$$\sin. u : x :: \sin. (u-P) : a$$

Whence, $$x=\frac{a \sin. u}{\sin. (u-P)} \qquad (1)$$

Similarly, $$y=\frac{b \sin. v}{\sin. (v-P)} \qquad (2)$$

The area of the triangle $PhH=\frac{1}{2}ax \sin. P$

Also, $PkK=\frac{1}{2}by \sin. P$

Whence, $$b \sin. Py-a \sin. Px=2d \qquad (3)$$

By substituting the values of x and y, taken from (1) and (2), we have,

$$b^2\frac{\text{sin. } P \text{ sin. } v}{\text{sin. } (v-P)}-a^2\frac{\text{sin. } P \text{ sin. } u}{\text{sin. } (u-P)}=2d \qquad (4)$$

Equation (4) contains only one unknown quantity P, the value of P, or the angle HPh can therefore be deduced.

$$b^2\left(\frac{\text{sin. } P \text{ sin. } v}{\text{sin. } v \text{ cos. } P-\text{cos. } v \text{ sin. } P}\right)-a^2\left(\frac{\text{sin. } P \text{ sin. } u}{\text{sin. } u \text{ cos. } P-\text{cos. } u \text{ sin. } P}\right)$$
$$=2d.$$

Dividing the numerator and denominator of the first fraction by (sin. P sin. v), and of the second fraction by (sin. P sin. u.), recollecting that *cosine* divided by sine gives cotangent. Thus we shall obtain

$$b^2\left(\frac{1}{\text{cot. } P-\text{cot. } v}\right)-a^2\left(\frac{1}{\text{cot. } P-\text{cot. } u}\right)=2d \qquad (5)$$

This last equation shows the surveyor that if he can make it convenient to run his *random* line from P, perpendicular to one of the sides, his equation will be less complex. For instance, if $AHP=90°$, its cotangent will be 0, and cot. u would then $=0$, and equation (5) would become.

$$\frac{b^2}{\text{cot. } P-\text{cot. } v}-\frac{a^2}{\text{cot. } P}=2d \qquad (6)$$

For the sake of convenience put cot. $P=z$, and cot. $v=c$.

Then

$$\frac{b^2}{z-c}-\frac{a^2}{z}=2d$$

Or,
$$z^2+\left(\frac{a^2-b^2}{2d}-c\right)z=\frac{a^2c}{2d}$$

The numerical value of z will be the numerical value of cot. P; its logarithm taken, and 10 added to the index will be logarithmic cot. in our table. The same remarks will apply to cot. v or c.

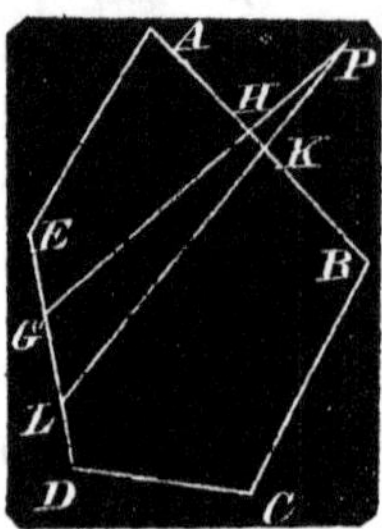

Case 3. When the given point is without the polygon.

Let $ABCDE$ be the polygon, and P the given point without it.

From the last case we learn that the surveyor had better run his random line perpendicular to one of the sides, therefore let PHG be the random line, perpendicular to AB.

As before, compute the area, $AEGH$, subtract it from mc, the difference is the difference between the triangles PGL and PHK. Draw PKL the line that divides the polygon as required.

Put $PH=a$, $PG=b$, $PK=x$, $PL=y$, angle $H=90°$, angle $PGE=u$, and the angle at P, designated by P.

The triangle $PHK=\frac{1}{2}ax \sin. P$

" $PGL=\frac{1}{2}by \sin. P$

Whence, $$b \sin. Py-a \sin. P\, x=2d. \qquad (1)$$

Here (2) represents a similiar quantity as in the last case.

In the triangle PHK, we have

$$1 : x :: \cos. P\, a, \text{ or } x=\frac{a}{\cos. P} \qquad (2)$$

In PGL, $\sin. u : y :: \sin. (u-P) : b$.

$$y=\frac{b \sin. u}{\sin.(u-P)} \qquad (3)$$

When the values of x and y are substituted in (1) we have

$$b^2\frac{\sin. P \sin. u}{\sin. (u-P)}-a^2\frac{\sin. P}{\cos. P}=2d \qquad (4)$$

$$\text{Or,}\quad b^2\left(\frac{\sin. P \sin. u}{\sin. u \cos. P-\cos. u \sin. P}\right)-a^2\frac{\sin. P}{\cos. P}=2d \qquad (5)$$

$$\text{Or,}\quad \frac{b^2}{\cot. P-\cot. u}-\frac{a^2}{\cot. P}=2d \qquad (6)$$

This equation is exactly similar to equation (6) of the last case, and it is reduced in the same manner.

PROBLEM VI.

To divide a polygon into three or more parts, having a given ratio, m, n, p, q, by lines passing through a given point.

This problem admits of three cases.

CASE 1. When the given point is on one side of the polygon.

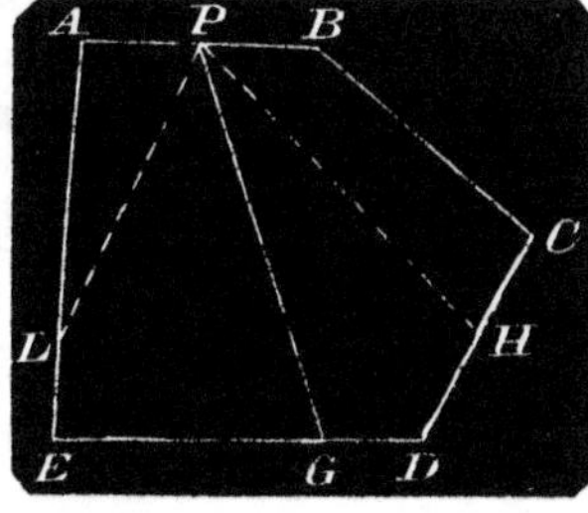

Divide the numerical area of the whole into parts, mc, nc, pc, qc, corresponding to the given ratio. Unite these into *two parts* $(mc+nc)$ and $(pc+qc)$.

From the given point P, draw PG, by Case 1, Problem V, so as to divide the polygon into the two parts $(mc+nc)$ and $(pc+qc)$.

We have now to divide the polygon $PAEG$ into two parts, mc, nc, by the line PL, and the polygon $PBCDG$ into two parts, pc and qc, by the line PH.

CASE 2. When the given point is within the polygon.

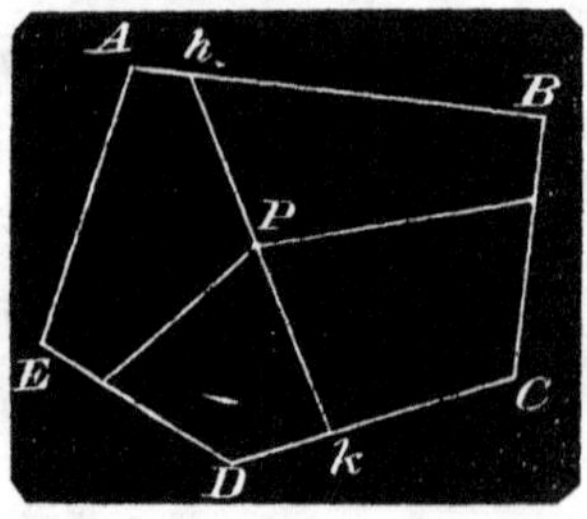

Let $ABCDE$ be the given polygon and P the given point within.

Draw hk through P, by Case 2, Problem V, so that the area $AhkDE$ shall equal mc, and the area $hkCB$ shall equal $(nc+pc)$, when the whole is required to be divided into *three parts* in the ratio of m, n, p.

When the whole is to be divided into *four parts*, in the ratio of m, n, p, q, then draw hk, so that one portion shall be $(mc+nc)$ and the other $(pc+qc)$.

Then we have P *as a given point*, in one side of the polygon, $AhkDE$, to divide it into two parts, in the ratio m to n, and P a given point on one side of the polygon, $hkCB$, to divide it into two parts, in the ratio of p to q, and this is done by Case 1, Problem V.

CASE 3. When the given point is without the polygon.

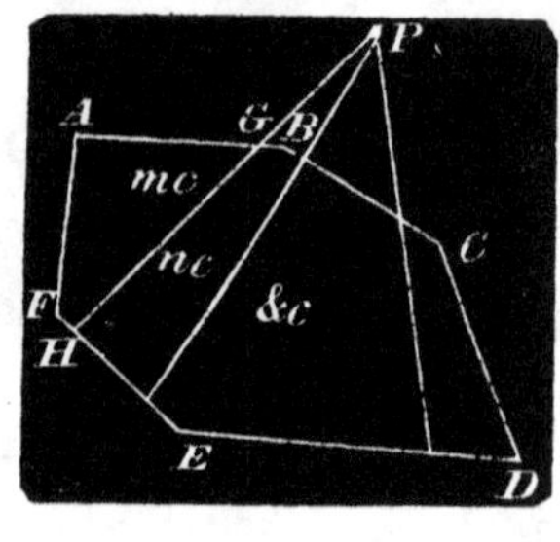

Let $ABCDEF$ be the given polygon, and P the given point without it.

Divide the numeral area into the required proportional parts, mc, nc, pc, &c., as many as required.

From the point P draw the line PH, as directed in Case 3, Problem V, dividing the polygon into *two* parts, mc and $(nc+pc+$&c.$)$.

Then divide the polygon, $GHEDCB$, into *two* parts, one of which is nc, and the other $(pc+qc$, &c.$)$, and thus we can proceed and cut off one portion after another, as many as may be required.

The application of the foregoing principles will meet *any case* that can occur in the division of lands ; and we now close this subject with the following practical

EXAMPLES.

1. *A triangular field, whose sides are 20, 18, and 16 chains, is to have a piece of 4 acres in content fenced off from it, by a right line drawn from the most obtuse angle to the opposite side. Required the length of the dividing line, and its distance from either extremity of the line on which it falls?*

Ans. Length of the dividing line, 13 chains, 95 links, if run nearest the side 16. Distance it strikes the base from the next most obtuse angle is 5.55 chains.

2. *The three sides of a triangle are 5, 12, and 13. If two-thirds of this triangle be cut off by a line drawn parallel to the longest side, it is required to find the length of the dividing line, and the distance of its two extremities from the extremities of the longest side.*

Ans. Distance from the extremity on 5, is $5(\sqrt{3}-\sqrt{2})$; on the side of 12, it is $12(\sqrt{3}-\sqrt{2})$; both divided by $\sqrt{3}$.

The division line is $13\sqrt{\frac{2}{3}}$.

3. *It is required to find the length and position of the shortest possible line, which shall divide, into two equal parts, a triangle whose sides are 25, 24, and 7 respectively.*

Remark.—It is obvious that the division line must cut the sides 25 and 24, and to make it the shortest line possible, the triangle cut off must be Isosceles.

Ans. The division line makes an angle with the sides 25 and 24 of $81° 52' 31''$, and its length is 4.896.

4. *The sides of a triangle are 6, 8, and 10. It is required to cut off nine-sixteenths of it, by a line that shall pass through the center of its inscribed circle.*

Ans. The division line cuts the side of 10, at the distance of 7.5 from the most acute angle, and on the side of 8, at the distance 6 from the most acute angle.

5. *Two sides of a triangle, which include an angle of 70°, are 14 and 17 respectively. It is required to divide it into three equal parts, by lines drawn parallel to its longest side.*

Ans. The first division line on the side 17, cuts that side at the distance $\frac{17}{\sqrt{2}}$; the second division line $\frac{17\sqrt{2}}{\sqrt{3}}$. The side 14 is cut at $\frac{14}{\sqrt{3}}$ and $\frac{14\sqrt{2}}{\sqrt{3}}$.

6. *Three sides of a triangle are* 1751, 1257.5, *and* 2364.5. *The most acute angle is* 31° 17′ 19″. *This triangle is to be divided into three equal parts by lines drawn from the angular points to some point within. Required the lengths of these lines.*

Ans. The line from the most acute angle is 1322.42, and from the next most acute angle 1119

7. *The legs of a right-angled triangle are* 28 *and* 45. *Required the lengths of lines drawn from the middle of the hypotenuse, to divide it into four equal parts.*

Ans. A line drawn from the middle of the hypotenuse to the right angle, divides the triangle into two equal parts.

8. *In the last example, suppose the given point on the hypotenuse at the distance of* 13 *from the most acute angle, whereabouts on the other sides will the division lines fall to divide the triangle into three equal parts?*

N. B. The sine of an acute angle to any right-angled triangle is equal to the side opposite that angle divided by the hypotenuse.

Ans. Both division lines fall on the side 28, distance of the first from the acute angle $12\frac{11}{30}$, or the second $24\frac{22}{30}$

9. *There is a farm containing* 64 *acres, commencing at its south westerly corner, the first course is North* 15° *E., distance* 12 *chains; the second is N.* 80° *E.* (*distance lost*), *the third S.* (*distance lost*), *the fourth is N.* 82° *W.* (*distance lost*), *to the place of beginning. It is required to determine the distances lost.*

OBSERVATION.— Extend the northern and southern boundary westward, and thus form a triangle on the west side of 12.

Ans. The 2nd side is 35.816 ch. 3rd, 23.21 ch. 4th, 38.76 ch.

The two following problems are from GUMMERE's *Surveying*, and are considered very difficult.

1. *There is a piece of land bounded as follows:*

Beginning at the south-west corner; thence,

1. *N.* 14° 00′ *W.*,	Distance	15.20 chains	$=a$;
2. *N.* 70° 30′ *E.*,	"	20.43 "	$=b$;
3. *S.* 6° 00′ *E.*,	"	22.79 "	$=c$;
4. *N.* 86° 30′ *W.*,	"	18.00 "	$=d$.

Within this lot there is a spring; the course to it from the second corner is S. 75° *E., distance* 7.90 *chains. It is required to cut off ten acres from the west side of this lot, by a line running through the spring. Where will this line meet the fourth side, that is, how far from the first corner?* Ans. 4.6357 chains.

First make a plot of the field. It is as here represented.

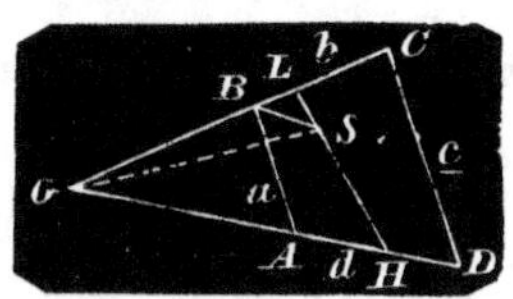

Produce the sides b and d, the second and fourth, until they meet at G. Let S be the position of the spring, and join SG. We may or may not find the contents of the field*: it is not necessary for the location of the line LSH.

It is necessary to find the area of the triangle ABG. Conceive a meridian line run through B; then we perceive that the angle $ABG=14°+70°\ 30'=84°\ 30'$. Conceive also a meridian line to be run through A, and then we perceive that the angle $BAG=86°\ 30'-14°=72°\ 30'$; whence $AGB=23°$. With the angles and the side $AB=15.20$, we readily find $AG=38.72$, $BG=37.10$, and the area $AGB=280.65$ square chains.

It is necessary to find the line GS and the angle BGS. From the given direction of the lines BC and BS, we find the angle $GBS=145°\ 30'$; and then from the triangle GBS, we find $BGS=5°\ 51'\ 30''$, and $GS=43.83$. Also we have the angle $SGA=17°\ 8'\ 30'$.

To the area of the triangle $AGB=280.65$, add 10 acres, or 100 square chains: then the area of the triangle GLH must equal 380.65

* When we have *four sides only*, and all the angles, as in this field, the best method of finding the contents is by conceiving it to be two triangles. Thus in this case the area is represented by

$$\tfrac{1}{2}\, ab \sin. ABC+\tfrac{1}{2}\, dc \sin. CDA.$$

square chains ; but GL and GH are both unknown. Put $GL=y$, $GH=x$: then we shall have the equation.

$$xy \sin. 23^\circ = 2(380.65). \qquad (1)$$

It is obvious that the sum of the two triangles LGS, SGH is equal to the triangle GLH.

But $GS=m=43.83$, $\sin. 23^\circ=P$, $\sin. (17^\circ\ 8'\ 30'')=Q$, $\sin. (5^\circ\ 51'\ 30'')=R$, and $2(380.65)$ or $761.3=a$: then we have

$$Pxy=a, \qquad (1)$$

and

$$Rmy+Qmx=a, \qquad (2)$$

From (1), $y=\frac{a}{Px}$. This value put in (2), gives

$$\frac{aRm}{Px}+Qmx=a\ ; \qquad (3)$$

whence,

$$x^2-\frac{a}{mQ}x=-\frac{a\,R}{PQ}. \qquad (4)$$

We now find the numerical values of $\frac{a}{mQ}$ and $\frac{a\,R}{PQ}$ by logarithms, as follows :

As our radius is unity, we diminish the indices of the logarithmic sines by 10.

	log. a, 2.881556		log. a, 2.881556
log. m, 1.641771			log. R,—1.008880
log. Q,—1.469437			
			1.790436
1.111208	1.111208		
58.932	1.770348	log. P,—1.591878	
		log. Q,—1.469437	
		—1.061315	—1.061315
674.72	- - - - - -		2.829121

Equation (4) now becomes

$$x^2-58.932x=-674.7\ ;$$

whence $x=GH=43.366$: from which subtract $GA=38.72$, and we have AH the distance required $=4.646$, which differs from the given answer about *one link* of the chain.

LEMMA. *Find the point in any trapezoid, through which any straight line which meets the parallel sides will divide the trapezoid into two equal parts.*

Let $ABCD$ be the trapezoid.

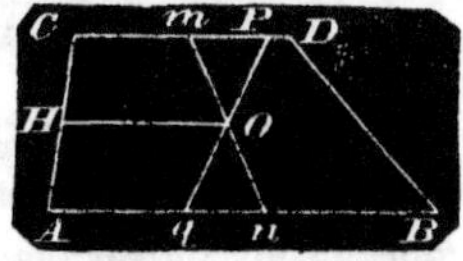

Bisect the parallel sides AB and CD in the points n and m. Join mn, and bisect mn in O, and O is the point required.

Any line meeting the parallel sides, and passing through O, will divide the trapezoid into two equal trapezoids. It is obvious that the line mn divides the figure into two equal parts, because the sum of the parallel sides is the same in each. Now draw any other line through O, as pOq: the trapezoid $pqBD = mnBD$; because the triangle $Omp = qOn$, and one triangle is cut off and the other is put on at the same time. The triangles are equal, because $mO = On$, the angle $pmO = Onq$, and the opposite angles at O are equal: therefore $pm = nq$; and whatever *more* than Cm is taken on one side, the equal quantity qn *less* than An is taken on the other side.

Another method of finding the point O, is to bisect AC in H, and draw HO parallel to AB or CD, and equal to one-fourth the sum of AB and CD.

2. *There is a piece of land bounded as follows:*

Beginning at the westernmost point of the field; thence,

1. *N.* 35° 15′ *E.*, 23.00 chains;
2. *N.* 75° 30′ *E.*, 30.50 "
3. *S.* 3° 15′ *E.*, 46.49 "
4. *N.* 66° 15′ *W.*, 49.64 "

It is required to divide this field into four equal parts, *by two lines, one running parallel to the third side, the other cutting the* first *and* third *sides. Find the distance of the parallel line from the first corner measured on the fourth side, and the bearing of the other line.*

Ans. Distance to the parallel, 32.50 chains; Bearing of the other side, *S.* 79° 48′ *E.*

CHAPTER VII.

TRIANGULAR SURVEYING: THE PLANE TABLE—ITS DESCRIPTION AND USES: MAPPING: MARINE SURVEYING.

TRIANGULAR SURVEYING, as here understood, requires the actual measurement of only one line, and all other lines can be deduced from this by means of observed angles forming triangles, of which this measured line forms a base of the first triangle in the series or chain of triangles.

Some of the other lines however should be measured after being computed, as a test to the accuracy or inaccuracy of the operations.

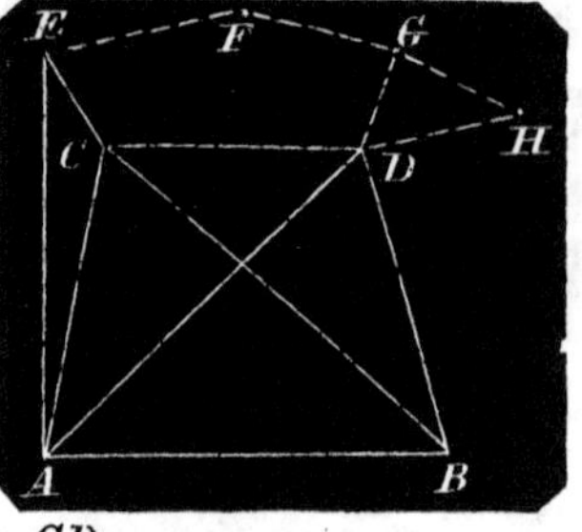

Let *AB* represent a base line which must be very accurately measured, for any error on *AB* will cause a proportional error in every other line.

If at *A* we measure the angles *BAC*, *BAD*, and at *B* we measure *or observe* the angles *ABC*, *ABD*, we then have sufficient data to determine the points *C* and *D*, and the line *CD*.

With equal facility that we determine the point *C*, we can determine the point *E*, or *F*, or *G*, or any other *visible* point.

Thus we may determine all the sides and angles of the figure *CEFGHD*, or any *visible* part of it, by triangulating from the base *AB*.

The lines forming the triangles are not drawn, except those to the points *C* and *D*; we omitted to draw others to *avoid confusion*.

After any line, as *FG*, has been computed, it is well to measure it, and if the measurement corresponds with computation, or nearly so, we may have full confidence of the accuracy of the work as far as it has been carried.

We may take *CD* as the base, and determine any visible number of points, as *A*, *B*, *H*, *F*, *G*, &c., trace any figure and determine its area, or show the relative positions and distances of objects from each other, such as buildings, monuments, trees, &c.

But to make the computation, triangle after triangle, for the sake of making a map, would be very tedious, and to measure every side and angle would be as tedious, and to facilitate this kind of operation we may have an instrument called the

PLANE TABLE.

The plane table is exactly what the name indicates ; it is a plane board table, about two feet long, and twenty inches wide, resting on a tripod, to which it is firmly screwed, yet capable of an easy motion onits center, having a ball and socket like a compass staff.

Directly under the table is a brass plate, in which *four milled screws* are worked, for the purpose of adjusting the table, the screws pressing against the table.

To level the table, a small detached spirit level may be used. The level being placed on the table over two of the screws, the screws are turned contrary ways, until the level is horizontal ; after which it is placed over the other two screws, and made horizontal in the same manner.

The table has a clamp screw, to hold it firmly during observations, and also a tangent screw, to turn it minutely and gently, after the manner of the theodolite.

The upper side of the table is bordered by four brass plates, about an inch wide, and the center of the table is marked by a pin.

About this center, and tangent to the corners of the table, *conceive* a circle to be described. Suppose the circumference of this circle to be divided into degrees and parts of a degree, and radii to be drawn through the center, and each point of division.

The points in which these radii intersect the outer edge of the brass border, are marked by lines on the brass plates ; these lines of course show degrees and parts of degrees ; they are marked from right to left, from 0 to 180° on both sides, but on some tables the numbers run all the way round, from 0 to 360°.

Near the two ends of the table are two grooves, into which are fitted brass plates, which are drawn down into their places by screws coming up from the under side. The object of these grooves and corresponding plates, is to hold down paper firmly and closely to the table.

The paper before being put on, should be moistened to expand it, then carefully drawn over the table, and fastened down by the plates that fit into the grooves; on drying, it will fit closely to the table.

A delicate fine edged ruler is used with the plane table, it has vertical sights; the hairs of the sights are in the same vertical plane as the edge of the ruler.

A compass is sometimes attached to the table, to show the bearings of the lines; but the most practical mathematicians prefer each instrument by itself.

The plane table may be used to advantage for three distinct objects.

1. For the measurement of horizontal angles.

2. For the determination of the shorter lines of a survey, both as to extent and position.

3. For the purpose of mapping down localities, harbors, water-courses, &c.

1. *To measure a horizontal angle.*

Place the center of the table over the angular point, by means of a plumb. Level the table, then place the fine edge of the ruler against the small pin at the center; direct the sights to one object, and note the degree on the brass plate; then turn the ruler to the other object, and note the degree as before.

The difference of the degrees thus noted, is the angle sought.

If the ruler passed over 0, in turning from one object to the other, subtract the larger angle from 180°, and to the remainder add the smaller angle, for the angle sought.

2. *To determine lines in extent and position.*

Let *CD* be a base line; having the paper on the table, all dried and ready for use. Place the table over *C*, so that the point on the table, where we wish *C* to be represented, shall fall directly over *C*; and place the position of the table so that *CD* shall take the desired direction on the table.

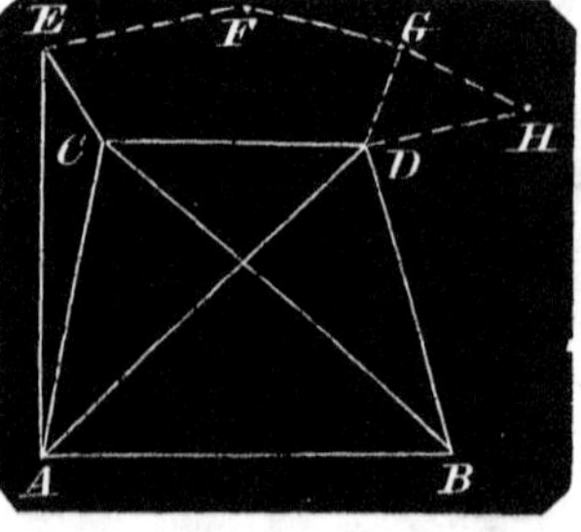

Now level the instrument, and clamp it fast; it is then ready for use.

Sight to the other end of the base line, and mark it along the fine edge of the ruler.

In the same manner sight along the direction of *CE*, and mark that direction in a fine lead line, that can be easily rubbed out, the point *E* is *somewhere* in that line.

Sight in the direction of *F*, and mark the line on the paper; *F* is somewhere in that line. In this manner sight to as many objects as desired, as *G*, *H*, *B*, *A*, &c.

Now the base *on the paper*, may be as long, or as short as we please; suppose the *real base* on the ground to be 1200 feet; this may be represented on the table by 3, 4, 5, 10, or 12 inches, more or less; suppose we represent it by 6 inches, then one inch on the paper, will correspond with 200 feet on the ground (*horizontally*).

Take *CD* six inches, and place a pin at *D*, remove the instrument to the other end of the base, and place *D* of the table right over the end of the base, by the aid of a plumb, and give the table such a position as will cause *CD* on the table, to correspond with the direction of the base.

Level the table and clamp it. Now, if *CD* on the table, does not exactly correspond with the direction of the base on the ground, make it correspond by means of the tangent screw.

Now from *D*, by means of the ruler and its sight vanes, draw lines on the paper, in the direction of the points *E*, *F*, *G*, *H*, *B*, *A*, &c.; and where these lines intersect those from the other end of the base, to the same points, is the real localities of those points, *in proportion* to the base line. Lines drawn from point to point, where these lines intersect, as *EF*, *FG*, *GH*, &c., will determine the distances from point to point, at the rate of 200 feet to the inch.

Lines drawn from the center of the table, parallel to *FE* and *FG*, will determine the angle *EFG*, in case the angle is required. After the points *E*, *F*, *G*, &c., have been determined, the light pencil lines to them, from the ends of the base, may be rubbed out, except those that we may wish to retain.

Here we perceive the utility of the plane table; we have a multitude of results, as soon as the observations are made.

The plane table will give us at once, the relative distances of buildings from the base, and from each other, and if we are careful and particular, we can obtain the magnitudes of the buildings, as is obvious by the adjoining figure.

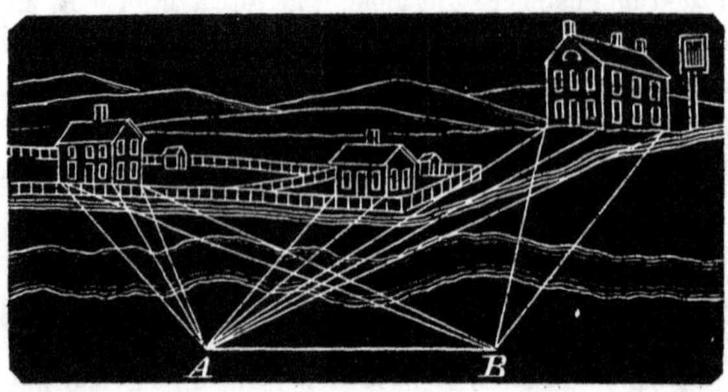

This is most useful to an officer or a spy, who wishes as exact knowledge of an enemy's locality as possible. Or from a distant place *AB*, we may examine and measure any objects whatever, on the other side of a river, or give a correct delineation of the river itself.

The plane table can be made very useful to civil engineers, for mapping the localities through which a canal or railroad passes. Take, for example, a railroad line, *ABCDE*, represented in the next figure. The lines being all measured and marked in distances of 100 or 200 feet, the bases are all ready where the line is straight.

Set the table as at *A*, and draw lines to all objects that you wish to appear on the map, both to the right and to the left,—then move the instrument to *B*, drawing lines to the same object *from a corresponding base on the paper*, and also draw lines to other objects further in advance on the line that may be seen from another base.

The intersections of the lines from the extremeties of any base to to the same object will locate the object.

When all the objects are thus located, both to the right and to the left, we pass on to new objects, taking care to keep at least two of the old objects in sight, to connect one new observation with those previously taken. We now commence a series of observations from a new base, which base must take its proper relative position on the

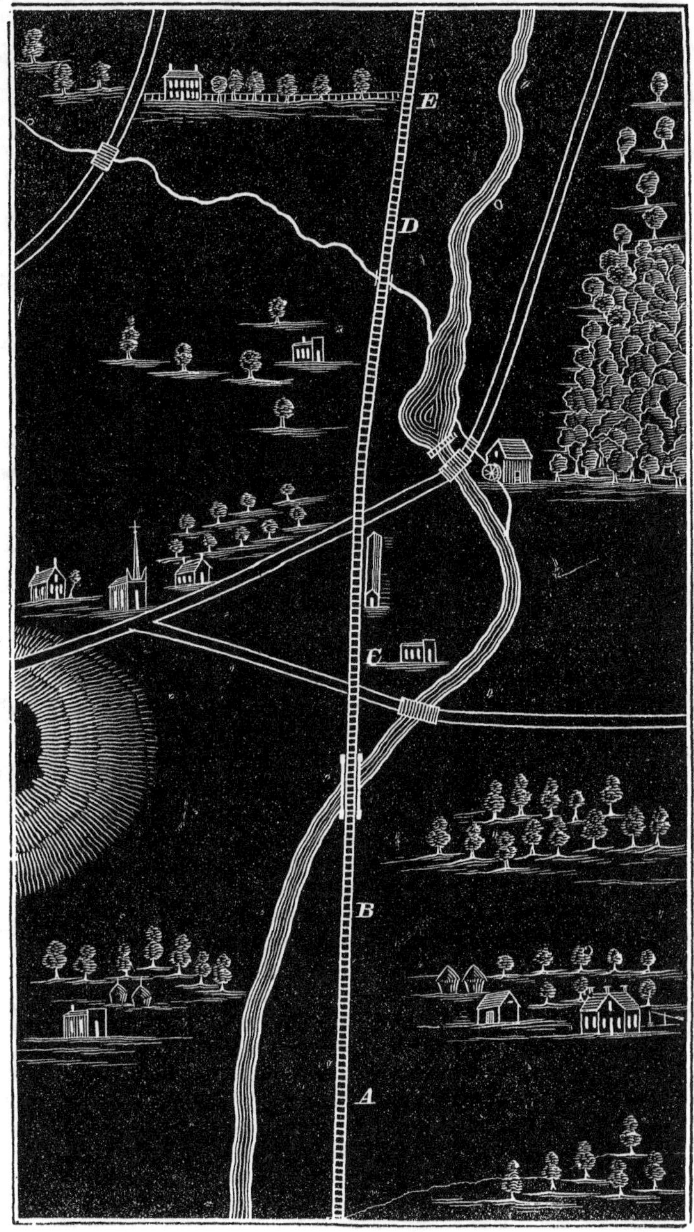
E
D
C
B
A

paper, and if the paper on the table is not large enough, it must be taken off and new paper put on, and two of the objects on the old paper must appear on the new, and then these two papers can be put together so that the objects which are on both papers will coincide, and then the two papers will be the same as one, and thus we may put any number of papers together and form as large a map as we please.

If the different bases are not in the same direction, the objects which are on two different papers on being put together will show it, and several papers put together may make a very inconvenient figure; but they must be put together and then a square sheet of tissue paper put over the whole, and the map taken off. From the tissue paper the map can be put on any other paper.

The engineers of Napoleon's army frequently made maps of the localities they were about to pass; indeed it is a military principle never to go into an unknown locality, except in cases of absolute necessity.

This subject naturally leads us to

MARINE SURVEYING.

Marine Surveying is too extensive a subject to be fully investigated in any work like this. We shall only explain how to find shoals, rocks, and turning points in a channel, by ranging to objects on shore.

In trigonometrical surveying, on shore, the observer is supposed to take his angles from the extremities of a base line, but in trigonometrical surveying on water, the observer can take his angles only from single points which may be connected together by distant base lines on the shore.

Important points along the shore are determined by taking latitude and longitude, and intermediate places, by regular land surveying.

The localities of rocks and shoals are also determined by astronomical observations, establishing latitude and longitude, in case no land is in sight, or they are far from the shore; but in the vicinity of the land, the determination of a point is commonly effected by the *three point problem.*

The three point problem is the determination of any point from observations taken at that point, on *three* other distant points where the distances of these three points from each other are known.

It is immaterial how those points are situated, provided the three points and the observer are not in the same right line, the middle one may be nearest or most remote from the observer, or two of them may be in one right line with the observer, or all three may be in one right line, provided the observer be not in that line. The following example will illustrate the principle.

Coming from sea, at the point *D*, I observed two headlands, *A* and *B*, and inland *C*, a steeple, which appeared between the headlands. I found, from a map, that the headlands were 5.35 miles from each other; that the distance from *A* to the steeple was 2.8 miles, and from *B* to the steeple 3.47 miles; and I found with a sextant, that the angle *ADC* was 12° 15′, and the angle *BDC* 15° 30′. Required my distance from each of the headlands, and from the steeple.

If the direction of *AB* is known, the direction of *AC* is equally well known.

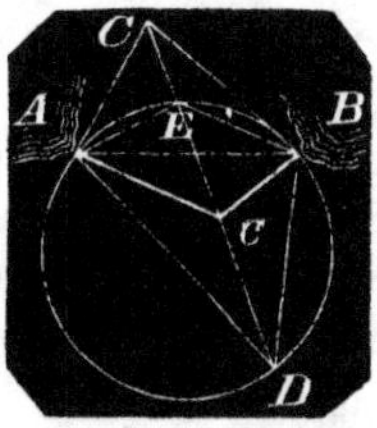

The case in which the three objects, *A*, *C*, and *B*, are in one right line may require illustration.

At the point *A*, make the angle *BAE*= the observed angle *CDB*, and at *B*, make the angle *CBE*= the observed angle *ADC*.

Describe a circle about the triangle *ABE*, join *E* and *C*, and produce that line to the circumference in *D*, which is the point of observation. Join *AD*, *BD*. The angle *ADB* is the sum of the observed angles, and *AEB* added to it must make 180°.

The Trigonometrical Analysis.—In the triangle *ABE*, we have the side *AB* and all the angles, *AE* and *EB* can therefore be computed.

In the triangle *AEC*, we now have *AC*, *AE*, and the angle *CAE*, from which we can compute *ACE*, then we know *ACD*.

Now in the triangle *ACD*, we have *AC* and all the angles, whence we can find *AD* and *CD*.

When the bearing of the base line on shore is known, as it generally is, and the bearings to its extremities, or even to one extremity, are taken, the triangle is known at once.

A pilot guides a vessel in and out of a port by ranging lines on the shore, minutely or approximately, as the case may require.

We will illustrate this by a figure. Let the deep shaded lines represent the shores, *A* a light house on a rocky promontory. *B* another prominent object on the opposite shore; the position of the arrow indicates the direction north and south.

The faint dotted lines represent the boundaries of shoal water, over which it would not be prudent, if even possible, to conduct a vessel. The line *a, b, d, E,* the center of the ship channel. All the pilots know that a line from *A* to *a,* which is nearly east south east, runs safe to the open sea, after passing the *shoal coast* near *n.*

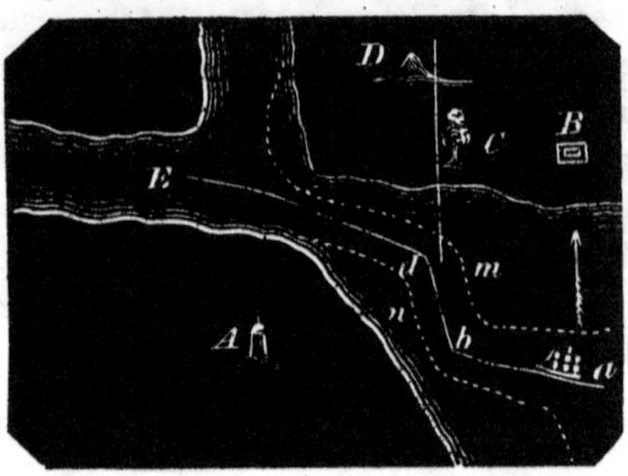

Now suppose a pilot boards a ship coming in from sea, sufficiently far from the coast, he directs her sailing so as to bring the light house at *A* to bear west north-west. He then sails toward the light house until he finds the object *B* bearing due north of him. He then knows that the ship must be near *a,* the mouth of the channel.

He continues the same course, and knows when he is about half way from *a* to *b* by the two objects, *C* and *D,* appearing in the same line. When *C* and *D* becomes fairly open, and *C* nearly north, and *B* not quite north-west, he is then at the turning point *b* of the channel. His course is then north, a little to the west, until the ship is nearly in a line between the two objects *A* and *B.* From thence, west north-west takes the ship directly through the proper channel into the harbor.

CHAPTER VIII.

LEVELING.

Two or more points are said to be on a level, when they are equally distant from the center of the earth, or when they are equally distant from a tranquil fluid, situated immediately below them. A level surface on the earth, is nearly spherical, *and is not a plane;* it is everywhere *perpendicular to a plumb line.*

Any small portion of a true level surface, cannot be distinguished from a plane; and, therefore, when observations are taken in respect to level, within short distances of each other, the spherical form of the earth is disregarded, and the level treated as a plane. But when any considerable portion of surface is taken into account, the curvature of the earth's surface must be considered.

The apparent level, at any point on the earth, is a tangent plane, touching the earth at that point only, and the true level is below this, and the distance below, depends on the distance from the tangent point.

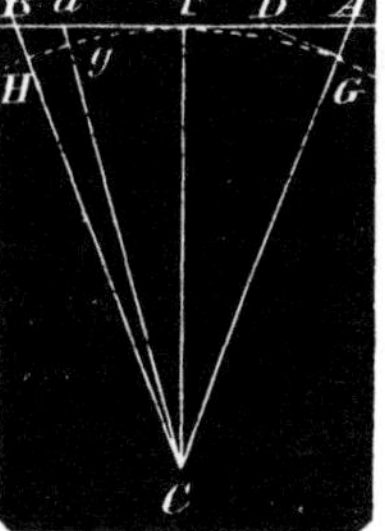

Let T be any point on the surface of the earth, at right angles to the plumb line from this point is the plane or apparent level ATB; but the true level, or the surface of standing water, would be the curved surface GTH.

The distance AG, depends on the distance AT, and the radius of the earth CG or CT.

From G, draw GD, at right angles to AC; then the two triangles ATC, AGD, are equiangular and similar, and give the proportion

$$CT : TA :: DG : GA.$$

Practically.—TA is a very short distance compared to CT, for TA, the distance within which we can take observations, is never more than two or three miles, while the distance CT is near 4000 miles; therefore, CT is nearly equal to CA, consequently, DA is nearly equal to DG, so near that we shall call it equal.

Observe that $TD=DG$; hence $DG=\frac{1}{2}TA$.

Now, in the preceding proportion, in the place of DG, put its equal $\frac{1}{2}TA$, and we shall have,

$$CT : TA :: \tfrac{1}{2}TA : GA.$$

Whence, $$GA=\frac{\overline{TA}^2}{2CT} \qquad \text{Also, } ga=\frac{\overline{Ta}^2}{2CT} \qquad (1)$$

That is, *The square of the distance, divided by the diameter of the earth, is the distance between the apparent and the true level.*

We can arrive at the same result by the direct application of the 36th proposition of the third book of Euclid, or by the application of theorem 18, third book of Robinson's Geometry.

Because A is a point without a circle, and AT touches the circle, we must have

$$AG\times(2CG+AG)=\overline{AT}^2$$

But $2CG$, which is the diameter of the earth, cannot be *essentially* or *appreciably* increased, by the addition of AG, which is at most, but a few feet, therefore, AG within the parentheses, may be suppressed without making any appreciable error. Then divide by $2CG$.

Whence, $$AG=\frac{\overline{AT}^2}{2CG}, \text{ the same as before.}$$

If we take one mile for the distance TA, the value of GA will be $\frac{1}{7918}$=8.001 inches.

By comparing equations (1) we perceive that,

$$GA : ga :: \overline{TA}^2 : \overline{Ta}^2$$

That is, *The corrections for apparent levels, are in proportion to the squares of the distances.*

The correction for one mile is 8.001 *inches; what is it for* 10 *miles?*

Ans. It is x inches; then we have the following proportion,

$$8.001 : x :: 1^2 : 10^2 \qquad x=800.1 \text{ inches.}$$

We have seen above, that the correction for *one mile* or 80 chains distance, on an apparent level, is 8.001 inches, what is the correction for the distance of 20 chains?

Let x=the correction sought, and the solution is thus,

$$8.001 : x :: (80)^2 : (20)^2$$
$$:: 4^2 : 1^2 \qquad x=0.500 \text{ inches.}$$

In this manner, the following table was computed.

Table showing the differences in inches, between the true and apparent level, for distances between 1 *and* 100 *chains.*

Ch's.	In's.	Ch's.	In's.	Ch's.	In's.	Ch's.	In's.
1	.001	26	.845	51	3.255	76	7.221
2	.005	27	.911	52	3.380	77	7.412
3	.011	28	.981	53	3.511	78	7.605
4	.020	29	1.051	54	3.645	79	7.802
5	.031	30	1.125	55	3.785	80	8.001
6	.045	31	1.201	56	3.925	81	8.202
7	.061	32	1.280	57	4.061	82	8.406
8	.080	33	1.360	58	4.205	83	8.612
9	.101	34	1.446	59	4.351	84	8.832
10	.125	35	1.531	60	4.500	85	9.042
11	.151	36	1.620	61	4.654	86	9.246
12	.180	37	1.711	62	4.805	87	9.462
13	.211	38	1.805	63	4.968	88	9.681
14	.245	39	1.901	64	5.120	89	9.902
15	.281	40	2.003	65	5.281	90	10.126
16	.320	41	2.101	66	5.443	91	10.351
17	.361	42	2.208	67	5.612	92	10.587
18	.405	43	2.311	68	5.787	93	10.812
19	.451	44	2.420	69	5.955	94	11.046
20	.500	45	2.531	70	6.125	95	11.233
21	.552	46	2.646	71	6.302	96	11.521
22	.605	47	2.761	72	6.480	97	11.763
23	.661	48	2.880	73	6.662	98	12.017
24	.720	49	3.004	74	6.846	99	12.246
25	.781	50	3.125	75	7.032	100	12.502

This table is of *little* or no practical use, for levelers rarely take sight to a greater distance than 10 chains, and at that distance the correction is only *one-eighth* of an inch, and if they put the level midway between two stations, they *annihilate* the corrections altogether.

Suppose a level to be placed at T, midway between A and B; the instrument will show them to be on the *same level*, as so they really are, for they are at equal distances from the *center of the earth*; but if the observations were taken in reference to A and a, the apparent level would not show equal distances from the center of the earth, and a correction must be applied, if the difference of distances is more than four or five chains.

To comprehend the whole subject, we must now describe the modern

SPIRIT LEVEL.

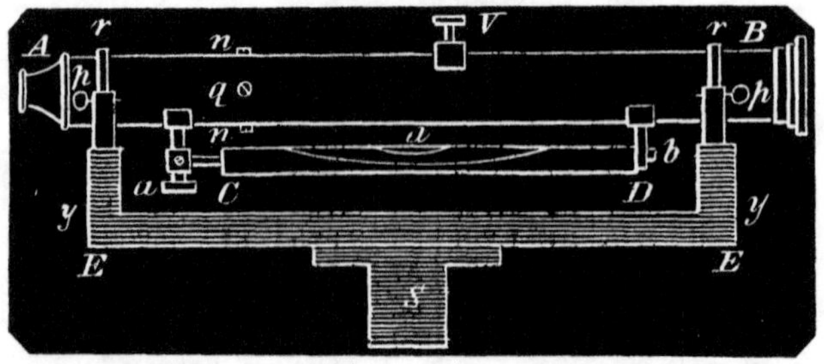

The figure before us represents this useful instrument, apart from its tripod.

Its principal parts are the telescope *AB*, to which is attached the leveling tube *CD*; the telescope rests in a bed, which is supported by posts *yy*, called the *y's*; *EE* is a firm bar, supporting the *y's*. In *S* is a socket, which receives the central pivot of the tripod (which is not here represented).

When the instrument is put upon its tripod, the tube *S* can be clasped on the outside, and held firmly by a clamp screw, it can then be moved horizontally, as minutely and readily as desired, by means of a tangent screw.

The tripod contains a pair of brass plates, to the lower one the legs of the tripod are firmly attached, the other plate moves in all directions on its center, and is worked by four screws; these are called the leveling screws; these plates are purposely made small as a greater surety against bending: the four leveling screws are placed at the four quadrant points of the circle, and, with the center, form diameters at right angles.

The eye-glass of the telescope is at *A*, the object glass at *B*. The screw *V* runs out the tube which holds the object glass, to adjust it to different distances.

The telescope is fastened into the *y's*, by the loops *r r*, which are fastened by the pins *p p*. The telescope can be reversed in the *y's*, by taking out the pins *p p*; opening the loops *r r*; taking up the tube, turning it round, and again placing it in the *y's*; then *A* will take the position of *B*, and *B* of *A*. The necessity of this construction will appear when we describe the adjustment.

At *n n* are two small screws that are attached to a ring inside of the tube; this ring holds a horizontal spider line; the object of the screw is to elevate and depress that spider line.

At *q q* (only one *q* can be seen in the figure), are two screws that work another ring, which holds a perpendicular spider line, which can be moved to the right and left by the screws *q q*. The two spider lines show a perpendicular and horizontal cross at the focus of the telescope.

Before using the instrument it must be adjusted. The adjustment consists:

1st. *In making the center of the eye-glass and the intersection of the spider's lines coincide with the axis of the telescope*, and this line is called the line of *collimation.*

2nd. *In making the axis of the attached level, CD, parallel to the line of collimation, in respect to elevation.*

3rd. *In making the attached level lie exactly in the same direction as the line of collimation.*

To make the *first* adjustment, the telescope is made to revolve in the *y's*.

To make the *second* adjustment, there is a screw *a*, which serves to elevate and depress the end of the leveling tube at *C*.

To make the *third* adjustment, there is a side screw *b*, which drives the end of the tube *D* to the right and left, as the case may require.

First Adjustment.— Plant the tripod, place the instrument upon it, and direct the telescope to some well defined and distant object.

Draw out the eye-glass at *A*, until the spider's lines are distinctly seen, then run out the object glass by the screw *V* to its proper focus, when the object and the spider's lines will be distinct. Now note the precise point covered by the horizontal spider's line. Having done this, revolve the telescope in the *y's* half round, when the attached level will be on the upper side. See if in this position the horizontal spider's line appears *above* or *below* the same object.

If this line should appear exactly in the same point of the object as before, this spider's line is already in adjustment, but if it appears above or below, bring it half way to the same point by means of the screws *n n*, loosening the one and tightening the other.

Carry back the telescope to its original position, and repeat the observation, and continue to repeat it until the telescope will revolve half round without causing the horizontal line to rise or fall. This will show that the horizontal line is a *diameter* of the circle of revolution, and not a *chord* of it. Make the same adjustment in respect to the vertical hair, and the line of collimation is then adjusted.

Second Adjustment.— That is, to make the tube *CD* horizontally parallel to the line of collimation. Place the instrument properly on its tripod, and bring the horizontal bar *EE* directly over two of the leveling screws; turn these screws until the bubble *d* rests in the center of the tube.

Now *CD* is on a level, but we are not able to say that the line of sight through the telescope, that is the line of collimation, is on a level also. To test this, take out the pins *p p*, open the loops *r r*, and take out the telescope with its attached level, and turn it end for end, put it back in its bed, and put the loops over and pin them down. If the bubble now rests in the middle, no adjustment is required; if not, the bubble will run to the elevated end. In that case the bubble must be brought half way back by the leveling screws, and the other half by the screw *a*.

Repeat the operation until the bubble will settle in the middle of the tube after reversing the telescope.

Third Adjustment.— The second adjustment being completed, revolve the telescope in the *y's*, and if the bubble continues in the middle, the axis of the telescope and the axis of the tube *CD* lie in the same direction, or in the same vertical plane; and if they be not in the same vertical plane the bubble will run to one end or the other; in that case the side screw *b* will remedy the defect.

The three adjustments are now made *approximately*, no one of them can be made perfectly while the instrument is greatly out of adjustment in relation to the others; therefore commence anew.— Bring the bar *EE* over two of the leveling screws, and level the instrument; then turn it over the two other screws and level it in that direction also. Now, if we can turn the instrument quite round without removing the bubble from the center it is in pretty good adjustment, but if otherwise, as is to be expected, *make all these adjustments over again;* they can now be made with much less difficulty.

It is important that a level should be in as perfect adjustment as possible, but perfection in all respects is almost, yea, quite impossible. Yet, with a level considerably out of adjustment, we can obtain the *relative* elevation of any two points, provided we can set the level midway between them.

To illustrate this, suppose the instrument placed at *D*, midway between two perpendicular rods *Aa Bb*.

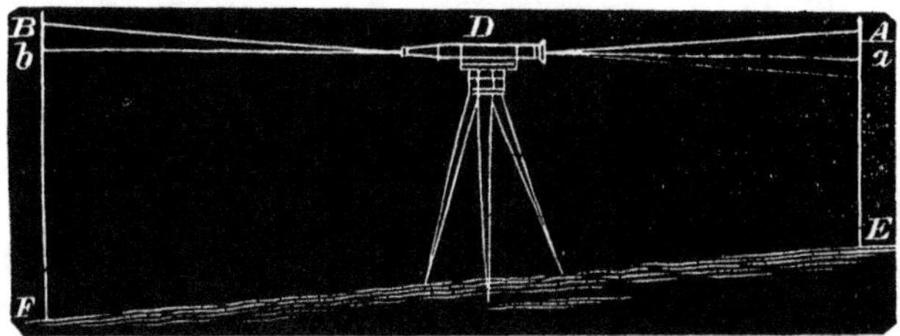

Let *ab* represent the true horizontal line, but suppose that the instrument is so imperfect, or out of adjustment, that when the leveling tube *CD* is horizontal, the telescope would point out the *rising line DA*, and the rise would be *Aa*. On turning the instrument round and sighting to *B*, the rise must be the same as in the opposite direction: *for the distance is the same,* therefore *A* and *B* are as truly on a level with each other as *a* and *b*.

By this problem, practical men complete the second adjustment of the instrument. They make the three adjustments as just explained, as accurately as possible. They then measure, very carefully the distance between two stations, as *E* and *F*, and set the instrument exactly midway between them as represented in the last figure.

They then level the instrument (that is the tube *CD*), and find the difference of the levels between *E* and *F* (two pegs driven into the ground).

Now, suppose *AE* measures on the rod, -	4.752 feet.
And *BF* " " " - -	6.327 feet.
Then *E* is above *F* - - - - - -	1.575 feet.

They now bring the level near to one of the stations as E, and level it very accurately, and sight to the rod AE.

Now, suppose the target stands at - - -	5.137 feet.
To this add - - - - - - -	1.575 feet.
	6.712

The rod man now goes to the station F, puts his target on the rod exactly at 6.712, and the telescope is turned upon it, and the horizontal spider's line ought to just coincide with the target, and will, if the instrument is in perfect adjustment; if it is not, the error is taken out by the screws n n. If the error was but slight, as in such cases it always is with good instruments, the adjustment is as complete as it can be made.

With the level there must be

A ROD.

The rod is commonly ten feet long, and divided into tenths and hundredths, some have also a vernier scale which in effect subdivides to *thousandths*. The target slides up and down the rod, and carries the vernier on the back of the rod; the target has equal alternate portions painted black and white for contrast.

A party to take the necessary levels on the line of a railroad or canal, after the stations are measured off, should consist of a leveler, and assistant leveler, a rod man, and an axe man.

The leveler and assistant leveler both keep book, and sometimes the rod man also. If there is no assistant leveler, the rod man will have an abundance of time to keep book, and under such circumcumstances always does so. Under all circumstances, two persons keep book, to have a check on each other and guard against mistakes.

In the field the aim is to put the level midway between the two stations, but they are not particular about it if the instrument is in good adjustment; they rather take the most advantageous spot to sight from.

When the ground admits of it, that is, sufficiently level, two or three intermediate stations are observed, as well as the extreme back and fore stations. The extreme back and fore stations are called *changing stations;* at these stations pegs are driven into the ground by the axe man for the rod man to plant his rod, so as to secure the same point for both the fore and back sight. At the intermediate stations they have no pegs, and are *not particular in any respect*, for all errors cancel each other.

The common railroad chain is 100 feet, divided into 100 links; each link is therefore *one foot*. Levels are commonly taken at intervals of 200 feet, oftener if the ground is very uneven, but a station is considered 200 feet, and the number of the station multiplied by 200 gives the number of feet from the commencement.

The field book is kept thus:

B. S. means back sight, F. S. fore sight, A. ascent, D. descent, T. total elevation above a common base.

N. B.—When the back sight is less than the fore sight, the ground is descending, and the converse.

Sta.	B. S.	F. S.	A.	D.	Total. 100
0	4.32	7.21		2.89	97.11
1	5.52	8.17		2.65	94.46
2	*9.18	6.27	2.91		99.37
3	6.27	6.12	0.15		97.52
4	6.12	3.76*	2.36		99.88
5	9.81	11.62		1.81	98.07
6	8.47	9.02		0.55	97.52
7	2.64	8.91		6.27	91.25
8	1.07	7.38		6.31	84.94
9	4.29	5.32		1.03	83.91
10	5.32	4.85	0.47		84.38
11	4.85	3.17	1.68		86.06
12	8.22	1.53	7.31		93.37

Thus we go through the whole line. We commenced with the *constant* 100, but this is arbitrary; the object of taking a constant is to avoid the minus sign, that is, getting below our ruled paper in making a profile of the vertical section.

Where the line is to be generally ascending, we assume a small constant, where generally descending a large one; taking care in all cases to have it so large as not to run it out.

At each and every section we know exactly how much we are above or below the constant base, and the exact ascent or descent from any one station to any other.

The following diagram represents a vertical section of the ground where these levels were taken, with the exception of the exaggeration of the roughness caused by the difference of scales for the base and perpendicular lines. From one station to another is 200 feet; we have made 10 feet occupy more space in the perpendicular direction than 400 feet does in a horizontal direction. We do this to show more clearly where any particular grade will enter the ground, and how much it is necessary to cut or fill at any particular point.

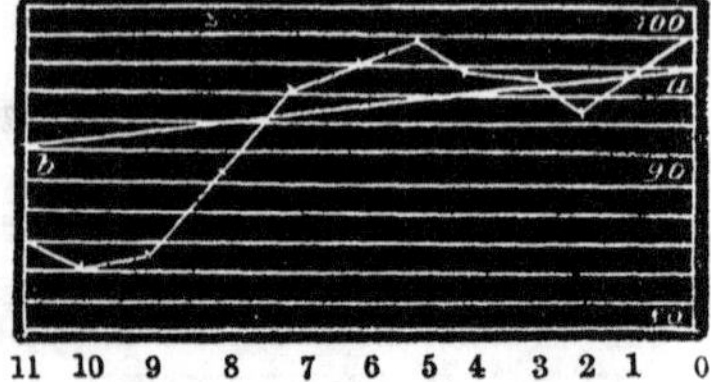

The zigzag line from 100 of altitude to 86, represents the surface of ground, and suppose that we wished to reduce it to a regular grade so as to remove as little earth as possible. By the mere exercise of judgment, we conclude to run the grade between 0 station and 11, from 98 to 92 (but the propriety of this conclusion would depend on the contour of the ground on each side of these stations). The direct line *a b* drawn, shows that the grade runs out of the ground at station 1, we must fill in about $2\frac{1}{2}$ feet at station 2, the grade runs into the ground again at about 80 feet before we come to station 3. At section 4 the cutting must be a little over 2 feet, at station 5 a little over 5 feet, at 7, 2 feet, and runs out of the ground midway between stations 7 and 8.

At stations 9 and 10 we must fill in about 8 feet, and so on; the depth of cutting or filling is obvious at every station.

If the contour of the ground beyond 11 was generally level or descending, we would change the grade at station 7 and render it more descending, so as to make less filling up at stations 9, 10, and 11. In the adoption of grades for a railroad, an engineer has great scope to exercise his judgment.

In the representation here made, *ab* appears like a steep grade, but it is not; it could scarcely appear on the ground other than a level, for the difference is only 6 feet in 2200 feet of distance, which is at the rate of $14\frac{4}{10}$ feet to the mile.

An engineer can freely vary his grade, while it keeps under 18 or 20 feet to the mile; but they submit to a great deal of cutting and filling, before they establish a grade over 40 feet to the mile.

CONTOUR OF GROUND.

Contour of ground is shown on maps, by marking where parallel planes run out on the surface.

We shall give only the general principle.

Let A be the top of a hill, whose contour we wish to delineate; measure any convenient line as AB, up or down the hill, and by the level or theodolite, ascertain the relative elevations of a, b, c, d, &c., as many planes as we wish to represent.

At a, place the level or theodolite, and level it ready for observation; measure the height of the instrument, and put the target on the rod at that height.

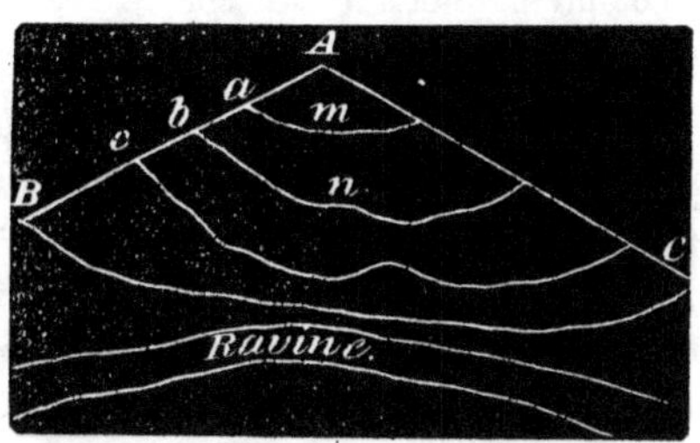

Send the rod-man and axe-man round the hill, on the same level as the instrument. Let the rod-man set the rod; the leveler will sight to it through the telescope, and if the target is below the level, he will motion the rod-man up the hill, if too high, down the hill; at length he will get the same level, and there the axe-man will drive a stake. In the same manner we will establish another stake further on; and thus proceed from point to point. To get round the hill, it may be necessary to move the instrument several times. The plane thus established, is represented by the curve am.

In the same manner, by placing the instrument at b, we can establish the next plane bn.

Then the next, and the next, as many as we please. Where the hill is more steep, two of these parallel planes will be nearer together in the figure; where less steep, they will appear at a greater distance asunder, and this, with the proper shading, will give a true representation of the ground.

ELEVATIONS DETERMINED BY ATMOSPHERIC PRESSURE, AS INDICATED BY THE BAROMETER.

The higher we ascend above the level of the sea, the less is the atmospheric pressure (other circumstances being the same), and therefore we can determine the ascent, provided we can accurately measure the pressure, and know the law of its decrease.

As this work is designed to be educational as well as practical, we shall here make an effort to explain the philosophy of the problem, in such a manner, as to *force* it upon the comprehension of the learner.

The pressure of the atmosphere at any place, is measured by the height of a column of mercury it sustains in the barometer tube.

It is found *by experiment*, that air is compressible, and the amount of compression is always in proportion to the amount of the compressing force.

Now, suppose the atmosphere to be divided into an indefinite number of strata, of the *same thickness*, and so small that the density of each stratum may be considered as uniform.

Commence at an indefinite distance above the surface of the earth, as at A, and let w represent the weight of the whole column of atmosphere resting on A. Let the small and indefinite distances between AB, BC, CD, &c., be equal to each other, and we shall call them *units* of some unknown magnitude.

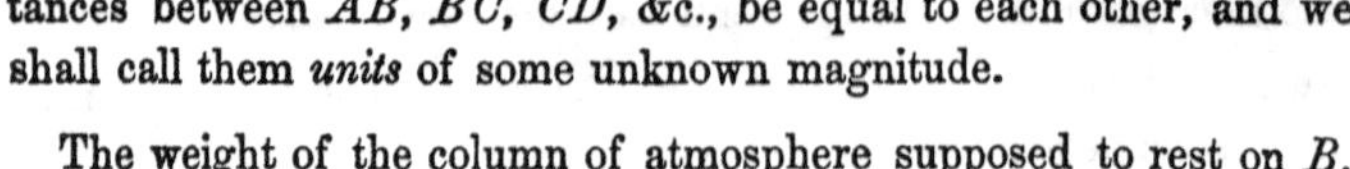

The weight of the column of atmosphere supposed to rest on B, is greater than w, by some indefinite part of w, say the nth part. Then the weight on B, must be expressed by $\left(w+\frac{w}{n}\right)$ or $\left(\frac{n+1}{n}\right)w$.

In the same manner, the weight or pressure resting on C, must be the weight above B, increased by its nth part; that is, it must be $\left(\frac{n+1}{n}\right)w+\left(\frac{n+1}{n^2}\right)w$, which by addition is $\left(\frac{n+1}{n^2}\right)^2 w$.

In the same manner, we find that the weight resting on D, must be $\frac{(n+1)^3}{n^3}w$, and so on. For the sake of perspicuity, we recapitulate.

The pressure on A is	w.	Units from A 0
" " on B is	$\left(\frac{n+1}{n}\right)w$	" " 1
" " on C is	$\frac{(n+1)^2}{n^2}w$	" " 2
" " on D is	$\frac{(n+1)^3}{n^3}w$	" " 3
" " on E is	$\frac{(n+1)^4}{n^4}w$	" " 4
" " on F is	$\frac{(n+1)^5}{n^5}w$	" " 5
	&c. &c.	&c. &c.

Here we observe the series which represents the pressure of atmosphere, at the different points A, B, C, &c., is a series in *geometrical progression*, and it corresponds with another series in *arithmetical progression;* therefore, by the nature of logarithms, the numbers in the arithmetical series, may be taken as the logarithms of the numbers in the geometrical series.

But this system of logarithms, may not be *hyperbolic* nor tabular, indeed it is neither ; the *base of this system* is as yet unknown, but our investigations will soon lead to its discovery.

Now, let the number of units from A to S (the surface of the sea), or *to the lower of two stations*, be represented by x, then the expression for the pressure of the air would be $\left(\frac{n+1}{n}\right)^x w$, but this is neither more nor less than the weight of the column of mercury in the barometer, which is sustained by this pressure.

By calling this b, and designating the logarithms of this unknown system by L', we shall have

$$L'b = x \qquad (1)$$

Taking y to represent the number of units from A to V, and $b_{,}$ to represent the pressure of the air at that point, we shall have

$$L'b_{,} = y \qquad (2)$$

Subtracting (2) from (1), we shall have

$$L'b - L'b_{,} = x - y = SV$$

This is, a certain *peculiar logarithm* of the barometer column at the lower station, diminished by the logarithm of the barometer at the upper station will give the difference of levels between the two stations. But still all is indefinite and unknown, because we know nothing of these logarithms.

In algebra, we learn that the logarithms of one system can be converted into another by multiplying them by a constant multiplier called the *modulus* of the system, therefore,

Assume Z to be the modulus or constant that will convert common logarithms into these peculiar logarithms.

Then, $$Z(\log. b - \log. b_{,}) = SV \qquad (3)$$

Here, log. b denotes the common logarithms of the barometer column.

Equation (3) is general, and determines nothing until we know SV in some particular case.

Taking SV some *known elevation*, and observing the altitude of the barometer column at both stations, and then equation (3) will give Z once for all.

Putting h to represent the *known elevation*, and we have, in general,

$$Z = \frac{h}{\log. b - \log. b_{,}} \qquad (4)$$

Example.— At the bottom and top of a tower, whose height was 200 feet, the mercury stood in the barometer as follows.

At the bottom, - - - - 29.96 inches $= b$

At the top, - - - - - 29.74 inches $= b_{,}$

the temperature of the air being 49° of Fahrenheit's thermometer.

Whence, $$Z = \frac{200}{\log. 29.96 - \log. 29.74} = \frac{200}{0.003201} = 62500 \text{ nearly.}$$

" But this multiplier is constant only when the mean temperature of the air at the two stations is the same; and for a lower temperature the multiplier is less, and for a higher it is greater. *A correction, however, may be applied for any deviation from an assumed temperature, by increasing or diminishing (according as the temperature is higher or lower) the approximate height by its 449th part for each degree of Fahrenheit's thermometer.* We can moreover change the multiplier to a more convenient one by assuming such a temperature as shall reduce this number to 60000 instead of 62500. Now 62500 exceeds 60000 by its 25th part; and, since 1° causes a change of one 449th part, the proportion

$$\tfrac{1}{449} : 1^{\circ} :: \tfrac{1}{25} : 17.9,$$

gives 18° nearly for the reduction to be made in the temperature of the air at the time of the above observations, in order to change the constant multiplier from 62500 to 60000, or to 10000, by calling the height fathoms instead of feet. Thus, instead of the thermometer standing at 49°, we may suppose it to stand at 49°—18° or 31°; and then, we take 10000 as the multiplier, and apply a correction additive for the 18° excess of temperature."

The same observations, for example, being given, to find the height of the tower.

29.96 - -	log. - -	1.47654
29.74 - - -	log. - - -	1.47334
	Diff. of log. - -	0.00320
	Multiplier - - -	10000
	Product - - -	32

Then the height of the tower is 32 fathoms, or $32 \times 6 = 192$ feet, on the supposition that the temperature of the air was 31° in place of 49°. But it being 49°, we must increase 192 by its $\frac{1}{449}$ part for each degree above 31°, that is, by $\frac{18}{449}$ or $\frac{1}{25}$ nearly of its approximate height, which gives nearly 8 feet to add to 192, making 200 feet for the height of the tower.

The same method is applicable to other cases whatever be the temperature of the air at the two stations, provided it be the same or nearly the same at both stations, or provided we take the mean temperature of the two stations. We can find the difference of levels of two stations to considerable accuracy by the following

RULE.— 1st. *Take the difference of the logarithms of the two barometer columns, and remove the decimal point four places to the right. This is the approximate difference of levels in fathoms.*

2d. *Add $\frac{1}{449}$ of the approximate height for each degree of temperature above* 31°, *and* SUBTRACT *the same for each degree below* 31°; *the result cannot be far from the truth.*

EXAMPLES.

1. The barometer at the base of a mountain stood at 29.47 inches, and taking it to the top, it fell to 28.93 inches.

The mean temperature of the air was 51°. What was the height of the mountain in feet? Ans. 503.34 feet.

29.47 log. - - -	1.469380
28.93 log. - - -	1.461348
	0.008032

Approximate height in fathoms, 80.32.

Correction.—Add $\frac{20}{449}$ of 80.32 to itself, that is, add 3.57.

Height, in fathoms, - -	83.89
Multiply by - - -	6
Height in feet, - -	503.34

The average height of the barometer, at the level of the sea, is 30.09 inches; and now if we know the average height of the barometer at any other place, and the average temperature, it is equivalent to knowing the elevation of the latter place above the level of the sea.

For example, the mean height of the barometer at Albany Academy is 29.96, and the mean temperature is 49°. How high is the academy above tide water?

Ans. 117.3 feet.

30.09 log.	1.478422	49°
29.96 log.	1.476542	31
	0.001880	18°
Approximate height	18.80	fathoms.
Add $\frac{18}{449}$ or $\frac{1}{25}$	75	
	19.55×6=117.3 feet.	

2. The average height of the barometer at Amenia Seminary in Duchess, Co., New York, is stated in the Regents' report at 29.81 inches: average temperature 49°. What is the height of that point above tide water?

Ans. 253.32 feet.

3. The mean height of the barometer at Pompey Academy, Onondagua Co., New York, is 28.13, corrected and reduced to 32° Fahr. What then is the elevation of that locality?

Ans. 1755 feet.

Others have made it 1745 feet.

4. From various observations on the summit of Mount Washington, in New Hampshire, the mean height of the barometer there is 24.20, mean temperature at the times of observation was about 65° Fahr.

Now admitting that the mean temperature at the surface of the sea, in the same latitude must have been 75°, which would make the mean temperature between the two stations 70°, what then is the height of Mount Washington above the sea?

Ans. 6170 feet, nearly.

By some observations the elevation is estimated at 6496 feet, by others at 6290 feet.

5. Lieutenant* Fremont, in his narrative of the exploring expedition across the Rocky Mountains, page 45, under date of July 13th, 1842, states his latitude at 41° 8′ 31″, longitude 104° 39′ 37″, height of the barometer 24.86 inches, *attached* thermometer 68°; what was his elevation above the sea? Ans. 5389.2 feet.

REMARK.—The author states his elevation at 5449. feet. As he does not state the temperature of the air by the *detached* thermometer, we know not what correction he made. These *solitary* barometrical observations are more or less valuable, according to the settled or unsettled state of the weather. A person of experience and good judgment in such matters, like Fremont, would not of course record the inapplicable observations.

6. Lieut. Fremont, in his journal, page 104, under date of August 15th, 1842, when, as he supposed, on the highest point of the Rocky Mountains, observed the barometer to stand at 18.29 inches, and the attached thermometer at 44°†; what was the elevation above the level of the sea? Ans. 13522 feet.

Fremont estimates the elevation at 13570 feet.

This shows that he estimated the mean temperature above 50°, and no doubt a similar cause made the difference in the result of the previous example.

7. On page 140, of Fremont's Journal, under date of July 12th, 1843, he says; "The evening was tolerably clear, with a temperature at sunset, of 63°. Elevation of the camp, 7300 feet."

Taking the mean temperature of the two stations, the sea and his

---

* Lieutenant was his proper title at this time.

† If the sea were at the base of the mountain, the temperature at the lower station would no doubt be as high as 60°. Making this supposition, the mean temperature of the two stations would be 50°. We therefore take 50° as the mean temperature.

place of observation, at 67°, what must have been the height of his barometer ? Ans. 23.21 inches.

Represent the approximate elevation by y, then

$$y+\frac{36y}{449}=7300 \qquad \text{Or, } y=6738.14$$

Divide y by 6, which gives 1126.35. Divide this by 10000. Then, let x represent the altitude of the barometer column.

Whence, $1.478422-\log. x=0.112635$

Therefore, $\log. x=1.365787$

In the preceding examples we could only be general and approximate, we had only the observations at one station referred to general observations at the other; but our results cannot be far from the truth.

When the difference of temperatures at the two stations is considerable, the result must be affected by it.

When the upper station is the coldest, which is generally the case, the mercurial column will be shorter than it otherwise would be, and consequently indicate too great a height.

If the temperature of the upper station is taken for the temperature of the lower, the mercurial column at the lower station would not be high enough, and the deduced result would be too small, as is the case in example 5.

The contraction of mercury being about one 10000th part for each degree of cold, or, 0.0025 in a column of 25 in., it would require 4° difference of temperature, to produce an effect amounting to one division on the scale of a common barometer, where the graduation is to hundredths of an inch.

This correction is combined with the former in the following equation, in which $t\,t'$ represents the temperature of the air at the two stations; t at the lower station, q and q' the temperature of the mercury, as indicated by the *attached* thermometer.

The fraction 0.00223, is equal to $\frac{1}{449}$ nearly; h=the height sought, b and $b_{,}$ represent the observed height of the mercurial column.

$$h=10000\left\{1+0.00223\left(\frac{t+t'}{2}-31\right)\right\}\log.\frac{b}{b_{,}(1+0.001(q-q')}$$

Beside the corrections previously considered, regard is sometimes had to the effect of the variation of gravity in different latitudes, and at different elevations above the earth's surface. The latter, however, is too small to require any notice in an elementary work. The former may be found by multiplying the approximate height by 0.0028371 × cos. 2 lat. It is additive, when the latitude is less than 45°, and subtractive when greater. Or it may be taken from the following table.

Latitude.	Correction.
0°	$+\frac{1}{352}$ of the app. height.
5°	$+\frac{1}{358}$
10°	$+\frac{1}{375}$
15°	$+\frac{1}{407}$
20°	$+\frac{1}{460}$
25°	$+\frac{1}{548}$
30°	$+\frac{1}{705}$
35°	$+\frac{1}{1030}$
40°	$+\frac{1}{2030}$
45°	0
50°	$-\frac{1}{2030}$
55°	$-\frac{1}{1030}$
60°	$-\frac{1}{705}$
65°	$-\frac{1}{548}$
70°	$-\frac{1}{460}$
75°	$-\frac{1}{407}$
80°	$-\frac{1}{375}$
85°	$-\frac{1}{358}$
90°	$-\frac{1}{352}$

Given, the pressure of the atmosphere at the bottom of a mountain, equal to 29.68 in. of mercury, and that at its summit, equal to 25.28 in., the mean temperature being 50°, to find the elevation.

Ans. 727.2 fathoms, or 4363.2 feet.

The following observations being taken at the foot and summit of a mountain, namely,

at the foot,	bar. 29.862	attach. therm. 78°	detach. therm.	71°
at the summit,	" 26.137	" 63°	"	55°

to find the elevation.

Ans. 618.9 fathoms, or 3713.4 feet.

It is required to find the height of a mountain in latitude 30°, the observations with the barometer and thermometer being as follows; namely,

at the foot,	bar. 29.40	attach. therm. 50°	detach. therm.*	43°
at the summit,	" 25.19	" 46°	"	39°

Ans. 683.27 fathoms, or 4099.62 feet.

If we assume any temperature, for instance 45°, and the height of the barometer at the level of the sea, at 30.09 inches; we can compute the elevation of the point, where it would be 29.99, 29.89, 29.79, 29.69, &c., inches; and thus we might form a table, showing the elevations that would correspond to any assumed height of the barometer at that temperature. It will be found, that the first fall of $\frac{1}{10}$ of an inch will correspond to about 88 feet in elevation, but every subsequent tenth, will require a greater and greater elevation.

* The attached thermometer measures the temperature of the mercury in the barometer, and the detached thermometer, that of the surrounding air.

NAVIGATION.

CHAPTER I.

INTRODUCTION.

Navigation is the art of conducting a ship from one place to another.

In most works this art is mixed up with seamanship and elementary science. In this work, navigation will stand by itself — alone; and we shall presume that the reader is properly prepared in elementary science.

This being the case, it will not be necessary to take up time and space in giving definitions of latitude, longitude, meridian, horizon, &c., &c., the previous indispensible knowledge of geography necessarily gives a knowledge of all these terms.

Navigation, rightly understood, requires an accurate knowledge of the geography of the seas — the winds and currents that here and there prevail, and also a good general knowledge of astronomy.

Running a line in surveying and running a course at sea, are mathematically the same thing, except that the latter is on a larger scale than the former, without its accuracy, and it is for a different object.

In surveying we take no account of the magnitude and figure of the earth. In navigation we are compelled to do so, unless the limits of the operation be very small.

There are two methods of finding the position of a ship.

1st. By tracing her courses and distances, as in a survey. This is called *dead* reckoning.

2d. By deducing latitude and longitude from observations on the heavenly bodies. This is called *nautical astronomy*.

No one expects accuracy from *dead reckoning*, and as a general thing it is only used from day to day, between observations; or to keep the approximate run during cloudy weather or until observations can again give a new starting point.

Some navigators keep a continuous dead reckoning from port to port, which enables them to judge of drifts, currents, and unknown causes of error.

The earth is so near a sphere that for the practical purposes of navigation it is taken as precisely so. Its magnitude corresponds to 69¼ English miles to one degree of the circumference, but in the early days of navigation 60 miles were supposed to be about a degree, which for the sake of convenience is still retained.

The *sixtieth* part of a degree is called a *nautical mile*, and it is, of course, larger than an English mile.

In an English mile there are - - - 5280 feet.
In a nautical mile there are - - - 6079 feet.

The rate which a ship sails is measured by a line running off of a reel, called the *log line.*

The log is nothing more than a piece of thin board in the form of a sector, of about six inches radius, the circular part is loaded with lead to make it stand perpendicular in the water.

The line is so attached to it that the flat side of the log is kept toward the ship, that the resistance of the water against the face of the log may prevent it, as much as possible, from being dragged after the ship by the weight of the line or the friction of the reel.

The time which is usually occupied in determining a ship's rate is half a minute, and the experiment for the purpose is generally made at the end of every hour, but in common merchantmen at the end of every second hour. As the time of operating is half a minute, or the hundred and twentieth part of an hour, if the line were divided into 120ths of a nautical mile, whatever number of those parts a ship might run in a half minute she would, at the same rate of sailing, run exactly a like number of miles in an hour. The 120th part of a mile is by seamen called a *knot*, and the knot is generally subdivided into smaller parts, called *fathoms*. Sometimes (and it is the most convenient method of division) the knot is divided into ten parts, more frequently perhaps into eight; but in either case the subdivision is called a *fathom*.

We shall consider a fathom the tenth of a knot, and as the nautical mile is 6079 feet, the 120th part of it is 50.66, the length of a knot on the line, and a little over 5 feet is the length of a fathom.

The operation of ascertaining the rate of sailing is called by seamen *heaving the log.*

At the end of an hour the *loaded chip*, or log, is thrown over the stern into the sea; a quantity of the line, called the *stray line*, is allowed to run off, then the glass is turned and the number of knots that runs off the reel during the *half* minute is the rate of the ship's motion.

The log is then hauled in and the same operation is repeated at the end of the next hour.

The officer of the watch, who has been on deck during the hour, will mark on the slate or board, called the log board, the number of miles and parts of a mile which the ship has sailed during the last hour, according to the best of his *judgment;* the log was thrown only to help make up that judgment, for the rate at the time the log was thrown may have been considerably more or less than the average motion during the hour.

The course of a ship is marked by the mariner's compass.

The mariner's compass differs from the surveyor's compass only by its construction, that is, the magnetic needle is the motive power in both. In consequence of the motion of a ship at sea, the mariner's compass is suspended in a double box, moving on a double axis, one at right angles with the other, the whole balanced by a central weight which keeps the compass card nearly steady and horizontal, whatever be the motion of the vessel.

The card is attached to the needle, and is moved by the needle. The card is divided into 32 equal parts, called points, and to read over these points in order, is called by seamen, *boxing the compass,* and to know the north star and box the compass is too often the amount of the common sailor's knowledge of navigation.

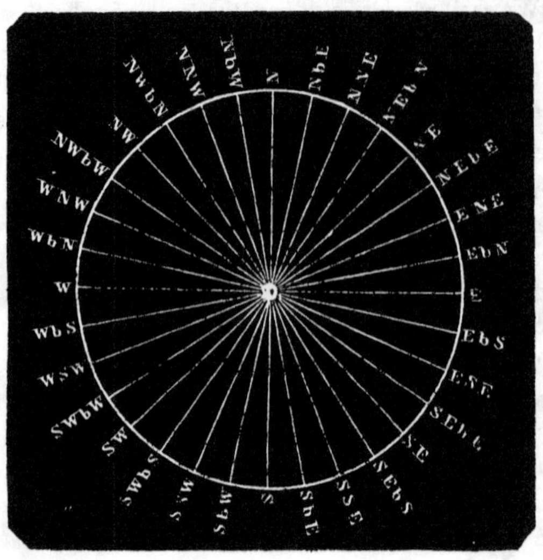

The figure before us represents the card of the mariner's compass. The four quadrant points are marked by a single letter as *N.* for north, *E.* for east. The midway points between these by two letters, as *N. E.* for north-east, *N. W.* for north-west, &c. One point either way from any one of these eight points is marked by the word *by.* Thus, *N.* by *E.* is one point from the north toward the east, and it is read north by east; *S. E.* by *S.* indicates one point from the south-east more to the south, and it is read *south-east by south; W.* by *N.* means west one point toward the north.

To box the compass we begin at any point, as north, and mention every point in order all the way round, thus:

North; north by east; north north-east; north-east by north; north-east; north-east by east; east north-east; east by north; east, &c.

A point of the compass is 11° 15′, which is subdivided into four equal parts. Mariners never take into account a less angle than a quarter point, in running a course.

When the mariner sets his course, he makes allowance for the variation of the needle, and his magnetic courses he reduces to true courses, by the following

Rule. — *Make a representation of the compass card on paper, and draw a line through the compass course.*

Now, conceive the compass card turned equal to the variation to the eastward, if variation is west, and VICE VERSA.

The line will now pass over the true course.

In the following examples, the true courses are required.

	Compass Course.	Variation.	Answer. True Course.
1.	*S. S. E.* ¾ *E.*	2 ¼ *W.*	*S. E.* b *E.*
2.	*E.* ½ *N.*	3 *E.*	*E. S. E.* ½ *S.*
3.	*N. W.* b *W.*	3¾ *E.*	*N.* b *W.* ¼ *W.*
4.	*W. S. W.* ½ *S.*	4 *W.*	*S.* b *W.* ½ *W.*
5.	*S. S. W.*	1½ *W.*	*S.* ½ *W.*
6.	*N.*	5 *E.*	*N. E.* b *E.*
7.	*E.* b *S.*	2¼ *E.*	*S. E.* ¾ *E.*
8.	*S.* 60°*E.*	18°*W.*	*S.* 78°*E.*
9.	*N.* 24 *W.*	36 *E.*	*N.* 12 *E.*
10.	*S.* 16 *W.*	40 *E.*	*S.* 56 *W.*

LEEWAY.

The angle included between the direction of the fore and aft line of a ship, and that in which she moves through the water, is called the *leeway.*

When the wind is on the right hand side of a ship, she is said to be on the *starboard* tack; and when on the left hand side, she is said to be on the *larboard* tack; and when she sails as near the wind as she will lie, she is said to be *close hauled.* Few large vessels will lie within less than six points to the wind, though small ones will sometimes lie within about five points, or even less; but, under such circumstances, the real course of a ship is seldom precisely in the direction of her head; for a considerable portion of the force of the wind is then exerted in driving her to leeward, and hence her course through the water, is in general found to be leeward of that on which she is steered by the compass. Therefore, to determine the point toward which a ship is actually moving, the leeway must be allowed *from the wind,* or toward the *right* of her apparent course, when she is on the *larboard* tack; but toward the *left,* when she is on the *starboard* tack.

It is only when a ship crowds to the wind, that leeway is made.

It is seldom that two ships on the same course make precisely the same leeway; and it not unfrequently happens, that the same ship makes a different leeway on each tack. It is the duty of the officer of the watch, to exercise his best skill in determining, or estimating, how much this deviation from the apparent course amounts to; and *in the dark*, the chief reliance must be placed on the judgment of the experienced mariner.

CHAPTER II.

PLANE SAILING.

In plane sailing, the earth is considered as a plane, the meridians as parallel straight lines, and the parallels of latitude as lines cutting the meridians at right angles. And though it is not strictly correct to consider any part of the earth's surface as a plane, yet when the operations to be performed are confined within the distance of a few miles, no material error will arise, from considering them as performed on a plane surface. And, as we have already seen, in all questions where the *nautical distance, difference of latitude, departure, and course*, are the objects of consideration, the results will be the same, whether the lines are considered as curves drawn on the surface of the globe, or as equal straight lines drawn on a plane.

In all maps, and charts, and constructions, when it is not otherwise stated, it is customary to consider the top of the page as pointing toward the north, the lower part as the south, the right side as the east, and the left as the west. The meridians therefore, in any construction, will be represented by vertical lines, and the parallels of latitude, by horizontal ones.

Hence, in constructing a figure for the solution of any case in plane sailing, the difference of latitude will be represented by a vertical line, the departure by a horizontal one, and the distance by the hypotenusal line, which forms, with the difference of latitude and departure, a right-angled triangle; and the course will be the angle included between the difference of latitude and distance.

With this understanding, the solution of any case that can arise from varying the data in plane sailing will present no difficulty.

EXAMPLES.

If a ship sail from Cape St. Vincent, Portugal (Lat. 37° 2′ 54″ north), *S. W.* ½ *S.* 148 miles; required her latitude in, and the departure which she has made?

By Construction.—Draw the vertical line *AB*, to represent the meridian; from the point *A*, make the angle *BAC*=3½ points, the given course; and from a scale of equal parts, take *AC*=148 miles, the given distance; from *C* on *AB*, draw the perpendicular *CB*, then *AB* will be the difference of latitude, and *BC* the required *departure*, and measured on the scale from which *AC* was taken, *AB* will be found 114.4, and *BC* 93.9.

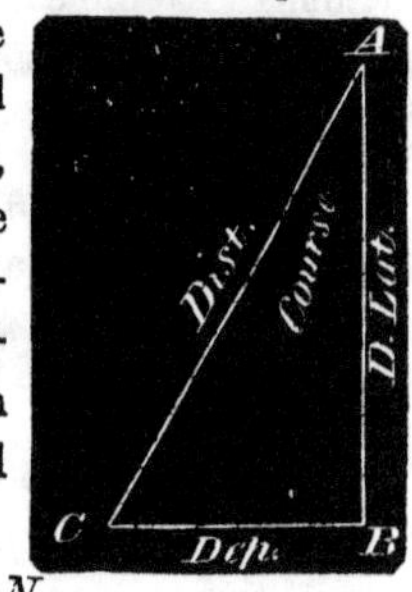

Lat. left - - -	37° 2′ 54″ *N.*	
Diff. lat. - - - -	1 54 24 *N.*	
Lat. in - - - -	35° 8′ 30″ *N.*	Dep. 93.9

2. If a ship sail from Oporto Bar, in Lat. 41° 9′ north, *N. W.* ¼ *W.* 315 miles; required her departure and the latitude arrived at?

Ans. Dep. 233.4 miles *W.*; Lat. 44° 41′ *N.*

3. If a ship sail from lat. 55° 1′ *N.*, *S. E.* by *S.*, till her departure is 45 miles; required the distance she has sailed and her latitude? Ans. Dep. 81 miles; Lat. 53° 54′ *N.*

4. A ship from lat. 36° 12′ *N.*, sails south-westward, until she arrives at lat. 35° 1′ *N.*, having made 76 miles departure; required her course and distance.

Ans. Course *S.* 46° 57′ *W.*; Distance 104 miles.

5. If a ship sail from Halifax, in lat. 44° 44′ *N.*, *S. E.* ½ *E.*, until her departure is 128 miles; required her latitude and distance sailed.

Ans. Lat. 42° 51′ *N.*, and dis. sailed 165.6 miles.

6. A ship leaving Charleston light, in latitude 32° 43′ 30″ north, sails *N.* eastward 128 miles, and is then, by observation, found 93 miles north of the light; required her course, latitude, and departure. Ans. Lat. 33° 22′ 30″ *N.*; Course *N.* 72° 16′ *E.*; Dep. 122 miles.

7. A ship from Cape St. Roque, Brazil, in latitude 5° 10′ south, sails *N. E.* ½ *N.*, 7 miles an hour, from 3 P. M. until 10 A. M.; required her distance, departure, and latitude in.

Ans. Dis. 133. miles; Dep. 84.4 miles; Lat. in, 3° 27′ south.

8. A ship from latitude 41° 2′ *N.*, sails *N. N. W.* $\frac{3}{4}$ *W.* 5$\frac{1}{2}$ miles an hour, for 2$\frac{1}{2}$ days; required her distance, departure, and latitude arrived at.

Ans. Dis. 330 miles; Dep. 169.7 miles; Lat. 45° 45′ *N.*

Similar examples, might be given without end, but these are sufficient, for they only involve the principles of the solution of a plane right-angled triangle.

In the preceding examples it will be observed that we traced latitude from latitude, and the distances east and west we called departure—*not difference of longitude. It now remains to determine difference of longitude from departure.*

On the equator, 60 miles of departure are equal to one degree of longitude, and the further we are from the equator, north or south, that is, the greater the latitude we are in, the same departure will cover more longitude.

To discover the *law* for changing departure to difference of longitude, we adduce the following figure.

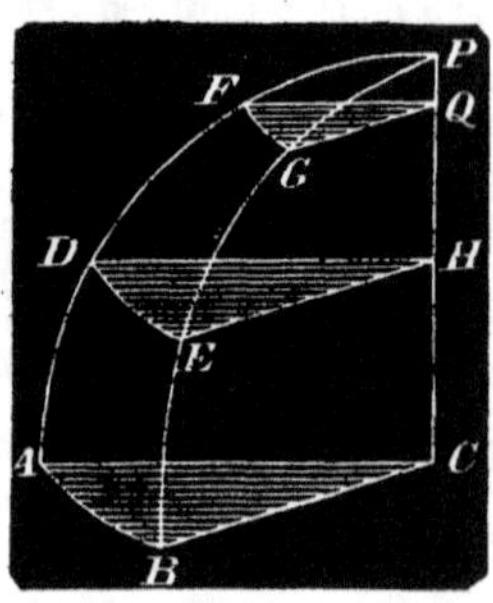

Let C be the center of the earth, P the pole, PC a portion of the earth's axis.

The plane PCB is the plane of one meridian, and PCA the plane of another meridian.

AB is a portion of the equator between the two meridians. ACB is a sector in the plane of the equator, and DEH and FGQ are sectors similar to ACB.

Observe that DE, FG, &c., are parallels of latitude, that is, they are parallel to AB, the plane of the equator.

The magnitude of DE is called *departure,* and it corresponds to the difference of *longitude* AB.

Also, FG is departure corresponding to the same difference of longitude AB. The difference of longitude is always greater than any corresponding departure, that is, AB is obviously greater *than any other parallel distance* between the same two meridians.

Because the two sectors ACB DEH are similar, we have the proportion.

$$AC : DH : : AB : DE \qquad (1)$$

Observe that AC is the radius of the sphere, DH is the sine of the arc PD, or the cosine of DA, which is the cosine of the latitude of the point D.

Therefore the preceding proportion becomes,

rad. : cos. lat. : : dif. lon. : dep.

Or, cos. lat. : rad. : : dep. : dif. lon.

In words, *The cosine of the latitude is to the radius so is the departure to the difference of longitude.*

These words are indelibly engraved on the memory of every navigator, and they embrace all the rules that can be made for changing departure into longitude, or longitude into departure.

When a ship sails east or west, the distance sailed is called departure, and is reduced to longitude by the preceding proportion.

This is called parallel sailing.

EXAMPLES.

1. What difference of longitude corresponds to 47 miles departure in the latitude of 37° 23′? Ans. 59.15 miles.

Let x= the difference of longitude required.

Then $\cos. 37° 53' : \text{rad.} : : 47 : x = \dfrac{47 \times \text{rad.}}{\cos. 37° 23'}$

By log.			
	47 R	- - -	11.672098
	Cos. 37° 23	- - -	9.900144
	Diff. Lon. 59.15	- - -	1.771954

2. How many miles, or how much departure corresponds to a degree in longitude on the parallel of 42° of latitude?

Ans. 44.59 miles.

Here the longitude of one degree is given.

$$R. : \cos. 42° : : 60 : x = \frac{60 \cos. 42°}{R.}$$

By log.			
	60	- - - -	1.778151
	cos. 42°	- - - -	9.871073
	44.59	- - - -	1.649224

N. B.— In this manner the length of a degree in longitude corresponding to each degree of latitude has been found and put in a table.

3. A ship sails east from Cape Race, 212 miles; required her longitude. The latitude of the cape is 46° 40′ *N.*, longitude 53° 3′ 15″ west. Ans. lon. 47° 54′ west.

4. Two places in lat. 50° 12′ differ in longitude 34° 48′; required their distance asunder in miles. Ans. 1336.

5. How far must a ship sail *W.* from the Cape of Good Hope, that her course to Jamestown, St. Helena, may be due north?

Ans. 1193 miles.

Cape { lat. 34° 29′ *S.* / lon. 18° 23′ *E.* Jamestown { lat. 15° 55′ *S.* / lon. 5° 43′ 30″ *W.*

6. How far must a ship sail *E.* from Cape Horn to reach the meridian of the Cape of Good Hope? The latitude of Cape Horn being 55° 58′ 30″ *S.*, lon. 67° 21′ *W.*, and the latitude and longitude of the Cape of Good Hope being as stated in the last example.

Ans. 2878 miles.

7. In what latitude will the difference of longitude be three times its corresponding departure? In other words in what latitude will *FG* be one-third of *AB*? (See last figure.)

Ans. lat. 70° 31′ 44″.

8. Take the 2d example in plane sailing (page 181), the departure made, as stated in the answer, is 233.4 miles. What is the corresponding difference of longitude? Ans. 5° 18′ 24″.

This inquiry now arises. To what latitude does this departure correspond? Is it to the latitude left, 41° 9′, or to the latitude arrived at, 44° 41′? Or does it correspond to the mean latitude between the two?

If we suppose the departure corresponds to lat. 41° 9′, then the difference of longitude by the preceding rule must be 5° 10′, and if the departure corresponds to lat. 44° 41′, then the difference of longitude is 5° 28′; the mean of these is 5° 19′, and if we take the departure to correspond with the mean latitude 42° 55′, then the difference of longitude would be 5° 18′ 24″.

In the examples under Plane Sailing we have supposed the earth a plane, and the course a ship sails a straight line, but neither supposition is strictly true.

Meridians are *not parallel* with each other, and therefore when a ship sails by the compass, and cuts all the meridians at the same angle, the line that the ship sails will not be a right line: it will be a curve line peculiar to itself, called a *rhumb line.*

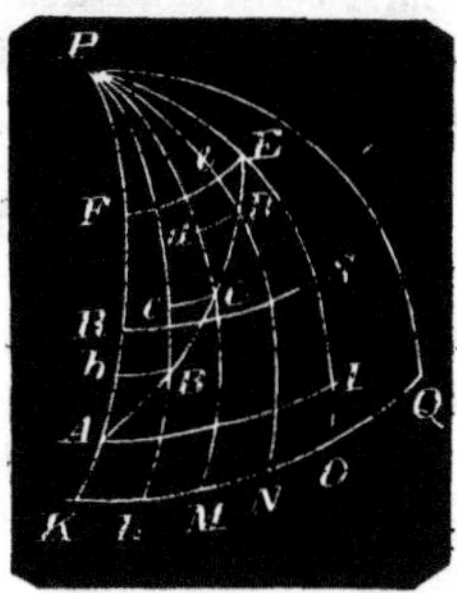

For the sake of illustration, let us suppose that in the annexed figure, P is the north pole, KQ the equator, or a great circle, every part of which is a quadrant distance from P; PK, PL, PM, &c., great circles passing through P, and of course cutting the equator at right angles; AI, bB, RS, &c., arcs of smaller circles parallel to the equator, and therefore cutting the meridians at right angles; AE a curve cutting every meridian which it meets, as PK, PL, PM, &c., at the same angle. Then PK, PL, &c., produced till they meet at the opposite pole, are called meridians; AI, bB, RS, &c., continued round the globe, are called parallels of latitude; AE is called the rhumb line, passing through A and E; the length of AE is called the nautical distance from A to E; and the angle BAb, or any of its equals, cBC, dCD, &c., is called the course from A to E.

If a ship sail from A to E, EF will be her meridian distance; but if she sail from E to A, AI will be her meridian distance.

If AB, BC, CD, &c., be conceived to be equal, and indefinitely small, and their number indefinitely great; then the triangles ABb, BcC, &c., may be considered as indefinitely small right-angled plane triangles. And as the angles BAb, CBc, &c., are equal, and the right angles AbB, BcC, &c., are equal, the remaining angles ABb, BCc, &c., are equal; and as the sides AB, BC, &c., are also equal, these elementary triangles ABb, BCc, CDd, &c., will be all identical triangles; therefore AE will be the same multiple of AB, that the sum of Ab, Bc, Cd, &c., is of Ab; and that the sum of Bb, Cc, Dd, &c., is of Bb.

It is obvious that

$$Ab+Bc+Cd+\&c.=AF$$

the whole difference of latitude. And,

$$Bb+Cc+Dd+Ee+\&c.=\text{the whole departure.}$$

16

But this departure is neither EF nor AI, it is greater than EF and less than AI; because bB is less than its corresponding portion on AI and greater than its corresponding portion on FE. The same may be said of cC, Dd, &c. Therefore the departure on any course corresponds to *neither of the extreme latitudes*, but to *some* mean between the two, and it is so near the arithmetical mean, that the arithmetical mean is taken as the true.

Therefore in all those examples in Plane Sailing, on page 181, we can take the departures and find the corresponding differences of longitude, provided we take the *middle latitude* and consider the departure run on that parallel.

This method of connecting the change in longitude with a ship's change of place is called

MIDDLE LATITUDE SAILING.

But in reality there is no such thing as middle latitude sailing; the cosine of the middle latitude is compared to the radius, as *the ratio* between the departure made and the corresponding difference of longitude, but the departure made may be made on one course or on several courses. When a ship sails on several courses before the run is summed up, the summing up and finding the result in one course and distance is called *working a traverse*, and sailing from one point to another by several courses is called

TRAVERSE SAILING.

With adverse winds or crooked channels, vessels are obliged to run a traverse. Going round a survey and keeping an account of our course and distance from the starting point is *working a traverse*, and the operation is the same on sea or land, except on land we aim at coming round to the same point again, but on sea we wish to make some other point.

With this explanation it is obvious that we must make a table as in a survey, and compute the course and distance from the starting point, and this is called the *course and distance made good.*

To work a *traverse* we use the *traverse* table of course; that table is made to every half degree, and the column in the table nearest to the course is sufficiently exact.

The following table gives the degree and parts of a degree corresponding to every point and quarter point of the compass.

Points.	Deg.	Points.	Deg.
$\frac{1}{4}$	2° 48′ 45″	$4\frac{1}{4}$	47° 48′ 45″
$\frac{1}{2}$	5° 37′ 30″	$4\frac{1}{2}$	50° 37′ 30″
$\frac{3}{4}$	8° 26′ 15″	$4\frac{3}{4}$	53° 26′ 15″
1	11° 15′	5	56° 15′
$1\frac{1}{4}$	14° 3′ 45″	$5\frac{1}{4}$	59° 3′ 45″
$1\frac{1}{2}$	16° 52′ 30″	$5\frac{1}{2}$	61° 52′ 30″
$1\frac{3}{4}$	19° 41′ 15″	$5\frac{3}{4}$	64° 41′ 15″
2	22° 30′	6	67° 30′
$2\frac{1}{4}$	25° 18′ 45″	$6\frac{1}{4}$	70° 18′ 45″
$2\frac{1}{2}$	28° 7′ 30″	$6\frac{1}{2}$	73° 7′ 30″
$2\frac{3}{4}$	30° 56′ 15″	$6\frac{3}{4}$	75° 56′ 15′
3	33° 45′	7	78° 45′
$3\frac{1}{4}$	36° 33′ 45″	$7\frac{1}{4}$	81° 33′ 45″
$3\frac{1}{2}$	39° 22′ 30″	$7\frac{1}{2}$	84° 22′ 30″
$3\frac{3}{4}$	42° 11′ 15″	$7\frac{3}{4}$	87° 11′ 15″
4	45° 0′	8	90° 0′

In works exclusively designed for practical navigation, the traverse table is adapted exactly to the points and quarter points of the compass, but the table in this work is sufficient for the purpose.

The use of this table is to find the degree corresponding to any given course, thus: *N.* by *E.*, *N.* by *W.*, *S.* by *E.*, *S.* by *W.*, each correspond to 1 point or 11° 15′. In using our traverse table for 1 point we should take a mean result between 11° and 11° 30′, which *mean* result can be taken by the eye.

Again *S.E.* by *E.* $\frac{1}{4}$ *E.* is $5\frac{1}{4}$ points, or *S.* 59° *E.* nearly, and so on for any other course that may be named.

The student is now fully prepared to work the following examples in traverse sailing.

EXAMPLES.

1. A ship from Cape Clear, Ireland, in lat. 51° 25′ *N.* and longitude 9° 29′ *W.*, sails as follows:

S. S. E. $\frac{1}{4}$ *E.* 16 miles, *E. S. E.* 23 miles, *S. W.* by *W.* $\frac{1}{2}$ *W.* 36 miles, *W.* $\frac{3}{4}$ *N.* 12 miles, and *S. E.* by *E.* $\frac{1}{4}$ *E.* 41 miles.

Required her course and distance made good, the departure, latitude and longitude of the ship.

By Construction.—Take A for the place sailed from, and draw the vertical line $NASC$, to represent the meridian. About A, as a center, with the chord of 60°, describe a circle, cutting NC in N and S; then N and S will represent the north and south points of the compass. Take $2\frac{1}{4}$ points from the line of chords,* and apply it from S to a, join Aa, and on it take $AD=16$ from a line of equal parts. Then D will be the place of the ship at the end of the first course. From S, set off $Sb=6$ points from the line of chords; join Ab, and through D draw DE, parallel to Ab, and make it equal to 23 from the same scale of equal parts that AD was taken from. Then E will be the place of the ship, at the end of the second course. Make $Sc=5\frac{1}{2}$ points, Nd $7\frac{1}{4}$ points, and Se $5\frac{1}{4}$ points, taken from the line of chords. Through E draw EF, parallel to Ac, and make it equal to 36, from the scale of equal parts; through F draw FG parallel to Ad, and make it equal to 12, from the scale of equal parts; through G draw GB parallel to Ae, and make it equal to 21, from the scale of equal parts. Demit BC a perpendicular, on the meridian NC, and join AB; then B will be the place of the ship, AB her distance from the place which she left, AC her difference of latitude, BC her departure, and BAC the course which she has made on the whole. Now, AB, AC, and BC being measured on the scale of equal parts from which the distances were taken, we have $AB=62.7$, $AC=59.6$, and $BC=19.6$ miles. And the arc included by AC and BC, if measured on the line of chords, gives about 18° for the measure of the course BAC.

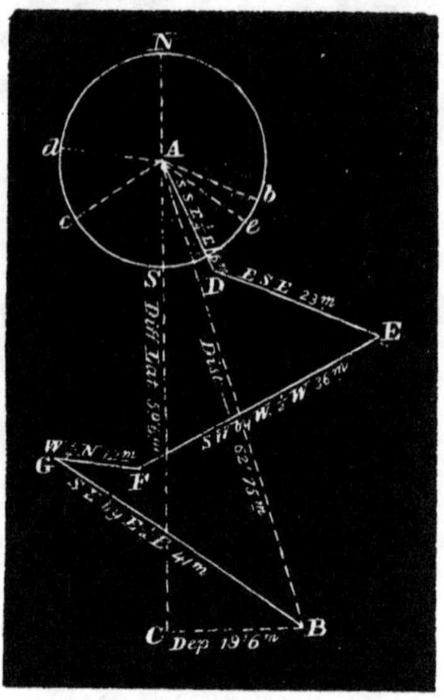

* Two and ¼ points is 25° 18′ 45″, that is, take the chord of 25° 18′ in the dividers, and set it off from s to a, and so on, for other angles.

TRAVERSE TABLE.

Courses.	Points.	Dis.	Diff. Lat. N.	Diff. Lat. S.	Dep. E.	Dep. W.
S. S. E. ¼ *E.*	2¼	16		14.5	6.8	
E. S. E.	6	23		8.8	21.3	
S. W. by *W.* ½ *W.*	5½	36		17.9		31.8
W. ¾ *N.*	7¼	12	1.8			11.9
S. E. by *E.* ¼ *E.*	5¼	41		21.1	35.2	
			1.8	61.4	63.3	45.7
				1.8	43.7	
Result				59.6	19.6	

Lat. left - -	51° 25′ *N.*	
diff. lat. - - -	1 00 *S.*	
Lat. in - -	50 25 *N.*	Mid. lat. 50° 55′

To find the course and distance, by trigonometry,

As dif. lat. 59.6 miles	1.775246
: radius 90°	10.000000
: : dep. 19.6 miles	1.292256
: tan. course 18° 12′	9.517010
As sin. course 18° 12′	9.484621
: dep. 19.6 miles	1.292256
: Radius	10.000000
: Dis. 62.75 miles	1.797635

To find the dif. of longitude.

As cos. 50° 55′	9.799651
: Radius	10.000000
: : dep. 19.6	1.292256
: diff. lon. 31.09 miles	1.492605
Longitude left	9° 29′ west
diff. lon.	31 east
Lon. in	8° 58′ west

Thus, we have found the course 18° 12′; Distance 62.75 miles; diff. longitude 31′ *E.*; lat. in 50.25 *N.*; lon. 8° 58′ *W.*

If these be the distances run in a day, from noon to noon again, then the preceding operation is called working a day's work; otherwise it is called working a traverse, as we have mentioned before,

But this is not the *seaman's* way of working a day's work, he does it all by inspection, in the traverse table. For example, taking the result of the traverse 59.6 south, and 19.6 east, which shows that the resulting course is between the south and the east, and with these numbers he enters the traverse table, and finds, as near as possible, 59.6 and 19.6, standing as latitude and departure; and they are found nearly under the angle of 18°, and opposite the distance 63. nearly.

Hence, he takes his course as *S.* 18° *E.*, and dis. 63. To find the difference of longitude, he takes the *middle* latitude as a course, and the departure as difference of latitude, *then the distance in the table is difference of longitude.*

In this instance, we take 51° as a course, and in the difference of latitude column we find 19.5, and the distance opposite to it is 31., which we take for difference of longitude.

The reason for this is as follows:

For the longitude we have,

$$\text{cos. mid. } L : R :: \text{dep.} : \text{diff. lon.} \qquad (1)$$

In the construction of the traverse table, we have,

$$\text{cos. course} : R. :: \text{diff. lat.} : \text{dist.} \qquad (2)$$

Now, in proportion (2), if we take the middle latitude for a course, and the dep. for difference of latitude; it necessarily follows, that the last term of proportion (2) must be diff. of longitude; for proportion (2) would then be transformed into proportion (1).

2. A ship sails from Cape Clear, as follows; *S.* by *W.* 23 miles; *W. S. W.* 40 miles; *S. W.* $\frac{3}{4}$ *W.* 18 miles; *W.* $\frac{1}{2}$ *N.* 28 miles; *S.* by *E.* 12 miles; *S. S. E.* $\frac{3}{4}$ *E.* 16 miles.

Required the course and distance made good, and the latitude and longitude arrived at.

Ans. Course *S.* 45° 47′ *W.*; dis. 102.4 miles.
Latitude of ship 50° 14′ *N.*; Lon. 11° 25′ *W.*

3. A ship at noon, on a certain day, was in lat. 41° 12′ *N.*, and longitude 37° 21′ *W.*, she then sailed as follows:

S. W. by *W.* 21 m.; *S. W.* $\frac{1}{2}$ *S.* 31 m.; *W. S. W.* $\frac{1}{2}$ *S.* 16 m.; *S.* $\frac{3}{4}$ *E.* 18 m.; *S. W.* $\frac{1}{4}$ *W.* 14, and *W.* $\frac{1}{2}$ *N.* 30 miles.

Required her course, distance, latitude, and longitude.

Ans. course *S.* 52° 49′ *W.*; dis. 111.7; lat. 40° 5′ *N.*; lon. 39° 18′ *W.*

4. Last noon we were in latitude 28° 46′ south, and longitude 32° 20′ west; since then we have sailed by the log:

*S. W.*¾ *W.* 62 m.; *S.* by *W.* 16 m.; *W.* ¼ *S.* 40 m.; *S.W.* ¾ *W.* 29 m.; *S.* by *E.* 30 m.; and *S.* ¾ *E.* 14 miles. Required the direct course and distance, and our present latitude and longitude.

Ans. Course *S.* 43° 14′ *W.*; Dis. 158 m.; lat. 30° 41′ *S.*; lon. 34° 24′ *W.*

5. A ship from Toulon, lat. 43° 7′ *N.*, lon. 5° 56, *E.*, sailed *S. S. W.* 48 m.; *S.* by *E.* 34 m.; *S. W.* ¼ *W.* 26 m.; and *E.* 17 miles. Required her course and distance to Port Mahon, Lat. 39° 52′ *N.*, and longitude 4° 18 30″ east.

Ans. Lat. of ship 41° 32′ *N.*; lon. 5° 37′ east. Course to Port M. *S.* 31° *W.* nearly, and distance 117.5 miles.

6. On leaving the Cape of Good Hope, for St. Helena, we took our departure from Cape Town, bearing *S. E.* by *S.* 12 miles, after running *N. W.* 36 m., and *N. W.* by *W.* 140 miles. Required our latitude and longitude, and the course and distance made.

N. B. Lat. of Cape Town 33° 56′ *S.* Lon. 18° 23′ *E.*
Lat. of St. Helena 15° 55′ *S.* Lon. 5° 43′ 30″ *W.*

Ans. Lat. 32° 3′ *S.*; Lon. 18° 25′ *E.*; course *N.* 52° 41′ *W.*; dis. 187 miles.

SAILING IN CURRENTS.

If a ship at *B*, sailing in the direction *BA*, were in a current which would carry her from *B* to *C*, in the same time that in still water she would sail from *B* to *A*, then, by the joint action of the current and the wind, she would in the same time, describe the diagonal *BD* of the parallelogram *ABCD*. For her being carried by the current in a direction parallel to *BC*, would neither alter the force of the wind, nor the position of the ship, nor the sails, with respect to it; the wind would therefore continue to propel the ship in a direction parallel to *AB*, in the manner as if the current had no existence. Hence, as she would be swept to the line *CD*, by the independent action of the current, in the same time that she reached the line *AD*, by the independent action of the wind on her sails, she would be found at *D*, the point of intersection of the lines *AD* and *CD*, having moved along the diagonal *BD*.

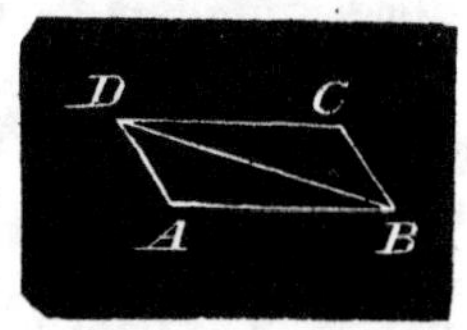

Now the log heaved from the ship in the ordinary way, can give no imitation of a current; for the line withdrawn from the reel is only the measure of what the ship sails *from the log;* and, consequently, as the log itself, as well as the ship, will move with the current, the distance shown by the log in a current, is merely what it would have been if the ship had been in still water.

If the ship sail in the direction of the current, the whole effect of the current will be to increase the distance; but if she sail against the current, the difference between the rate of sailing given by the log and drift of the current, will be the distance which the ship actually goes; and she will move forward, if her rate of sailing be greater than the drift of the current, but otherwise, her motion will be retrograde, or she will be carried backwards, in the direction of the current.

Problems relating to the oblique action of a current upon a ship, may be resolved by the solution of an oblique-angled plane triangle, such as *ABD*, in the preceding figure, where if *AB* represent the distance which a ship would sail in still water, and *AD* the drift of the current in the same time, *BD* will be the actual distance sailed, and *ABD* the change in the course produced by the current.

A great variety of problems might be proposed relative to currents, but the chief ones of any practical importance, are the following:

1. To determine a ship's actual course and distance in a current, when her course and distance by the compass and the log, and the setting and drift of the current, are given.

2. To find the course to be steered through a known current, the required course in still water, and the ship's rate of sailing, being known.

3. To find the setting and drift of a current, from a ship's actual place, compared with that deduced from the compass and the log.

The first of these cases may be conveniently resolved, by considering the ship as having performed a traverse, the setting and drift of the current being taken as a separate course and distance.

EXAMPLES.

1. If a ship sail *W.* 28 miles in a current, which in the same time carries her *N. N. W.* 8 miles, required her true course and distance.

N. B. Conceive the current to be one course and distance, and with the other courses find the course and distance made good.

Thus, by the traverse table:

Course.	Dis.	Diff. Lat N.	Diff. Lat S.	Dep. E.	Dep. W.
W.	28				28
N. N. W.	8				
		7.39			2.06
		7.39			31.06

As 7.39 : rad. : : 31.06 : tan. 76° 36′, the course,
cos. 76° 31′ : *R* : : 31.06 : 31.93, the distance.

2. If a ship sail *E.* 7 miles an hour by the log, in a current setting *E. N. E.* 2.5 miles per hour; required her true course, and hourly rate of sailing.

Ans. Course *N.* 84° 8′ *E.*, and rate 9.358 per hour.

3. A ship has made by the reckoning *N.* ½ *W.* 20 miles, but by observation it is found, that, owing to a current, she has actually gone *N. N. E.* 28 miles. Required the setting and drift of the current in the time which the ship has been running.

Ans. Setting *N.* 64° 48′ *E.*, and drift 14.1 miles.

4. A ship's course to her port is *W. N. W.*, and she is running by the log 8 miles an hour, but meeting with a current setting *W.* ½ *S.* 4 miles an hour, what course must she steer in the current that her true course may be *W. N. W*?

Ans. Course *N.* 44° 39′ *W.*

5. In a tide running *N. W.* b *W.* 3 miles an hour, I wished to weather a point of land, which bore *N. E.* 14 miles. What course must I steer so as to clear the point, the ship sailing 7 miles an hour by the log, and what time shall I be in reaching the point?

Ans. Course *N.* 69° 51′ *E.*, and time 2 hours 25 minutes.

6. From a ship in a current, steering *W. S. W.* 6 miles an hour by the log, a rock was seen at 6 in the evening, bearing *S. W.* ½ *S.* 20 miles. The ship was lost on the rock at 11 P. M. Required the setting and drift of the current.

Ans. Setting *S.* 75° 10′ *E.*, and drift 3.11 miles per. hour.

CHAPTER III.

MERCATOR'S CHART AND MERCATOR'S SAILING.

In representing any small portion of the earth's surface, it is sufficiently accurate to represent the meridians as parallel; but if the portion of the earth is considerable, the representation will not be true unless the meridians are curved.

If we make a chart and draw all the meridians parallel with each other, the length of a degree of longitude in all places, *except on the equator,* will be greater on the chart than its true distance, but the true bearing of one place from another will be preserved, provided we increase the degrees of latitude in the *same ratio* as the degrees of longitude are increased.

Gerrard Mercator, a Fleming, in 1556, published a chart which seemed to embrace this idea, but he did not show its construction, nor were his degrees in their true proportion; but from this came the name of Mercator's Chart.

A Mr. Wright, an Englishman, in 1599, it is said, published the true sea chart, constructed on the following principles.

1. *The distance between two meridians at the equator, is to their distance in any parallel of latitude, as the radius is to the cosine of that latitude.*

2. *Any part of a parallel of latitude, is to a like part of the meridian, as the radius is to the secant of that parallel.*

We shall make an effort to illustrate these principles by the following figure.

Conceive the equator to be extended both ways parallel to the earth's axis, thus forming a cylinder, whose circumference is just equal to the circumference of the earth.

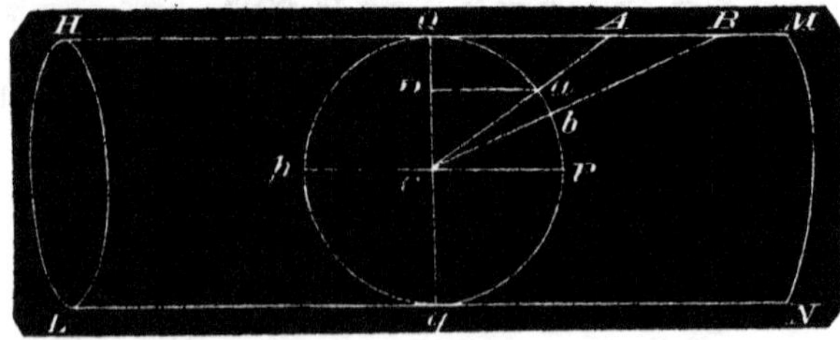

Let Qq be the plane of the equator, Pp the earth's axis; conceive a globe enclosed in the cylinder, *HLNM.*

Suppose there is an island on the earth at *a*, that island is projected on the cylinder at *A*. The surface of the earth at *b* is projected at *B*. Conceive this paper cylinder cut by a line at right angles to the equator and rolled out, it will then be a true representation of Mercator's chart.

The scale on the globe at *a* is, to the scale on the chart at *A*, as *Ca* to *CA*, that is, as radius to the secant of the latitude at *a*.

The scale on the chart at *A* is, to the scale on the chart at *B*, as *CA* is to *CB*, that is, the scale on the chart increases as the secants of the latitudes increase.

The poles of the earth, and places very near the poles, can never be represented on this chart.

The meridian distance of a degree on the globe, as at *a*, is 60 miles, on the chart at *A* it is 60, into *AC* the secant of the latitude, calling *Ca* unity.

If we commence at the equator at *Q*, and take one mile for unity. Then,

Mer. pts. of 1′= nat. sec. 1
Mer. pts. of 2′= nat. sec. 1+ nat. sec. 2
Mer. pts. of 3′= nat. sec. 1′+ nat. sec. 2′+ nat. sec. 3
Mer. pts. of 4′= nat. sec. 1′+ nat. sec. 2′+ nat. sec. 3′
+ nat. sec. 4′, &c., &c.

In this manner the table of *meridional parts* was originally constructed. It is Table IV of this work.

The following figures represent any problem than can arise in Mercator's sailing.

AC represents the true difference of latitude.

AD represents the meridional difference of latitude, which is always taken from the table.

CB represents the departure.

DE the difference of longitude.

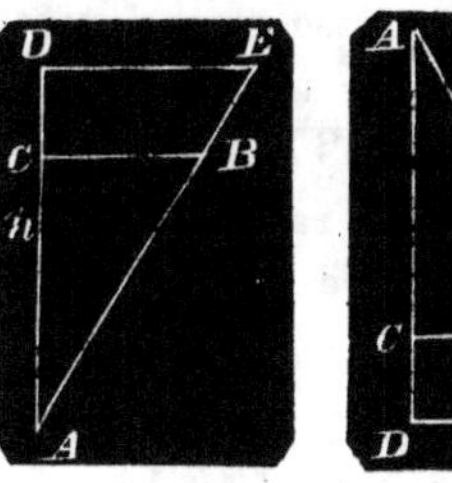

AB represents the distance.

A, the angle at *A*, represents the course.

Three of these *six* quantities must be given to solve a problem.

Observe that the difference of longitude *DE* is always greater than the departure *CB*, as it ought to be.

EXAMPLES.

1. A ship from Cape Finisterre, in lat. 42° 56′ *N.*, and longitude 8° 16′ *W.*, sailed *S. W.* ¼ *W.* till her difference of longitude is 134 miles; required the distance sailed and the latitude in.

By logarithms.	As radius - - - - - - -	10.000000
	: diff. lon. 134 miles - - - -	2.127105
	: : cot. course 4¼ points - - -	9.957295
	: mer. diff. lat. 121.5 miles - -	2.084400

Lat. Cape Finisterre 42° 56′ *N.* Mer. parts -	2858
Mer. diff. -	121
Lat. 41° 27′ *N.*, corresponding to - - -	2737 in table

As cosine course - - - - -	9.827085
: proper diff. lat. 89 miles - -	1.949390
: : radius - - - - - - - -	10.000000
: dis. 132.5 miles - - - - -	2.122305

2. A ship from lat. 40° 41′ *N.*, lon. 16° 37′ *W.*, sails in the *N. E.* quarter till she arrives in lat. 43° 57′ *N.*, and has made 248 miles departure; required her course, distance, and longitude in.

Ans. course *N.* 51° 41′ *E.*, dis. 316 miles, and lon. in 11° *W.*

3. How far must a ship sail *N. E.* ½ *E.* from lat. 44° 12′ *N.*, lon. 23° *W.*, to reach the parallel of 47° *N.*, and what from that point will be the bearing and distance of Ushant, which is in lat. 48° 28′ *N.* and lon. 5° 3′ *W.*?

Ans. She must sail 262 miles, and her course and distance to Ushant will then be *N.* 80° 32′ *E.*, and dis. 535 miles.

4. A ship from the Cape of Good Hope steers *E.* ½ *S.* 446 miles; required her place, and her course, and distance to Kerguelen's Land, in lat. 48° 41′ *S.*, and lon. 69° east.

Ans. lat. in 35° 13′ *S.*, lon. in 27° 21′ *E.*, course *S.* 66° 25′ *E.*, and distance 2018 miles.

5. By observation, a ship was found to be in lat. 41° 50′ *S.*, lon. 68° 14 *E.* She then sailed *N. E.* 140m, and *E* ½ *S.* 76m; required her place, and her course, and distance to the island of St. Paul, which is in lat. 38° 42′ *S.*, and in lon. 77° 18′ *E.*

Ans. lat 40° 18′ *S.*, lon. 72° 2′, course *N.* 68° 35′ *E*, and dis. 263 miles, nearly.

CELESTIAL OBSERVATIONS.

CHAPTER I.

We now come to the more scientific and essential parts of navigation, the determination of latitude and longitude by celestial observations.

We shall at present confine ourselves to latitude, first calling to mind the following necessary definitions and explanations:

1. Meridian. — The meridian of any place is the north and south line passing through that place, and it may be conceived to run along the ground or pass in the same direction in the heavens, through the point vertically over the place. The line on the earth is the terrestrial meridian, the line in the heavens is called the celestial meridian; they are both in one plane with the center of the earth.

2. Equator. — The equator is that circle around the earth over which the sun seems to pass when the days and nights are equal all over the earth.

3. Latitude. — The latitude of any place is the meridian distance of that place from the equator, measured by degrees and parts of a degree of arc.

4. Longitude. — The longitude of any place is the inclination of the plane of its meridian, with the plane of some other definite meridian from which reckoning is made. This inclination is measured on the equator by degrees, minutes, and seconds of *arc*, and it is either east or west.*

---

* The first meridian to reckon from may be arbitrarily chosen, and different nations have taken different meridians for the commencement of longitude, but custom and long association have pretty firmly fixed the meridian of Greenwich (England) as the first meridian for all who use the English language.

5. Declination. — The declination of a heavenly body is its meridian distance from the equator north or south.

6. Polar Distance. — The polar distance of a body is its declination added to, or subtracted from 90°. If both added and subtracted, we shall have the meridian distances from each pole.

The distance from the north pole, is called north polar distance, and from the south pole, south polar distance. The two polar distances must of course make 180°.

7. Zenith. — Zenith is the point in the heavens directly overhead.

8. Horizon. — The horizon is either apparent or real, or as commonly expressed, *sensible* or *rational.*

The *sensible* horizon is a plane conceived to touch the earth at any point at which an observer is situated.

The *rational* horizon is a plane parallel to the sensible one, passing through the center of the earth.

The zenith is the pole to the horizon.

9. Great Circles. — A great circle in the heavens is any circle whose plane passes through the center of the earth.

All great circles which pass through the zenith are perpendicular to the horizon, and such circles are called *vertical circles, azimuth circles,* or *circles of altitude.*

10. Azimuth. — The angle which the meridian makes with that vertical circle which passes through any object is said to be the azimuth of that object. Hence, azimuths may be reckoned from the north or south points of the horizon.

11. Altitude. — The altitude of any object is its angular distance from the horizon, measured on a vertical circle.*

Altitudes are very frequently measured at sea, several times in a day in fair weather; but altitudes observed from the surface of the earth, or above it, require several corrections before the true altitudes can be deduced from them.

---

* We do not pretend to give all the definitions of the sphere, but we suppose the reader is already acquainted with them, from his knowledge of Geography and Astronomy.

These corrections are for *semi-diameter, dip, refraction,* and *parallax.* The correction for semi-diameter is obvious.

At sea, the visible horizon (from which all observed altitudes are taken) is where the sea and sky apparently meet, and when the eye of the observer is above the water, this visible horizon is below the *sensible* horizon, and the amount of the depression is called the *dip of the horizon.* Its correction is always subtractive, and its amount is to be found in Table VI.

Refraction is to be found in Table V. It is always subtractive, and for the reason, see some treatise on natural philosophy.

Parallax is always additive. Conceive two lines drawn to a heavenly body; one from an observer at the circumference of the earth and the other from the center of the earth, the inclination of these two lines is parallax, and when the body is in the horizon its parallax is greatest, and it is then called horizontal parallax.

Parallax always tends to depress the object, but the parallax of any celestial object, except that of the moon, is so small, that we shall pay attention to lunar parallax only, but this is so important to navigation that we shall give it a full explanation.

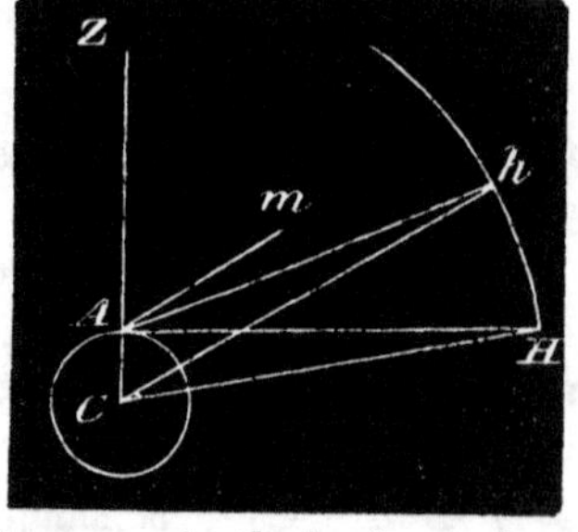

The moon's horizontal parallax is given in the Nautical Almanac for every noon and midnight of Greenwich time, and from the horizontal parallax we must deduce the parallax corresponding to any other altitude.

Let AC be the radius of the earth, A the position of an observer, Z his zenith, and suppose H to be the moon in the horizon; then the angle AHC* is the moon's horizontal parallax, and the angle AhC is the parallax corresponding to the apparent altitude hAH. Draw Am parallel to Ch, then mAH would be the *true altitude.*

---

* From this figure we draw the following definition for horizontal parallax. *The horizontal parallax of any body is the angle under which the semi-diameter of the earth would appear as seen from that body.* Of course then, when the body is at a great distance its horizontal parallax must be small, hence the sun and the remote planets have very little parallax, and the fixed stars none at all.

Let CH and Ch be each represented by R. Put $p=$ the horizontal parallax, and $x=$ the parallax in altitude, or the angle mAh or AhC.

Now in the triangle ACH, right-angled at A, we have

$$1 : \sin. p : : R : AC.$$

In the triangle ACh we have

$$\sin.\ CAh : \sin.\ x : : R : AC.$$

By comparing these two proportions, we perceive that

$$1 : \sin.\ p : : \sin.\ CAh : \sin.\ x$$

Whence, $\quad \sin. x = \sin. p. \sin.\ ACh$

But $\sin. CAh = \cos. hAH$, for the sine of any arc greater than 90° is equal to the cosine of the excess over 90°, hence,

$$\sin. x = \sin. p \cos. hAH$$

The lunar horizontal parallax is rarely over a degree, commonly less, and the sine of a degree does not materially differ from the arc itself, hence, the preceding equation becomes the following, without any essential error.

That is, $\quad x = p \cos.$ altitude.

Or, in words, *the parallax in altitude is equal to the horizontal parallax multiplied into the cosine of the apparent altitude* (radius being unity).

EXAMPLES.

1. The apparent altitude of the moon's center after being corrected for dip and refraction was 31° 25′; and its horizontal parallax at that time, taken from a nautical almanac, was 57′ 37″; what was the correction for parallax, and what was the true altitude as seen from the center of the earth?

$p=$57′ 37″=3457″ log.	- -	3.538699
31° 25′ cos.	- -	9.931152
$x=$49′ 10″=2950 log.	- -	3.469851

Ans. Cor. for parallax 49′ 10″
True altitude 32° 13′ 10″

2. The apparent altitude of the moon's center on a certain occasion was 42° 17′; and its horizontal parallax at the same time was 58′ 12″; what was the parallax in altitude, and what was the moon's true altitude? Ans. Parallax in alt. 43′ 4″
True alt. 43° 0′ 4″

No other examples of this kind are necessary, as they will incidentally occur in several places further on.

It now remains to describe the instrument used for taking angles at sea. We, therefore, give the following illustrations on the

QUADRANT AND SEXTANT.

The quadrant and sextant are essentially the same instrument, and the following is an explanation of the principle on which they are constructed.

Let ABC be a section of a reflecting surface, FB a ray of light falling upon it, and reflected again in the direction BE, and BD a perpendicular at the point of impact; then it is a well known optical fact, that the angles FBC and EBA are equal, and that FB, DB, and EB are in the same plane.

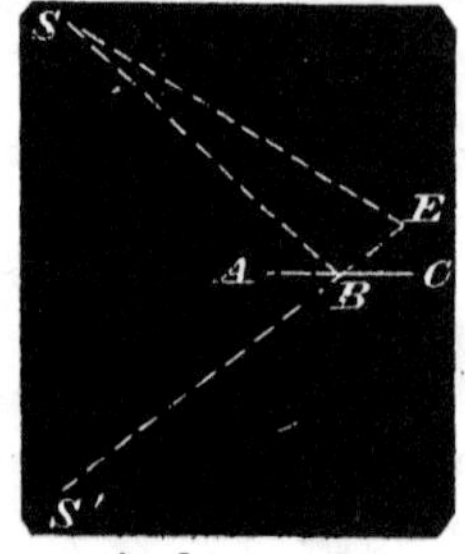

Again, if AC were a reflecting surface, and a ray of light, SB, from any celestial object S, were reflected to an eye at E, the image of the object would appear at S' on the other side of the plane, the angles SBA and ABS', as well as EBC, being equal; and if EB bear no sensible proportion to the distance of S, the angles SES' and SBS' may be considered as equal; for their difference, BSE, will be of no sensible magnitude.

Before we proceed to the direct description of the sextant, it is necessary to give the following important

LEMMA.

If the exterior angle of a triangle be bisected, and also one of the interior opposite angles, and the bisecting lines produced until they meet, the angle so formed will be half the other interior opposite angle.

Let ABC be the triangle, and bisect the exterior angle ACD by the line CE, and the angle B by the line BE.

The angle E will be half the angle A.

Let each of the angles ACE, ECD, be designated by x (as represented in the figure), and each of the equal parts of the angle B by y. Let A represent the angle A, and E the angle E.

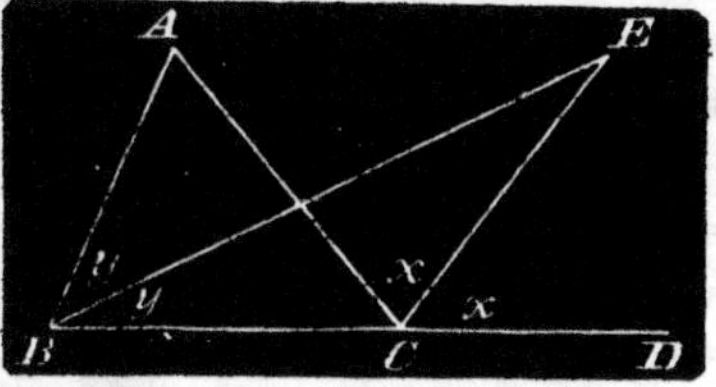

Now as the sum of the three angles of every plane triangle is equal to 180°; therefore, in the the triangle ABC, we have

$$A+2y+C=180° \qquad (1)$$

Also, in the triangle EBC, we have

$$E+y+C+x=180° \qquad (2)$$

Subtracting (2) from (1) gives us

$$A-E+y-x=0 \qquad (3)$$

Whence, $$A=E+(x-y) \qquad (4)$$

But because x is the exterior angle of the triangle ECB

$x=E+y$ (see Elementry Geometry.)

Or, $(x-y)=E$

This value of $(x-y)$ substituted in (4) gives

$$A=2E, \text{ or } E=\frac{A}{2} \qquad Q. E. D.$$

Another Demonstration.— The angle x being the half of ACD is equal to

$$\frac{A+2y}{2}$$

The angle x is also equal to $E+y$, because it is the exterior angle to the triangle EBC.

Therefore, by equality,

$$E+y=\frac{A+2y}{2}$$

Whence, $$E=\tfrac{1}{2}A \qquad Q. E. D.$$

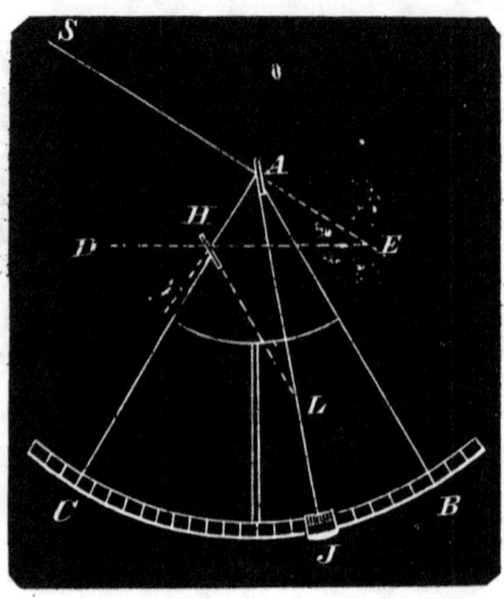

We are now prepared to show the construction of the sextant and quadrant.

The instrument represented by the annexed cut is a quadrant or a sextant, according as the arc contains 90° or 120°, but each actual degree of arc is graduated to 2°, and the space that covers 90° is really but 45°, and so on.

The reason why a half degree is counted and marked as a whole one, we are about to explain.

ABC is a firm plane sector, commonly made of metal or ebony; *AJ* is a revolving index bar, turning on the center *A*, to which is attached a vernier scale, revolving over the graduated arc.

The graduation commences at *B*. At *A* is a small plane mirror, perpendicular to the plane of the sector, it is attached to the index bar and revolves with it. This is called the index mirror or index glass.

At *H* is another small mirror, half silvered and the other half transparent. This is called the *horizon* glass; it might be called the *image* glass.

The horizon glass must be perpendicular to the instrument, *and parallel to AB.*

Now conceive a ray of light coming from an object *S*, striking the mirror *A*, the index and mirror being turned so as to throw the reflecting ray into the mirror *H*, this mirror agains reflects it toward *E*, and an eye anywhere in the line *DH* will see the image of the object behind the mirror *H*. Conceive the ray of light from *S* to pass right through the mirror at *A*, to meet the line *HE*; then, it is obvious that the angle *SED* measures the angle between the object *S* and its image *D*.

Now, in the triangle *AEH*, by a little inspection, it will be found that *HL* bisects the exterior angle, and *AJ*, the index, bisects one of the interior opposite angles ; therefore, by the *preceding lemma*, the angle HLA is half the angle at E, but as AB and the mirror H are parallel, the angle HLA is equal JAB. It is obvious that JAB is measured by the arc BJ, or it measures the angle at E, if half degrees on BC are counted as whole ones, which was to be shown.

A tube, and sometimes a small telescope, is attached to the bar AB, and placed in the direction of the line EH. This is called the *line of sight.*

THE ADJUSTMENT OF THE INSTRUMENT.

When this instrument is in adjustment, the two mirrors are perpendicular to the plane of the sector, and are parallel to each other when 0 on the vernier coincides with 0 on the arc. We therefore inquire : First

Is the index mirror perpendicular to the plane of the instrument?

The following experiment decides the question.

Put the index on about the middle of the arch, and look into the index mirror, and you will see part of the arch reflected, and the same part direct; and if the arch appears perfect, the mirror is in adjustment; but if the arch appears broken, the mirror is not in adjustment, and must be put so by a screw behind it, adapted to this purpose. Second,

Are the mirrors parallel when the index is at 0?

Place the index at 0, and clamp it fast; then look at some well-defined, distant object, like an even portion of the distant horizon, and see part of it in the mirror of the horizon glass, and the other part through the transparent part of the glass ; and, if the whole has a natural appearance, the same as without the instrument, the mirrors are parallel; but, if the object appears broken and distorted, the mirrors are not parallel, and must be made so, by means of the lever and screws attached to the *horizon glass.* Third,

Is the horizon glass perpendicular to the plane of the instrument?

The former adjustments being made, place the index at 0, and clamp it; look at some smooth line of the distant horizon, while holding the instrument perpendicular; a continued unbroken line will be seen in both parts of the horizon glass; and if, on turning the instrument from the perpendicular, the horizontal line *continues unbroken,* the horizon glass is in full adjustment; but, if a break in the line is observed, the glass is not perpendicular to the plane of the instrument, and must be made so, by the screw adapted to that purpose.

After an instrument has been examined according to these directions, it may be considered as in an approximate adjustment — a re-examination will render it more perfect — and, finally, we may find its *index error* as follows :— measure the sun's diameter both on and off the arch — that is, both ways from 0, and if it measures the same, there is *no index error;* but if there is a difference, half that difference will be the index error, *additive,* if the greater measure is off the arch, subtractive, if on the arch.

To measure the altitude of the sun at sea.

Turn down the proper screen or screens, to defend the eye. Put the index at 0, having it loose, look directly at the sun through the tube, and you will see its image in the silvered part of the horizon glass. Now move the index, and the image will drop; drop it to the horizon, and clamp the index.

Let the instrument slightly vibrate each side of the perpendicular, on the line of sight as a center, and the image of the sun will apparently sweep along the horizon in a circle. While thus sweeping, move the tangent screw,* so that the lower limb of the sun will just touch the horizon, without going below it. The reading of the index will be the altitude corresponding to that instant, provided there be no index error.

To measure the angular distance between two bodies as the sun and moon, or the moon and a star.

The most brilliant of the two objects is always reflected to the other. Loosen the index, place it at 0, and direct the line of sight to the brighter object, and catch a view of its image in the silvered part of the horizon glass.

Turn the plane of the instrument into the plane between the two objects; now move the index, keeping the eye on the image, and

---

* The screens, adjusting screws, clamp screw, and tangent screw, are not given in our description of the instrument, it is not necessary to describe them; should we attempt it, there is danger that the spirit and clearness of the description would be lost in the multitude of words.

bring it along to the other object; bring them as near as possible, then gently clamp the index.

Hold up the instrument again, in the plane between the two objects, and view one object through the transparent part of the horizon glass; and when the instrument is in the right position, the image of the other object will appear also in the same field of view, and then with the tangent screw, make the limb of the reflected object just touch the other, as it moves past it to and fro, by the gentle motion of the instrument. When the observer is satisfied that he has got the measure as near as he can, he cries out, *mark*, and his assistants mark the time by the watch, and the altitudes of the objects are also marked for the same time, if required, and observers are present to take them.

The first experiments in the use of this instrument, other than measuring a simple altitude, are generally failures, but a little practice will establish dexterity, skill, and confidence.

We are now prepared to give examples for finding latitude.

Let it be remembered, that latitude is the observer's zenith distance from the equator, and the nautical almanac gives the distance of all the heavenly bodies from the equator, under the name of Declination. We can therefore observe our zenith distance from any celestial object, and then apply its declination, and we shall have our zenith distance from the equator, *which is the latitude.*

EXAMPLES.

1. On a certain day, the meridian* altitude of the sun's lower limb was observed to be 31° 44′, bearing south. At that time its declination was 7° 25′ 8″ south, semi-diameter 16′ 9″, index error +2′ 12″, height of the eye 17 feet. What was the latitude?

Ans. 50° 48′ north.

---

* To obtain the meridian altitude of the sun, the observer commences observations before noon, while the sun is still rising; driving the index forward as fast as the image appears to rise, and there will come a time, a few minutes in succession, in which the image appears to rest on the horizon, neither rises nor falls, but at length the image will fall; then the observer knows that noon has passed, and the greatest apparent altitude will be shown by reading the index.

Semi-diameter	+16′ 9″	*N.A.*	Alt. ob.	31° 44
Index error	+ 2.12		Correction	12′ 46″
Refraction	— 1.31	Table.	Alt. ⊙'s center	31 56 46
Dip	— 4.04	Table.		90°
Sum	+12′46″		⊙'s zenith dis.	58° 3′ 14″
			Dec. south	7° 25′ 8″
			Latitude north	50° 48′ 6″

In this example, if the meridian altitude had been observed in the north, in place of the south, what then would have been the observer's latitude ? Ans. 65° 28′ 22″ south.

We may note the following

RULE. — *Subtract the corrected altitude from 90°. Then if the observer and the object are both on the same side of the equator, add the declination, but if on different sides, subtract the declination, and the sum or difference will be the latitude of the observer.*

Find the latitude from each of the following meridian observations:

	Object.	Alt. ob.	Direc.	S. D.	Height.	Declination.	Latitude.
1	Sun *L. L.*	45° 27′	South	16′ 15″	20 feet	17° 19′ 31″ *S.*	27° 2′ 51″ *N.*
2	" *L. L.*	81° 43′	South	15′ 47″	14 "	22° 13′ 7″ *N.*	30° 18′ 10″ *N.*
3	Jupiter	73° 17′	South		17 "	24° 10′ 13″ *S.*	7° 22′ 52″ *S*
4	Saturn	82° 12′	North		17 "	12° 9′ 6″ *N.*	4° 16′ 54″ *N.*
5	Sirius	75° 5′	North		18 "	16° 31′ *S.*	31° 30′ 26″ *S.*
6	Sun *U. L.*	40° 42′	North	16′ 17″	16 "	23° 22′ *S.*	73° 1′ 19″ *S.*
7	Sun *L. L.*	87° 29′	South	16′ 17″	16 "	22° 9′ *S.*	19° 50′ 12″ *S.*
8	Sun *L. L.*	15° 45′	South	16′ 0″	16 "	4° 43′ *N.*	69° 23′ 16″ *N*

In this table *L. L.* indicates lower limb, *U. L.* upper limb, *S. D.* semi-diameter, *N.* north, *S.* south, Direc. direction. In these examples, the instrument is supposed to have no index error.

Night observations at sea are of little value, for it is very seldom that the horizon can be defined, unless it is in bright moon-light, in the tropical climates.

For this reason, very few navigators attempt to find the latitude, by observations on the planets and fixed stars.

Occasionally, however, when one of the bright planets, or a conspicuous fixed star, comes to the meridian in the morning or evening, twilight observations can be made on them, and the latitude deduced.

Some navigators apply a summary correction to the sun's lower limb, for semi-diameter, dip, and refraction, such as is comprised in the following table.

Correction to be added *to the Observed Altitude of the Sun's Lower Limb, to find the True Altitude.*

Sun's Obs. Altitude.	Height of the Eye above the Sea in Feet.														
	6	8	10	12	14	16	18	20	22	24	26	28	30	32	34
°	′	′	′	′	′	′	′	′	′	′	′	′	′	′	′
5	3.8	3.5	3.1	2.8	2.5	2.3	2.1	1.8	1.6	1.4	1.2	1.0	0.8	0.6	0.5
6	5.3	4.9	4.6	4.3	4.0	3.7	3.5	3.3	3.0	2.8	2.6	2.4	2.2	2.1	1.9
7	6.4	6.0	5.7	5.4	5.1	4.8	4.6	4.4	4.1	3.9	3.7	3.5	3.3	3.2	3.0
8	7.2	6.8	6.5	6.2	5.9	5.7	5.4	5.3	5.0	4.8	4.6	4.4	4.2	4.0	3.9
9	7.9	7.5	7.2	6.9	6.6	6.4	6.1	5.9	5.7	5.5	5.3	5.1	4.9	4.7	4.5
10	8.5	8.1	7.8	7.5	7.2	6.9	6.7	6.5	6.2	6.0	5.8	5.6	5.4	5.3	5.1
11	8.9	8.6	8.2	7.9	7.6	7.4	7.2	6.9	6.7	6.5	6.3	6.1	5.9	5.7	5.6
12	9.3	9.0	8.7	8.3	8.0	7.8	7.6	7.3	7.1	6.9	6.7	6.5	6.3	6.2	6.0
14	9.9	9.6	9.2	8.9	8.7	8.4	8.2	7.9	7.7	7.5	7.3	7.1	6.9	6.8	6.6
16	10.4	10.1	9.7	9.4	9.1	8.9	8.7	8.4	8.2	8.0	7.8	7.0	7.4	7.2	7.1
18	10.8	10.4	10.1	9.8	9.5	9.3	9.0	8.8	8.6	8.4	8.2	8.0	7.8	7.6	7.5
20	11.1	10.7	10.4	10.1	9.8	9.6	9.3	9.1	8.9	8.7	8.5	8.2	8.1	7.9	7.7
22	11.4	11.0	10.7	10.4	10.1	9 8	9.6	9.4	9.1	8.9	8.7	8.5	8.3	8.2	8.0
26	11.7	11.4	11.0	10.7	10.5	10.2	10.0	9.7	9.5	9.3	9.1	8.9	8.7	8.6	8.4
30	12.0	11.7	11.3	11.0	10.8	10.5	10.3	10.0	9.8	9.6	9.4	9.2	9.0	8.9	8.7
35	12.3	11.9	11.6	11.3	11.0	10.7	10.6	10.3	10.1	9.9	9.7	9.4	9.2	9.2	9.0
40	12.5	12.2	11.8	11.5	11.3	11.0	10.8	10.5	10.3	10.1	9.9	9.7	9.5	9.4	9.2
45	12.7	12.4	12.0	11.7	11.5	11.2	11.0	10.7	10.5	10.2	10.1	9.8	9.7	9.6	9.4
50	12.8	12.5	12.2	11.9	11.6	11.3	11.1	10.9	10.6	10.4	10.3	10 0	9.8	9.7	9.5
55	13.0	12.6	12.3	12.0	11.7	11.5	11.2	11.0	10.7	10.5	10.3	10.1	9.9	9.8	9.6
60	13.1	12.7	12.4	12.1	11.8	11.6	11.3	11.1	10.9	10.6	10.4	10.2	10.1	9.9	9.7
65	13.2	12.8	12.5	12.2	11.9	11.7	11.4	11.2	11.0	10.7	10.5	10.3	10.1	10.0	9.8
70	13.3	12.9	12.6	12.3	12.0	11.8	11.5	11.3	11.0	10.8	10.6	10.4	10.2	10.1	9.9
75	13.4	13.1	12.7	12.4	12.1	11.9	11.7	11.4	11.2	11.0	10.8	10.6	10.4	10.2	10.1
80	13.6	13.2	12.9	12.6	12.3	12.0	11.8	11.6	11.3	11.1	10.9	10.7	10.5	10.4	10.2

Monthly Correction for Sun's Semi-diam.	Jan. +0′ 3	Feb. +0′.2	Mar. +0′.1	April, 0′0	May, —0′.1	June, —0′.2
	July, —0′.3	Aug. —0′.2	Sept. —0′.1	Oct. +0′.1	Nov. +0′.2	Dec. +0′.3

The most practical method of obtaining the latitude by observation, other than the meridian altitude of the sun, is by the meridian altitude of the moon; but to correct the observed altitude for semi-diameter, parallax, refraction, and dip, and do it to the utmost accuracy, requires more computation and attention than the mere practical navigator is disposed to give. Moreover, such like accuracy is not required in practical navigation. To know the latitude within a mile is all the ship master requires; and this can be done in a very summary manner, by observing the moon's meridian alititude and using, the following tables, according as the lower* or upper limb of the moon is observed.

* The bright limb, is the one observed, whether it be the upper or lower.

These tables make but one correction for semi-diameter, parallax and refraction

TABLE I.

Corrections to be added to the Observed Altitude of the Moon's lower limb.

Part 1st. Horizontal Parallax.

☾'s Alt.	53′	54′	55′	56′	57′	58′	59′	60′	61′
	° ′	° ′	° ′	° ′	° ′	° ′	° ′	° ′	° ′
6	0.59	1. 0	1. 1	1. 3	1. 4	1. 5	1. 6	1. 8	1. 9
8	1. 0	1. 2	1. 3	1. 4	1. 6	1. 7	1. 8	1. 9	1.11
10	1. 1	1. 3	1. 4	1. 5	1. 7	1. 8	1. 9	1.10	1.12
15	1. 2	1. 3	1. 5	1. 6	1. 7	1. 9	1.10	1.11	1.12
20	1. 2	1. 3	1. 4	1. 5	1. 6	1. 8	1. 9	1.10	1.11
25	1. 0	1. 2	1. 3	1. 4	1. 5	1. 6	1. 7	1. 9	1.10
30	0.59	1. 0	1. 1	1. 2	1. 3	1. 4	1. 5	1. 7	1. 8
35	0.57	0.59	0.59	1. 0	1. 1	1. 2	1. 3	1. 4	1. 5
40	0.55	0.55	0.56	0.57	0.58	0.59	1. 0	1. 1	1. 2
45	0.51	0.52	0.53	0.54	0.55	0.56	0.57	0.58	0.59
50	0.48	0.49	0.50	0.51	0.51	0.52	0.53	0.54	0.55
55	0.44	0.45	0.46	0.47	0.48	0.49	0.49	0.50	0.51
60	0.40	0.41	0.42	0.43	0.44	0.44	0.45	0.46	0.47
65	0.36	0.37	0.38	0.39	0.39	0.40	0.40	0.41	0.42
70	0.33	0.33	0.34	0.34	0.35	0.36	0.36	0.37	0.37
75	0.28	0.28	0.29	0.29	0.30	0.30	0.31	0.31	0.32
80	0.24	0.24	0.24	0.25	0.25	0.25	0 26	0.26	0.27
85	0.19	0.19	0.19	0.20	0.20	0.21	0.21	0.21	0.22

TABLE II.

Corrections to be applied to Observed Altitude of the Moon's upper limb

Part 2nd. Horizontal Parallax.

☾'s Alt.	53′	54′	55′	56′	57′	58′	59′	60′	61′
	° ′	° ′	° ′	° ′	° ′	° ′	° ′	° ′	° ′
10	+0.33	+0.33	+0.34	+0.34	+0.34	+0.36	+0.37	+0.37	+0.38
15	0.33	0.33	0.35	0.35	0.36	0.37	0.37	0.39	0.39
20	5.32	0.33	0.34	0.35	0.35	0.36	0.37	0.37	0.38
26	0.30	0.32	0.32	0.33	0.33	0.34	0.35	0.35	0.36
30	0.29	0.30	0.31	0.31	0.32	0.32	0.33	0.34	0.34
36	0.26	0.26	0.27	0.27	0.28	0.28	0.29	0.29	0.30
40	0.24	0.25	0.26	0.26	0.27	0.27	0.28	0.29	0.29
46	0.19	0.22	0.22	0.22	0.23	0.24	0.24	0.24	0.26
50	0.17	0.19	0.19	0.20	0.20	0.21	0.21	0.21	0.21
55	0.14	0.15	0.16	0.16	0.16	0.16	0.17	0.17	0.17
60	0.10	0.11	0.12	0.12	0.12	0.12	0.13	0.13	0.13
65	0. 6	0. 7	0. 7	0. 8	0. 8	0. 8	0. 8	0. 8	0. 9
70	0. 3	0. 3	0. 3	0. 3	0. 3	0. 3	0. 3	0. 3	0. 4
75	—0. 1	—0. 1	—0. 1	—0. 1	—0. 1	—0. 2	—0. 2	—0. 2	—0. 2
80	0. 6	0. 6	0. 6	0. 6	0. 6	0. 6	—0. 6	0. 7	0. 7
85	0.10	0.11	0.11	0.11	0.11	0.11	0.11	0.12	0.12
Height of the eye,					4ft.	9ft.	16ft.	25ft.	36ft.
Dip of the Horizon,					—2′	—3′	—4′	—5′	—6′

EXAMPLES.

1. In longitude about 45° west, on the 5th of January, 1852, at about 11h. in the evening, I observed the altitude of the moon's lower limb as she passed the meridian, and found it to be 68° 12′ from the south, height of the eye 16 feet. What was my latitude?

On the 5th of Jan., at 11h. evening, lon. 45 west, corresponds to 2 hours after midnight at Greenwich.

From the Nautical Almanac, we find, that,

At midnight of the 5th, the moon's horizontal parallax was - -	**57′ 40″**
At noon of the 6th, - - - - - - - - -	**58′ 1″**

Therefore, by proportion, the horizontal parallax at the time of observation, must have been 57′ 43″.

Moon's declination at midnight of the 5th, (N. Almanac),	**21° 47′ 53″ *N.***
" " noon of the 6th, - - - -	**22° 16′ 55″ *N***
Variation in 12 hours, - - - - - - - - -	**29′ 2″**
Therefore, the variation for 2 hours, was not far from -	**4′ 50″**
Hence the dec. at the time of observation was, - -	**22° 52′ 43″ *N.***

We enter table 1, and under the parallax, and opposite to the altitude *as near as we can find them,* we perceive that 37′ must be about the correction for the altitude.

Whence,	Observed alt. *L. L*	68° 12′
	Correction,	+ 37
		68 49
	Dip. always sub.	— 4
		68 45
		90
	Zenith dis. ☽	21 15
	☽ 's dec.	22 53
	Lat. in	44° 8′ North.

Find the true altitude of the moon's center, in each of the following examples. *L. L.* means lower limb ; *U. L.* upper limb.

	Observed Alt.		H. P.	Height of the eye.	Ans. True Alt.
1.	☽ *L. L.*	53° 23′	58′ 14″	14 feet	54° 10′ nearly.
2.	☽ *L. L.*	48° 58	60′ 27″	19 "	49° 48′ "
3.	☽ *U. L.*	57° 11′	54′ 30″	20 "	57° 19′ "
4.	☽ *L. L.*	63° 38′	55′ 29″	12 "	64° 14′ "
5.	☽ *U. L.*	20° 3′	54′ 14″	16 "	20° 32′ "
6.	☽ *L. L.*	16° 2′	59′ 38″	23 "	17° 12′ "

When the weather makes it doubtful whether meridian observations can be obtained, navigator's resort to double altitudes, or to the altitudes of two objects taken at the same time. We shall only show the principle on which this method is founded; it is the application of spherical trigonometry.

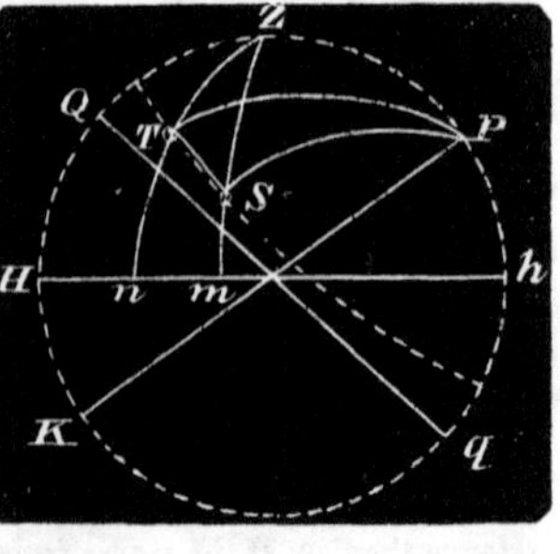

Let *Pp* be the earth's axis, *Qq* the equator. Suppose the sun to be the object, and let its position be *S* and *T* at two different times.

The elapsed time measures the angle *SPT*. In the triangle *PTS*, we have the two sides *PT*, *PS*, and the included angle, from which we compute the side *TS*, and the angle *TSP*.

Subtracting the altitudes *Sm* and *Tn* from 90°, we have *ZS*, and *ZT*, then we have all the sides of the triangle *ZTS*, from which we compute the angle *TSZ*. Subtracting this angle from *TSP*, gives us the angle *ZSP*. Now, in the triangle *ZSP*, we have the two sides *ZS*, *SP*, and their included angle, from which we compute *PZ* the complement of the latitude.

If the ship sails, during the interval between the observations, a correction will be required for the first altitude, and such corrections are found by the traverse table; a nautical mile in the direction of the sun, corresponds to one minute of a degree, to be applied to the altitude. When the proper correction is made, the result is equivalent to having both altitudes taken at the last station, and the deduced latitude is the latitude of that station.

CHAPTER II.

LONGITUDE.

LONGITUDE, from celestial observations, is measured by time. A place 15° west of another, will have noon one hour of absolute time later; if 30° west, the local time, noon will be two hours later, &c., &c.; 15° corresponding to an hour in time. Therefore, if we have any way of determining the times at two places, corresponding to the same absolute instant, the difference of such times will correspond to the difference of longitude between the two places at the rate of 15° to an hour, or 4 minutes to a degree.

A *perfect* time piece will keep the time at any *particular meridian,* and by carrying that *perfect* time piece with us, by it we can see the time at that particular meridian; and then if we can find the time at the place where we are, the comparison of these two times will give the difference of longitude, that is, the difference between our longitude and that of the particular meridian, to which the time piece refers.

For instance, a gentleman leaves Boston; his watch is a perfect time piece, and it is set to Boston time, he travels west on the railroads, his watch all the while shows Boston time; when it is twelve o'clock by his watch it is really so in Boston, but not so at the place where he is. The sun has arrived at the meridian of Boston, but not yet at the meridian of Albany, or Buffalo, or Detroit; and when the gentleman arrives at any of these places, or any intermediate place, the local time, compared with the time in Boston, will give the longitude of that locality from Boston, counting *one* degree for every four minutes in the difference of time.

Unfortunately, however, there is no such thing as a *perfect time piece,* but some do approximate toward perfection. Such ones, made with the greatest care and solely for accuracy in rate of motion, are called *chronometers*; they are supposed to keep time within certain known limits, and in the place of perfect time keepers, they are used at sea for finding longitude.

Chronometers show the time at the distant place, it then remains to find the time at ship, and this is done most accurately by spherical trigonometry, as will soon appear.

The sun's altitude is greatest just at apparent noon, but by observations we cannot define just the moment when that takes place; hence meridian observations, valuable as they are for latitudes, are worth nothing for time, when time is to be settled to anything like accuracy.

The best position of the sun (or any other celestial object) for an observation to find local time, is when it is nearly east or west, and its altitude more than ten degrees.

In such circumstances, an observer can find the local time

within 5 or 6 seconds, by taking an altitude of the sun, provided he at the same time knows his latitude and the sun's polar distance.

The operation is a beautiful application of spherical trignometry, and it is illustrated by the following figure.

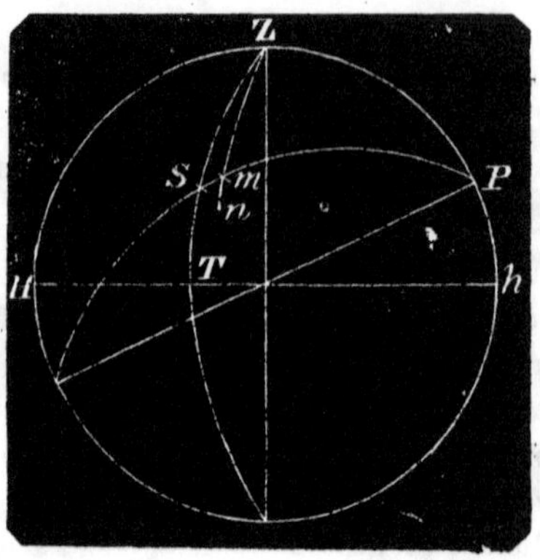

Let Z be the zenith of the observer, P the pole, S the position of the sun, and PS the sun's polar distance.*

When S comes on to the meridian, it is then apparent noon; and the angle ZPS of the triangle ZPS measures the interval from apparent noon, at the rate of four minutes to one degree.

The side PS is the polar distance, the side ZS is the co-altitude, and the side PZ is the co-latitude.

Now, in every treatise on spherical trigonometry, it is demonstrated as a fundamental principle, that

The cosine of any angle, of a spherical triangle, is equal to the cosine of its opposite side, diminished by the rectangle of the cosines of the adjacent sides, divided by the rectangle of the sines of the adjacent sides.

That is, $$\cos. P = \frac{\cos. ZS - \cos. PZ \cos. PS}{\sin. PZ \sin. PS}$$

Now, in place of cos. ZS, we take its equal, sin. ST, or the sine of the altitude, and in place of cos. PZ, we take its equal, the sine of the latitude.

In short, let $A=$ the altitude, $L=$ the latitude, and $D=$ the polar distance.

Then $$\cos. P = \frac{\sin. A - \sin. L \cos. D}{\cos. L \sin. D}$$

---

* When the observer is in the *northern* hemisphere, the polar distance is counted from the *north* pole; when in the southern hemisphere, from the south pole.

From a general equation, in plane trigonometry, we have

$$2 \sin.^2 \tfrac{1}{2} P = 1 - \cos. P$$

Substituting the value of cos. P, in this last equation, we have

$$2 \sin.^2 \tfrac{1}{2} P = 1 - \frac{\sin. A - \sin. L \cos. D}{\cos. L \sin. D}$$

$$= \frac{(\cos. L \sin. D + \sin. L \cos D) - \sin. A}{\cos. L \sin. D}$$

By comparing the quantity in parentheses with eq. (7), plane trigonometry, we perceive that

$$2 \sin.^2 \tfrac{1}{2} P = \frac{\sin. (L+D) - \sin. A}{\cos. L \sin. D}$$

Considering $(L+D)$ to be a single arc, and then applying equation (18), plane trigonometry, and dividing by 2, we shall have

$$\sin^2. \tfrac{1}{2} P = \frac{\cos. \left(\frac{L+D+A}{2}\right) \sin. \left(\frac{L+D-A}{2}\right)}{\cos. L \sin D.}$$

But $\frac{L+D-A}{2} = \frac{L+D+A}{2} - A$, and now if we put

$$S = \frac{L+D+A}{2} \text{ we shall have}$$

$$\sin^2. \tfrac{1}{2} P = \frac{\cos. S \sin. (S-A)}{\cos. L \sin. D}$$

Or, $$\sin. \tfrac{1}{2} P = \sqrt{\frac{\cos. S \sin. (S-A)}{\cos. L \sin. D}}$$

This is the final result when radius is unity, when it is R times greater, then the sin. $\frac{1}{2} P$ will be R times greater, and if R represents the radius of our tables, to correspond with these tables we must multiply the second member by R, and if we put it under the radical sign, we must multiply by R^2; in short we shall have,

$$\text{Sin. } \tfrac{1}{2} P = \sqrt{\left(\frac{R}{\cos. L}\right)\left(\frac{R}{\sin. D}\right) \cos. S \sin. (S-A)}$$

The right hand member of this equation, shows *four* distinct logarithms; thus, $\frac{R}{\cos. L}$ is the cosine of the latitude subtracted from 10, *which we shall call cosine complement.*

This equation furnishes the following rule for finding apparent local time, when the sun's altitude, its polar distance, and the latitude of the observer, are given.

The *altitude must be observed*, the latitude must be known, and the Nautical Almanac will furnish the polar distance.

RULE.—1. *Add together the altitude, latitude, and polar distance; take the half sum, and from the said half sum subtract the altitude, thus finding the remainder.*

2. *The logarithms. Find the cosine complement of the latitude, the sine complement of the polar distance, the cosine of the half sum, and the sine of the remainder.*

3. *Add these four logarithms together, and divide by 2, the logarithm thus found, is the sine of half the polar angle, or half the sun's meridian distance.*

4. *Take out the arc corresponding to this sine, and divide its double by 15 (as in compound division in arithmetic), and the quotient will be the hours, minutes, and seconds from apparent noon; and if the sun is east of the meridian, the hours, minutes, and seconds, must be subtracted from 12 hours, for the corresponding time of day.*

The time shown by a chronometer or a perfect clock, or rather graduation of clocks, is to mean and not to apparent time, and to convert apparent into mean time, the equation* of time is given in the Nautical Almanac for the noon of every day at Greenwich. The amount of it, reduced or modified to correspond to the time of observation, can be applied to apparent time, and the mean time of taking the observation will be determined. The difference between this time and the mean time at Greenwich, as determined by the chronometer, will be the longitude. The longitude will be *west*, if the time at Greenwich is *latest* in the day, otherwise it will be east.

If the observer is on land, without a sea horizon, and uses a reflecting instrument, he must have an artificial horizon. A proper artificial horizon, is a small dish of mercury, with a glass roof to put over it, to keep the mercury from being agitated by the wind. In place of the mercury, a plate of molasses will answer. In still calm weather any clear pool of water is a good artificial horizon.

In either of these, the reflected image of the object appears as much below the horizon as it is above it, and to measure the altitude,

* For the theory of equation of time, see works on astronomy.

the image reflected by the mirror of the instrument must be carried to the image in the artificial horizon; half of the angle shown by the index will be the apparent altitude. In using an artificial horizon there is no dip, other corrections are to be applied according to circumstances.

EXAMPLES UNDER THE PRECEDING RULE.

1. Being at sea, May 20th, 1823, in latitude 43° 30′ *N.*, and in longitude about 20° west, I observed the altitude of the sun's lower limb, and found it to be 32° 4′ rising, when an assistant marked the time per watch, at 7h. 43m. A. M.; height of the eye 16 feet. What was the true mean time?

Just before the observation, the watch was compared with the chronometer in the cabin*, and found to be 1 hour, 21 minutes, and 12 seconds slow of chronometer.

On the 8th of May, the chronometer was 3m. 7s. fast of Greenwich time, and gaining 1s.6 daily. What was the longitude?

				H. M. S.
☉ *S. D.*,	+ 15′ 49″			
Dip.,	— 3 56		Watch,	7 43 0
Ref.,	— 1 30		Diff.,	1 21 12
Correction,	+ 10 23		Face of chron. at ob.,	9 4 12
Observation,	32 4		Error 3*m.* 7*s.*, increase of error in 12 days 19*s.*, whole error,	— 3 26
Alt. ☉ center,	32° 14′ 23″		Greenwich time,	9 0 46

At noon on the 20th of May, 1823, the sun's declination, by the *N. A.*, was 19° 52′ 18″ north, increasing at the rate of 30″.6 per hour, and the time of taking the observation was 3 hours before noon at Greenwich; therefore, the declination must have been 19° 50′ 47″ *N.*

Altitude,	32° 14′ 23″		
Lat.,	43 30	cos. com.	.139435
P. D.,	70 9 13	sin. com.	.026603
	2)145 53 36		
S.	72 56 48	cosine	9.467253
	32 14 23		
(*S—A*)	40 42 25	sine	9.814363
			2)19.447657
	31° 58′ 8″	sin.	9.723828

* Chronometers should never be, and by careful persons, never are, taken out of their places during a voyage.

$$\begin{array}{l} 31^\circ\ 58'\ 8'' \\ \quad\quad\quad\ 2 \\ \hline 63^\circ\ 56'\ 16'' = 4h.\ 15m.\ 45s. \\ \quad\quad\quad\quad\quad\quad 12 \end{array}$$

Apparent time, - - -	7	44	15	A. M.
Equation of time *N. A.* -	—	3	51	
Mean time at ship, - -	7	40	24	
Watch, - - - - - -	7	43		
Watch too fast, - - -		2*m.*	36*s.*	
Time at Greenwich per chron.,	9*h.*	0*m.*	46*s.*	
Time at ship per observation,	7	40	24	
Diff., - - - - - -	1	20	22=20° 5′ west lon.	

2. August 10th, 1824, in latitude 54° 12′ north, at 2h 33m per watch, height of the eye 18 feet, I observed the altitude of the sun's upper limb 16° 50′ falling. My chronometer was 2h 20m 37s fast of the watch; and on the 7th, the same month, the chronometer was 40m 29.4s fast of Greenwich time, gaining $7\frac{5}{60}$ seconds daily. What was the error of the watch, and the longitude per chronometer?

Prepartion.

Time per watch -	5*h.*	33*m.*	0*s.*	P. M.
Diff. per chron. - -	2	20	37	
Face of chronometer -	7	53	37	P. M.
Chron. fast (whole error)		—40	52	
Greenwich, mean time,	7	12	45	P. M.

On the 10 of August, 1824, the sun's declination at noon, Greenwich time, was 15° 32′ 14″ north, decreasing at the rate of 45″ per hour, as given in the Nautical Almanac. The decrease for $7\frac{1}{5}$ hours must be 5′ 24″; whence, the declination at the time of observation, 15° 26′ 50″ *N.*, and the polar distance 74° 33′ 10″.

Observed altitude -	16° 50′ 00″	Equation of time, per N. A., Aug. 10, 1824, was - -	+5*m.* 2*s.*
Semi-diameter, N. A. -	—15 48	Hourly decrease $\frac{33}{100}s.$	—2
Dip and Ref. - -	—7 20		
True alt. center - -	16° 37′ 52″	Equation at ob. - -	5*m.* 0*s.*

We now leave the problem to be worked through by the pupil, giving only the answer.

Ans. Watch slow of local mean time, 3m 27s.
Longitude by chronometer, 24° 4′ 30″ west.

3. When it was 6h 0m 21s, P. M., mean time, at Greenwich, by my chronometer, I observed the altitude of the sun's lower limb to be 30° 17′, in the afternoon of January 12th, 1852. At noon our

latitude, by a meridian observation, was 21° 47′ north, and since that time we have made 11 miles of southing, by the log. The dip was 4′, and semi-diameter 16′ 17″. What was the longitude by chronometer?

Sun's Declination Jan. 12, '52, at noon, G. T. -	21° 44′ 10″ south.
Hourly decrease, per N. A., 25″, giving -	—2′ 30″
Declination at the time of observation - -	21° 41″40″ south
Equation of time at noon, Greenwich - -	+8*m*. 25*s*.
Hourly increase $\frac{96}{100}$ of a second, making - -	6*s*. nearly
Equation at time of observation (to add) -	+8*m*. 31*s*.

Were we sure that pupils would have access to nautical almanacs, we would give neither declination nor equation of time.

Ans. Lon. 45° 39′ west.

4. On the 16th of January, 1852, when my chronometer showed 11h 27m 41s, A. M., for the mean time at Greenwich, I observed the altitude of the sun's lower limb and found it 32° 21′ rising, height of the eye 16 feet, latitude 0° 41′ south. What was the longitude by chronometer?

By the N. A., the sun's declination at that time was 21° 2′ 36″ south, and the equation of time 9*m*. 53*s*. additive.

Ans. Lon. 46° 39′ west.

N. B. — Time at any place, is but the difference between the right ascension of the meridian and the right ascension of the sun; and to find the time from these two elements, we always subtract the right ascension of the sun from the right ascension of the meridian, increasing the latter by 24 hours, to render subtraction possible, when necessary.

The right ascensions of the stars are given, and the right ascension of the sun is given, in the Nautical Almanac, for the noon of every day in the year, Greenwich time. Now, if we can find the meridian distance of any known star, by observation, we can establish the right ascension of the meridian, and, consequently, the local time. Hence, we can find longitude by comparing the chronometer with the altitudes of the stars, as the following example will illustrate.

5. If on the 8th of March, 1852, when my chronometer showed the Greenwich time to be 7h 22m 3s, P. M., I found by observation,

that the true altitude of Sirius was 37° 52′ west of the meridian. My latitude was 32° 28′ south. What was the time at ship, and what was my longitude; the elements for computation being as follows?

1. Right ascension of the star - - 6*h*. 38*m*. 38*s*.
2. Declination of the star 16° 31′ south
3. Right ascension of the sun - - 23*h*. 17*m*. 25*s*.

By means of the triangle we find,

The meridian distance of the star -	3*h*.	40*m*.	58*s*.	
To which add ✱'s R. A., because ✱ is west	6	38	38	
Right ascension of the meridian - -	10	19	36	
Add - - - - - -	24			
	34	19	36	
Subtract the R. A. of the sun - -	23	17	25	
Diff. is apparent time at ship - -	11	2	11	P. M.
Equation of time, add - - - - -		10	48	
Mean time at ship - - - - -	11	12	59	
Time at Greenwich - - - - -	7	22	3	
Longitude in time - - - - -	3	50	56	
		=57° 44′ east		

N. B.—When the chronometer remains in the same place for a week or more, its rate can be determined by comparing it with the observed altitudes of the sun, taken from day to day. In different climates the same chronometer will have different rates, and on returning to its original station it will frequently resume its original rate.

For azimuths, and variations of the compass, see page 106.

CHAPTER III.

LUNAR OBSERVATIONS.

A GOOD and well-tried chronometer is a valuable and reliable instrument for finding the longitude at sea, during short runs; but still it is but an instrument, and is not one of the reliable works of nature. Near the end of a long voyage, the best of chronometers very frequently give false longitude, and in such cases, good navigators always resort to lunar observations, which from the hands of a good observer, can be relied upon to within 10 or 12 minutes of a degree, and they usually come within 5 or 6 miles, and sometimes even more exact, but that is accidental and unfrequent.

To comprehend the theory of lunars, we must call to mind the fact that the moon moves through the heavens, apparently among the stars, at the rate of more than 13° in a day, and any angular distance it may have from the sun or any star corresponds to some moment of Greenwich time.

About three days before and after the change of the moon, she is too near the sun to be visible, but at all other times, her distance from the sun, some of the larger planets, and certain bright fixed stars, called lunar stars,* which lie near her path, are computed and put down in the nautical almanac, for every third hour of mean Greenwich time commencing at noon. For any particular day, the distances are given to such objects only, east and west of her, as will be convenient to measure with the common instruments.

The distances put down in the nautical almanac, are such as would be seen if viewed from the center of the earth; but observers are always on the surface of the earth, and the distances thence observed, must always be reduced to equivalent distances seen from the center, and this reduction is called *working a lunar*, which is generally the highest scientific ambition of the young navigator.†

The true distance between the sun and moon, or between a star and the moon, can be deduced from the apparent distance by the application of spherical trigonometry.

The moon is never seen by an observer in its *true place*, unless the observer is in a line between the center of the earth and the moon, that is, unless the moon is in the zenith of the observer; in all other

---

* There are nine lunar stars, Arietis, Aldebaran, Pollax, Regulus, Spica, Antares, Aquilæ, Fomalhaut, and Pegasi.

† Many navigators, both old and young, direct all their efforts to knowing how to do, without attempting to comprehend the reasons for so doing; and this the world calls *practical*, — a complete perversion of the term. On the other hand, some men of the schools spend their energies in metaphysical nothings, splitting hairs in logic, and calling it scientific; this is equally a perversion.

positions, the moon is depressed by parallax, and appears nearer to those stars that are below her, and further from those stars that are above her, than would appear from the center of the earth. Therefore, the apparent altitudes of the two objects, must be taken at the same time that their distance asunder is measured. The altitudes must be corrected for parallax and refraction, thus obtaining the true altitudes.

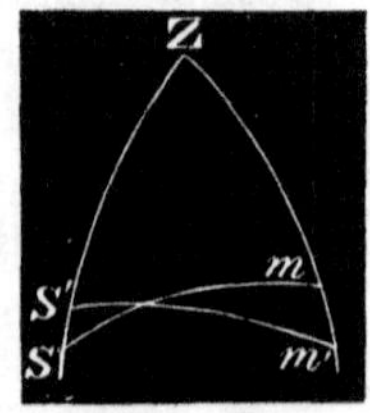

The annexed figure is a general representation of the triangles pertaining to a lunar observation.

Let Z be the zenith of an observer, S' the apparent place of the sun or star, and S its true place. Also, let m' be the apparent place of the moon, and m its true place as seen from the center of the earth.

Here are two distinct triangles, $ZS'm'$, and ZSm. The apparent altitudes subtracted from 90°, give ZS' and Zm', and $S'm'$ is the *apparent* distance; with these three sides, the angle Z can be found. Correcting the altitudes, and subtracting them from 90°, will give the sides ZS and Zm; these two sides, and their included angle at Z, will give the side Sm, which is the *true distance.*

The definite true distance must have a definite Greenwich time, which can be readily found; and this, compared with the local time deduced from an altitude of the sun, will of course give *the longitude.*

We shall now make a formula to clear the distance.

Let S'=the apparent altitude of the sun or star,
and S=the true altitude. Also,
Let m'=the apparent altitude of the moon,
and m =the true altitude.

Observe that the letters with the accent, indicate *apparent*, and without the accent, the *true* altitudes.

Put d to represent the apparent distance, and x to represent the true distance.

Bear in mind, that the sine of an altitude is the same as the cosine of its zenith distance, and conversely, the sine of a zenith distance is the same thing as the cosine of the corresponding altitude.

Now, by the fundamental equation of spherical trigonometry noted in the last chapter, we have

$\cos. Z = \frac{\cos. d - \sin. S' \sin. m'}{\cos. S' \cos. m'}$. Also $\cos. Z = \frac{\cos. x - \sin. S \sin. m}{\cos. S \cos. m}$.

Whence, $\frac{\cos. d - \sin. S' \sin. m'}{\cos. S' \cos. m'} = \frac{\cos. x - \sin. S \sin. m}{\cos. S \cos. m}$.

By adding unity to each member we have

$$1 + \frac{\cos. d - \sin. S' \sin. m'}{\cos. S' \cos. m'} = 1 + \frac{\cos. x - \sin. S \sin. m}{\cos. S \cos. m}.$$

$$\frac{(\cos. S' \cos. m' - \sin. S' \sin. m') + \cos. d}{\cos. S' \cos. m'} = \frac{(\cos. S \cos. m - \sin. S \sin. m) + \cos. x}{\cos. S \cos. m}.$$

By observing equation 9, plane trigonometry, we perceive that the preceding equation reduces to

$$\frac{\cos. (S' + m') + \cos. d}{\cos. S' \cos. m'} = \frac{\cos. (S + m) + \cos. x}{\cos. S \cos. m},$$

Whence $\cos. x = (\cos. (S' + m') + \cos. d) \frac{\cos. S \cos m}{\cos. S' \cos. m'} - \cos. (S + m)$.

It is here important to notice that the moon's horizontal parallax given in the Nautical Almanac, is the equatorial horizontal parallax; that is, it corresponds to the greatest radius of the earth. The diameter of the earth through any other latitude is less, and of course the corresponding parallax is less.

We therefore give the following table for the reduction of the equatorial horizontal parallax, to the horizontal parallax of any other latitude; it is computed on the supposition that the equatorial diameter is to its polar as 230 to 229. For example if the horizontal parallax in the Nautical Almanac is 55′—, in the latitude of 40° the reduction would be 6″, and the parallax reduced would be 54′ 54″, and if the parallax from the Nautical Almanac were 60′ the reduction would be 6″ 6, and reduced would be 59′ 53″.4.

The semi-diameter of the moon given in the Nautical Almanac is her horizontal semi-diameter, but when she is in the zenith she is nearer to us by the whole radius of the earth, about one-sixtieth part of her whole distance, consequently she must appear under a larger and larger angle as she rises from the horizon, and this is called the augmentation of the semi-diameter.

We give the reduction for the parallax; and the augmentation for the semi-diameter in the following tables:

Red. of ☾'s Eq. hor. parallax.		
Latitude.	Eq. par. 55′	Eq. par. 60′
20°	0″.9	1″
25	2 .8	3
30	3 .7	4
35	4 .6	5
40	6 .0	6 .6
45	7 .3	8
50	8 .6	9 .4
55	10 .1	11
60	11	12
65	11 .8	13
70	12 .8	14
75	13 .9	15
80	14 .6	16

Augmentation of the Moon's semi-diam.	
Ap. Alt.	Aug.
6°	2″
12	3
18	5
24	6
30	8
36	9
42	11
48	12
54	13
60	14
66	15
72	16
90	16

We now give an example showing all the details of finding the longitude by a lunar observation.

EXAMPLE.

Suppose that on the 25th of January, 1852, between three and four o'clock in the afternoon, local time, the observed distance between the nearest limbs of the sun and moon was 50° 3′ 20″, the altitude of the sun's lower limb was 20° 1′, and of the moon's lower limb 48° 57′, height of the eye 16 feet. The latitude corrected for the run from noon was 34° 12′ *N.*, and the supposed longitude about 65° west. What was the longitude ? (the Nautical Almanac being at hand.)

Preparation.

	H. M.
Supposed time at ship,	3 15 *P. M.*
Supposed longitude 65	4 20
Supposed time at Greenwich,	7 25 *P. M.*

On the 25th at noon the N. A. gives the ☽'s *S. D.* at 14′ 47″, and at midnight at 14′ 45″ 7 ; therefore at the time of observation we take it at 14′ 46″, by simple inspection. In the same summary manner we take the Equatorial horizontal parallax at 54′ 12″.

☽'s semi-diameter, - -	14′ 46″	☽'s Eq. hor. par. - -	54′ 12″
Aug for Alt. - - - -	12	Red. for lat. - - - -	4
☽'s true *S. D.* - - -	14′ 58″	Reduced hor. par. - -	54′ 8″

Observed distance, -	50° 3′ 20″	Alt. ☽'s *L L*	48° 57′
Sun's *S. D.* - -	16 16	☽'s *S. D.*	14′ 58″
Moon's *S. D.* - -	14 58	Dip	— 3 56
Apparent central dis.	50° 34′ 34″=*d*	☽'s app. alt.	49° 8′ 2″=*m.*′

Alt. ☉'s lower L	20° 1′	
Semi-diameter,	+16′ 16″	
Dip,	— 3′ 56″	
☉'s app. alt.	20° 13′ 12″=S′	
Refraction,	— 2′ 34″	
☉'s true alt.	20° 10′ 38″ S.	

N. B. To find the moon's parallax in altitude see problems on page 201.

☽'s app. Alt. 49° 8′	ocs. 9.815778
54′ 8″=3488′	log. 3.542576
36′ 2″=2282	3.358354

☽'s app. alt.	49° 8′ 2″
Parallax in alt.	36′ 2
Refraction,	—49
True alt.	49° 43′ 15″=m

$(S'+m')=69° 21' 14''$ $(S+m)=69° 53' 53''$.

We are now prepared to apply the equation to compute the true distance. The equation requires the use of natural sines and cosines.

$$\cos. x=(\cos. (S'+m')+\cos. d)\frac{\cos. S \cos. m}{\cos. S' \cos. m'}-\cos. (S+m)$$

$(S'+m')=69° 21' 14''$ N. cos. .35259
$d=50° 34' 34''$ N. cos.* .63449

.98708 log. —1.994350
$S=20° 10' 38''$ log. cos. 9.972496
$m=49° 43' 15''$ log. cos. 9.810578
$S'=20° 13' 12''$ cos. com. 0.017626
$m'=49° 8' 2''$ cos. com. 0.184228
Num. .97561 log. —1.989278
=sum less 20.†

N. cos. $(S+m)=69° 53' 53''$ —.34369
True distance, 50° 48′ 29″ cos.63192

In the Nautical Almanac, we find that at 6 P. M. mean Greenwich time, on said day, the true distance between the sun and moon was 49° 59′ 26″, and at 9 P. M., the distance was 51° 20′ 49″, showing a change of 1° 21′ 23″ in three hours of time. But the change

* When d is greater than 90° its cosine becomes *minus*, and its numerical value is then the natural sine of the excess over 90°. Thus if d were 105°, its cosine would be numerically equal to the sine of 15°, and must then be subtracted from the cosine of the sum of apparent altitudes. The result (cos. x) would then be the sine of the excess over 90°.

† Less 20 because the table of natural sines is to radius unity, and we used cos. S and cos. m to the radius of 10, making two tens to take away.

from 49° 59′ 26″ to 50° 48′ 29″ is 49′ 3″; and now on the supposition that the change is in proportion to the time (and it is very nearly), we have the following analogy

$$1° \ 21' \ 23'' : 49' \ 3'' :: 3h. : t$$

Or, $$4883 : 2943 :: 3 : 1h. \ 48m. \ 29s.$$

That is, the time that this observation was taken 1h 48m 29s after 6 at Greenwich. Or, 7h 48m 29s mean Greenwich time.

With the true altitude of the sun 20° 10′ 38″, the latitude 34° 12′, and the polar distance 109° 0′ 48″, we find the apparent time at ship 3h 10m 5s, to which we would add the equation of time, 12m 34, making the mean time 3h 22m 39s.

From the Greenwich time	*7h.*	*48m.*	*29s.*
Sub. time at ship - -	3	22	39
Giving lon. in time -	4	25	50=66° 27′ 38″ *W.*

West, because the time at Greenwich was later in the day.

If a lunar is taken with a star, or with the sun, when the sun is not in a proper position to depend upon its altitude for local time, the time must be noted by a watch, and the difference between the watch and true time made known, by a previous or subsequent observation on the sun, or some star which is nearly east or west of the observer.

The most material part of working a lunar is that of clearing the distance. We, therefore, give the following examples, without the little incidental details.

We show the working of one in which the distance is greater than 90°.

The apparent distance between the center of the sun and moon on a certain occasion, was 98° 12′; the apparent altitude of the sun's center was 74° 10′, and of the moon's 20° 37′; the moon's horizontal parallax at the same time was 57′ 12″. What was the true distance? Ans. 97° 34′ 27″

Horizontal par. 57′ 12″=3432	log.	3.532547
☽ Alt. 20° 37′	cos.	9.971256
Parallax in alt. 53° 31′=3211	log.	3.586803

☽'s app. alt.	20° 37′		☉ app. alt.	74° 10′ 0″
Refraction	—2 31		Refraction	16
Parallax	+53 31		☉'s true alt.	74° 9′ 44″
☽'s True alt.	21° 28′			

$(S'+m')=94° 47'$ $(S+m)=95° 37' 44''$

94° 47′ Nat. cos.—0.08339

d=98° 12′ Nat. cos.—0.14263

—0.22602 log. —1.356150*

S=74° 9′ 44″ cos. - - 9.436021

m=21° 28′ 00″ cos. - - 9.968777

S'=74° 10′ 00″ cos. complement 0.364092

m'=20° 37′ 00″ cos. complement 0.028744

—0.23007 log. —1.361784

cos. $(S+m)$=95° 37′ 66″+0.09826†

True dis. 97° 34′ 27″ $N.$ cos.—0.13181

EXAMPLES FOR PRACTICE.

No.	Ap. alt. of sun or a fixed star.	Moon's ap. altitude.	Apparent central distance.	Moon's hor. par.	True distances.
	° ′	° ′	° ′ ″	′ ″	° ′ ″
1	☉ 86 3	39 18	46 45 0	53 51	46 4 25
2	✱ 29 47	57 22	27 35 0	60 3	28 8 24
3	☉ 31 14	28 7	14 21 30	53 29	14 9 24
4	☉ 60 5	63 12	51 3 21	58 30	50 41 15
5	✱ 34 28	10 42	49 18 38	61 11	48 45 39
6	☉ 8 26	19 24	120 18 46	57 14	120 1 46
7	✱ 43 27	40 9	18 21 35	60 20	18 8 12
8	✱ 53 13	57 32	60 13 49	60 52	59 48 12
9	☉ 72 26	18 30	81 2 28	60 58	80 9 33
10	☉ 60 33	9 26	70 36 16	59 57	69 49 12

* The factor 0.22602 being minus renders the product minus.

† The cos. ($S+m$) in the equation is minus, but where the value of ($S+m$) is more than 90°, the *minus* necessarily becomes *plus*.

APPENDIX.

It is comparatively an easy matter, to conduct a survey, or navigate a vessel, when there are no important difficulties to be overcome; but the true test of knowledge or skill in any pursuit, is to be found only in real adversity. The mariner who successfully manages his ship, when every thing is provided, when all is in order, and the weather favorable, is deserving of little credit; but let the ship become disabled, and the storm terrific, and then there is scope for the exercise of every necessary acquirement, and its kindred talent.

So it is with the man of science; when every instrument is at hand, and all in order, it requires little skill, and but common knowledge, to make observations and experiments; but when we reverse the case; the tact, knowledge, and ingenuity of the man, may oft times more or less overcome the difficulties.

For instance, suppose it were necessary to find the altitude of the sun, for the purpose of finding the latitude of the place (on shore), or for the purpose of finding the time; and we had no sextant or quadrant, and in fact, no instrument to measure angles. It could be done approximately as follows:

Let a plumb line be suspended in water; have a knot in the line, and let the knot be at a known distance above the water. The knot will cast a shadow on the water; measure the distance of this shadow from the plumb line. The knot and its shadow, with the plumb line and water, will form a right angled triangle, and the angle at the base, computed by plane trigonometry, will be the altitude of the sun's *upper limb*, and this altitude may be used for any purpose, the same as if it were measured by a sextant, but the accuracy is not to be depended upon, for the want of delicacy in the instrument.

A person on shore having a good watch, and knowing his latitude, can regulate his watch, or at least determine its rate and error for a short period of time. Then, if he have a nautical almanac, the common tables of logarithms, and a knowledge of spherical trigonometry, and a corresponding knowledge of astronomy, he can find the longitude by a lunar observation, *without* a sextant, as follows:

By the means of his watch and a plumb line, he will be able to range off an *approximate* meridian line. He will then observe the transits of stars, and of the moon across that meridian, taking those stars which are at that time near the moon's meridian, some to the east and some to the west of the moon, and some more north, and others more south than the moon.

He will note the difference in time, between the transit of each star and the moon, across his approximate meridian, and by a combination, or rather comparison of these observations, he will be able to determine the moon's right ascension very nearly. By the moon's right ascension, and the aid of the nautical almanac, he can find the Greenwich time.

The Greenwich time, compared with the local time, will give the longitude.

When we can find the moon in a vertical plane with any two fixed stars, and it be at the time the moon changes her declination very slowly, so that we can depend upon a declination taken from the nautical almanac for the supposed time, we can then determine the moon's right ascension, and from thence the longitude as before, whether we are on land or sea.

Ship-wrecked mariners, and travelers similarly situated, have frequently resorted to these artifices to obtain their approximate localities.

We have frequently remarked in the course of this work, that the best position of a celestial object, at the time of taking its altitude, for the purpose of more exactly defining the time, is when the object is nearly east or west; we now propose to show this conclusively, and therefore give the following

INVESTIGATION.

To find under what circumstances, in a given latitude, a small mistake in observing or correcting the altitude of a celestial object, will produce the smallest error in the time computed from it.

Let *Z* be the zenith, *P* the pole, *r* the supposed place, and *m* the true place of the object. Let *ms* be a parallel of altitude, join the points *m* and *r*, and let *pq* be the arc of the equator contained between the meridians *Pm* and *Pr*.

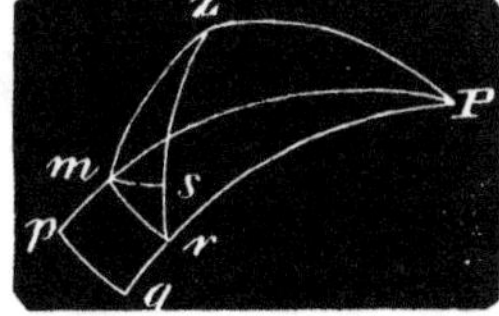

Then as *Pm* and *Pr* are equal, *mr* may be considered as a small portion of a parallel of declination *rs* will be the error in

altitude, and pq the measure of the required error in time. And as the sides of the triangle msr will necessarily be small, that triangle may be considered as a rectilinear one, right angled at s; and because the angle Prm is also a right angle, the angles smr and PrZ, being each the complement of mrs, are equal to each other

We now have,

$$rs : mr :: \sin. smr\ (ZrP) : \text{rad.} \qquad (1)$$

Also, $$mr : pq :: \cos. qr : \text{rad.} \qquad (2)$$

Multiplying these two proportions together, omiting the common factor mr, gives,

$$rs : pq :: \cos. qr \sin. (ZrP) : (\text{rad.})^2 \qquad (3)$$

But, $$\sin. rP \text{ or } \cos. qr : \sin. rZP : \sin. ZP : \sin. (ZrP) \qquad (4)$$

Whence, $$\cos. qr \sin. ZrP = \sin. rZP \sin. ZP \qquad (5)$$

The first member of equation (5), is the same as the third term in proportion (3); therefore, proportion (3) may be changed to the following,

$$rs : pq :: \sin. rZP \sin. ZP : (\text{rad.})^2$$

Whence, $$pq = \left(\frac{rs\ (\text{rad.})^2}{\sin. ZP} \right) \frac{1}{\sin. rZP}$$

Now, as the quantities in parentheses are supposed to be constant the value of pq, the error in time must vary as $\frac{1}{\sin. rZP}$ varies; and it is obvious that pq will be *least*, when $\sin. rZP$ is *greatest*, that is, when $rZP = 90°$, or the object due east or west.

LOGARITHMS OF NUMBERS

FROM

1 TO 10000.

N.	Log.	N.	Log.	N.	Log.	N.	Log.
1	0 000000	26	1 414973	51	1 707570	76	1 880814
2	0 301030	27	1 431364	52	1 716003	77	1 886491
3	0 477121	28	1 447158	53	1 724276	78	1 892095
4	0 602060	29	1 462398	54	1 732394	79	1 897627
5	0 698970	30	1 477121	55	1 740363	80	1 903090
6	0 778151	31	1 491362	56	1 748188	81	1 908485
7	0 845098	32	1 505150	57	1 755875	82	1 913814
8	0 903090	33	1 518514	58	1 763428	83	1 919078
9	0 954243	34	1 531479	59	1 770852	84	1 924279
10	1 000000	35	1 544068	60	1 778151	85	1 929419
11	1 041393	36	1 556303	61	1 785330	86	1 934498
12	1 079181	37	1 568202	62	1 792392	87	1 939519
13	1 113943	38	1 579784	63	1 799341	88	1 944483
14	1 146128	39	1 591065	64	1 806180	89	1 949390
15	1 176091	40	1 602060	65	1 812913	90	1 954243
16	1 204120	41	1 612784	66	1 819544	91	1 959041
17	1 230449	42	1 623249	67	1 826075	92	1 963788
18	1 255273	43	1 633468	68	1 832509	93	1 968483
19	1 278754	44	1 643453	69	1 838849	94	1 973128
20	1 301030	45	1 653213	70	1 845098	95	1 977724
21	1 322219	46	1 662578	71	1 851258	96	1 982271
22	1 342423	47	1 672098	72	1 857333	97	1 986772
23	1 361728	48	1 681241	73	1 863323	98	1 991226
24	1 380211	49	1 690196	74	1 869232	99	1 995635
25	1 397940	50	1 698970	75	1 875061	100	2 000000

N. B. In the following table, in the last nine columns of each page, where the first or leading figures change from 9's to 0's, points or dots are now introduced instead of the 0's through the rest of the line, to catch the eye, and to indicate that from thence the corresponding natural numbers in the first column stands in the *next lower line*, and its annexed first two figures of the Logarithms in the second column.

N.	0	1	2	3	4	5	6	7	8	9
100	000000	0434	0868	1301	1734	2166	2598	3029	3461	3891
101	4321	4750	5181	5609	6038	6466	6894	7321	7748	8174
102	8600	9026	9451	9876	.300	.724	1147	1570	1993	2415
103	012837	3259	3680	4100	4521	4940	5360	5779	6197	6616
104	7033	7451	7868	8284	8700	9116	9532	9947	.361	.775
105	021189	1603	2016	2428	2841	3252	3664	4075	4486	4896
106	5306	5715	6125	6533	6942	7350	7757	8164	8571	8978
107	9384	9789	.195	.600	1004	1408	1812	2216	2619	3021
108	033424	3826	4227	4628	5029	5430	5830	6230	6629	7028
109	7426	7825	8223	8620	9017	9414	9811	.207	.602	.998
110	041393	1787	2182	2576	2969	3362	3755	4148	4540	4932
111	5323	5714	6105	6495	6885	7275	7664	8053	8442	8830
112	9218	9606	9993	.380	.766	1153	1538	1924	2309	2694
113	053078	3463	3846	4230	4613	4996	5378	5760	6142	6524
114	6905	7286	7666	8046	8426	8805	9185	9563	9942	.320
115	060698	1075	1452	1829	2206	2582	2958	3333	3709	4083
116	4458	4832	5206	5580	5953	6326	6699	7071	7443	7815
117	8186	8557	8928	9298	9668	..38	.407	.776	1145	1514
118	071882	2250	2617	2985	3352	3718	4085	4451	4816	5182
119	5547	5912	6276	6640	7004	7368	7731	8094	8457	8819
120	9181	9543	9904	.266	.626	.987	1347	1707	2067	2426
121	082785	3144	3503	3861	4219	4576	4934	5291	5647	6004
122	6360	6716	7071	7426	7781	8136	8490	8845	9198	9552
123	9905	.258	.611	.963	1315	1667	2018	2370	2721	3071
124	093422	3772	4122	4471	4820	5169	5518	5866	6215	6562
125	6910	7257	7604	7951	8298	8644	8990	9335	9681	1026
126	100371	0715	1059	1403	1747	2091	2434	2777	3119	3462
127	3804	4146	4487	4828	5169	5510	5851	6191	6531	6871
128	7210	7549	7888	8227	8565	8903	9241	9579	9916	.253
129	110590	0926	1263	1599	1934	2270	2605	2940	3275	3609
130	3943	4277	4611	4944	5278	5611	5943	6276	6608	6940
131	7271	7603	7934	8265	8595	8926	9256	9586	9915	0245
132	120574	0903	1231	1560	1888	2216	2544	2871	3198	3525
133	3852	4178	4504	4830	5156	5481	5806	6131	6456	6781
134	7105	7429	7753	8076	8399	8722	9045	9368	9690	..12
135	130334	0655	0977	1298	1619	1939	2260	2580	2900	3219
136	3539	3858	4177	4496	4814	5133	5451	5769	6086	6403
137	6721	7037	7354	7671	7987	8303	8618	8934	9249	9564
138	9879	.194	.508	.822	1136	1450	1763	2076	2389	2702
139	143015	3327	3630	3951	4263	4574	4885	5196	5507	5818
140	6128	6438	6748	7058	7367	7676	7985	8294	8603	8911
141	9219	9527	9835	.142	.449	.756	1063	1370	1676	1982
142	152288	2594	2900	3205	3510	3815	4120	4424	4728	5032
143	5336	5640	5943	6246	6549	6852	7154	7457	7759	8061
144	8362	8664	8965	9266	9567	9868	.168	.469	.769	1068
145	161368	1667	1967	2266	2564	2863	3161	3460	3758	4055
146	4353	4650	4947	5244	5541	5838	6134	6430	6726	7022
147	7317	7613	7908	8203	8497	8792	9086	9380	9674	9968
148	170262	0555	0848	1141	1434	1726	2019	2311	2603	2895
149	3186	3478	3769	4060	4351	4641	4932	5222	5512	5802

N.	0	1	2	3	4	5	6	7	8	9
150	176091	6381	6670	6959	7248	7536	7825	8113	8401	8689
151	8977	9264	9552	9839	.126	.413	.699	.985	1272	1558
152	181844	2129	2415	2700	2985	3270	3555	3839	4123	4407
153	4691	4975	5259	5542	5825	6108	6391	6674	6956	7239
154	7521	7803	8084	8366	8647	8928	9209	9490	9771	..51
155	190332	0612	0892	1171	1451	1730	2010	2289	2567	2846
156	3125	3403	3681	3959	4237	4514	4792	5069	5346	5623
157	5899	6176	6453	6729	7005	7281	7556	7832	8107	8382
158	8657	.8932	9206	9481	9755	..29	.303	.577	.850	1124
159	201397	1670	1943	2216	2488	2761	3033	3305	3577	3848
160	4120	4391	4663	4934	5204	5475	5746	6016	6286	6556
161	6826	7096	7365	7634	7904	8173	8441	8710	8979	9247
162	9515	9783	..51	.319	.586	.853	1121	1388	1654	1921
163	212188	2454	2720	2986	3252	3518	3783	4049	4314	4579
164	4844	5109	5373	5638	5902	6166	6430	6694	6957	7221
165	7484	7747	8010	8273	8536	8798	9060	9323	9585	9846
166	220108	0370	0631	0892	1153	1414	1675	1936	2196	2456
167	2716	2976	3236	3496	3755	4015	4274	4533	4792	5051
168	5309	5568	5826	6084	6342	6600	6858	7115	7372	7630
169	7887	8144	8400	8657	8913	9170	9426	9682	9938	.193
170	230449	0704	0960	1215	1470	1724	1979	2234	2488	2742
171	2996	3250	3504	3757	4011	4264	4517	4770	5023	5276
172	5528	5781	6033	6285	6537	6789	7041	7292	7544	7795
173	8046	8297	8548	8799	9049	9299	9550	9800	..50	.300
174	240549	0799	1048	1297	1546	1795	2044	2293	2541	2790
175	3038	3286	3534	3782	4030	4277	4525	4772	5019	5266
176	5513	5759	6006	6252	6499	6745	6991	7237	7482	7728
177	7973	8219	8464	8709	8954	9198	9443	9687	9932	.176
178	250420	0664	0908	1151	1395	1638	1881	2125	2368	2610
179	2853	3096	3338	3580	3822	4064	4306	4548	4790	5031
180	5273	5514	5755	5996	6237	6477	6718	6958	7198	7439
181	7679	7918	8158	8398	8637	8877	9116	9355	9594	9833
182	260071	0310	0548	0787	1025	1263	1501	1739	1976	2214
183	2451	2688	2925	3162	3399	3636	3873	4109	4346	4582
184	4818	5054	5290	5525	5761	5996	6232	6467	6702	6937
185	7172	7406	7641	7875	8110	8344	8578	8812	9046	9279
186	9513	9746	9980	.213	.446	.679	.912	1144	1377	1609
187	271842	2074	2306	2538	2770	3001	3233	3464	3696	3927
188	4158	4389	4620	4850	5081	5311	5542	5772	6002	6232
189	6462	6692	6921	7151	7380	7609	7838	8067	8296	8525
190	8754	8982	9211	9439	9667	9895	.123	.351	.578	.806
191	281033	1261	1488	1715	1942	2169	2396	2622	2849	3075
192	3301	3527	3753	3979	4205	4431	4656	4882	5107	5332
193	5557	5782	6007	6232	6456	6681	6905	7130	7354	7578
194	7802	8026	8249	8473	8696	8920	9143	9366	9589	9812
195	290035	0257	0480	0702	0925	1147	1369	1591	1813	2034
196	2256	2478	2699	2920	3141	3363	3584	3804	4025	4246
197	4466	4687	4907	5127	5347	5567	5787	6007	6226	6446
198	6665	6884	7104	7323	7542	7761	7979	8198	8416	8635
199	8853	9071	9289	9507	9725	9943	.161	.378	.595	.813

N.	0	1	2	3	4	5	6	7	8	9
200	301030	1247	1464	1681	1898	2114	2331	2547	2764	2980
201	3196	3412	3628	3844	4059	4275	4491	4706	4921	5136
202	5351	5566	5781	5996	6211	6425	6639	6854	7068	7282
203	7496	7710	7924	8137	8351	8564	8778	8991	9204	9417
204	9630	9843	..56	.268	.481	.693	.906	1118	1330	1542
205	311754	1966	2177	2389	2600	2812	3023	3234	3445	3656
206	3867	4078	4289	4499	4710	4920	5130	5340	5551	5760
207	5970	6180	6390	6599	6809	7018	7227	7436	7646	7854
208	8063	8272	8481	8689	8898	9106	9314	9522	9730	9938
209	320146	0354	0562	0769	0977	1184	1391	1598	1805	2012
210	2219	2426	2633	2839	3046	3252	3458	3665	3871	4077
211	4282	4488	4694	4899	5105	5310	5516	5721	5926	6131
212	6336	6541	6745	6950	7155	7359	7563	7767	7972	8176
213	8380	8583	8787	8991	9194	9398	9601	9805	...8	.211
214	330414	0617	0819	1022	1225	1427	1630	1832	2034	2236
215	2438	2640	2842	3044	3246	3447	3649	3850	4051	4253
216	4454	4655	4856	5057	5257	5458	5658	5859	6059	6260
217	6460	6660	6860	7060	7260	7459	7659	7858	8058	8257
218	8456	8656	8855	9054	9253	9451	9650	9849	..47	.246
219	340444	0642	0841	1039	1237	1435	1632	1830	2028	2225
220	2423	2620	2817	3014	3212	3409	3606	3802	3999	4196
221	4392	4589	4785	4981	5178	5374	5570	5766	5962	6157
222	6353	6549	6744	6939	7135	7330	7525	7720	7915	8110
223	8305	8500	8694	8889	9083	9278	9472	9666	9860	..54
224	350248	0442	0636	0829	1023	1216	1410	1603	1796	1989
225	2183	2375	2568	2761	2954	3147	3339	3532	3724	3916
226	4108	4301	4493	4685	4876	5068	5260	5452	5643	5834
227	6026	6217	6408	6599	6790	6981	7172	7363	7554	7744
228	7935	8125	8316	8506	8696	8886	9076	9266	9456	9646
229	9835	..25	.215	.404	.593	.783	.972	1161	1350	1539
230	361728	1917	2105	2294	2482	2671	2859	3048	3236	3424
231	3612	3800	3988	4176	4363	4551	4739	4926	5113	5301
232	5488	5675	5862	6049	6236	6423	6610	6796	6983	7169
233	7356	7542	7729	7915	8101	8287	8473	8659	8845	9030
234	9216	9401	9587	9772	9958	.143	.328	.513	.698	.883
235	371068	1253	1437	1622	1806	1991	2175	2360	2544	2728
236	2912	3096	3280	3464	3647	3831	4015	4198	4382	4565
237	4748	4932	5115	5298	5481	5664	5846	6029	6212	6394
238	6577	6759	6942	7124	7306	7488	7670	7852	8034	8216
239	8398	8580	8761	8943	9124	9306	9487	9668	9849	..30
240	380211	0392	0573	0754	0934	1115	1296	1476	1656	1837
241	2017	2197	2377	2557	2737	2917	3097	3277	3456	3636
242	3815	3995	4174	4353	4533	4712	4891	5070	5249	5428
243	5606	5785	5964	6142	6321	6499	6677	6856	7034	7212
244	7390	7568	7746	7923	8101	8279	8456	8634	8811	8989
245	9166	9343	9520	9698	9875	..51	.228	.405	.582	.759
246	390935	1112	1288	1464	1641	1817	1993	2169	2345	2521
247	2697	2873	3048	3224	3400	3575	3751	3926	4101	4277
248	4452	4627	4802	4977	5152	5326	5501	5676	5850	6025
249	6199	6374	6548	6722	6896	7071	7245	7419	7592	7766

N.	0	1	2	3	4	5	6	7	8	9
250	397940	8114	8287	8461	8634	8808	8981	9154	9328	9501
251	9674	9847	..20	.192	.365	.538	.711	.883	1056	1228
252	401401	1573	1745	1917	2089	2261	2433	2605	2777	2949
253	3121	3292	3464	3635	3807	3978	4149	4320	4492	4663
254	4834	5005	5176	5346	5517	5688	5858	6029	6199	6370
255	6540	6710	6881	7051	7221	7391	7561	7731	7901	8070
256	8240	8410	8579	8749	8918	9087	9257	9426	9595	9764
257	9933	.102	.271	.440	.609	.777	.946	1114	1283	1451
258	411620	1788	1956	2124	2293	2461	2629	2796	2964	3132
259	3300	3467	3635	3803	3970	4137	4305	4472	4639	4806
260	4973	5140	5307	5474	5641	5808	5974	6141	6308	6474
261	6641	6807	6973	7139	7306	7472	7638	7804	7970	8135
262	8301	8467	8633	8798	8964	9129	9295	9460	9625	9791
263	9956	.121	.286	.451	.616	.781	.945	1110	1275	1439
264	421604	1788	1933	2097	2261	2426	2590	2754	2918	3082
265	3246	3410	3574	3737	3901	4065	4228	4392	4555	4718
266	4882	5045	5208	5371	5534	5697	5860	6023	6186	6349
267	6511	6674	6836	6999	7161	7324	7486	7648	7811	7973
268	8135	8297	8459	8621	8783	8944	9106	9268	9429	9591
269	9752	9914	..75	.236	.398	.559	.720	.881	1042	1203
270	431364	1525	1685	1846	2007	2167	2328	2488	2649	2809
271	2969	3130	3290	3450	3610	3770	3930	4090	4249	4409
272	4569	4729	4888	5048	5207	5367	5526	5685	5844	6004
273	6163	6322	6481	6640	6800	6957	7116	7275	7433	7592
274	7751	7909	8067	8226	8384	8542	8701	8859	9017	9175
275	9333	9491	9648	9806	9964	.122	.279	.437	.594	.752
276	440909	1066	1224	1381	1538	1695	1852	2009	2166	2323
277	2480	2637	2793	2950	3106	3263	3419	3576	3732	3889
278	4045	4201	4357	4513	4669	4825	4981	5137	5293	5449
279	5604	5760	5915	6071	6226	6382	6537	6692	6848	7003
280	7158	7313	7468	7623	7778	7933	8088	8242	8397	8552
281	8706	8861	9015	9170	9324	9478	9633	9787	9941	..95
282	450249	0403	0557	0711	0865	1018	1172	1326	1479	1633
283	1786	1940	2093	2247	2400	2553	2706	2859	3012	3165
284	3318	3471	3624	3777	3930	4082	4235	4387	4540	4692
285	4845	4997	5150	5302	5454	5606	5758	5910	6062	6214
286	6366	6518	6670	6821	6973	7125	7276	7428	7579	7731
287	7882	8033	8184	8336	8487	8638	8789	8940	9091	9242
288	9392	9543	9694	9845	9995	.146	.296	.447	.597	.748
289	460898	1048	1198	1348	1499	1649	1799	1948	2098	2248
290	2398	2548	2697	2847	2997	3146	3296	3445	3594	3744
291	3893	4042	4191	4340	4490	4639	4788	4936	5085	5234
292	5383	5532	5680	5829	5977	6126	6274	6423	6571	6719
293	6868	7016	7164	7312	7460	7608	7756	7904	8052	8200
294	8347	8495	8643	8790	8938	9085	9233	9380	9527	9675
295	9822	9969	.116	.263	.410	.557	.704	.851	.998	1145
296	471292	1438	1585	1732	1878	2025	2171	2318	2464	2610
297	2756	2903	3049	3195	3341	3487	3633	3779	3925	4071
298	4216	4362	4508	4653	4799	4944	5090	5235	5381	5526
299	5671	5816	5962	6107	6252	6397	6542	6687	6832	6976

N.	0	1	2	3	4	5	6	7	8	9
300	477121	7266	7411	7555	7700	7844	7989	8133	8278	8422
301	8566	8711	8855	8999	9143	9287	9481	9575	9719	9863
302	480007	0151	0294	0438	0582	0725	0869	1012	1156	1299
303	1443	1586	1729	1872	2016	2159	2302	2445	2588	2731
304	2874	3016	3159	3302	3445	3587	3730	3872	4015	4157
305	4300	4442	4585	4727	4869	5011	5153	5295	5437	5579
306	5721	5863	6005	6147	6289	6430	6572	6714	6855	6997
307	7138	7280	7421	7563	7704	7845	7986	8127	8269	8410
308	8551	8692	8833	8974	9114	9255	9396	9537	9667	9818
309	9959	..99	.239	.380	.520	.661	.801	.941	1081	1222
310	491362	1502	1642	1782	1922	2062	2201	2341	2481	2621
311	2760	2900	3040	3179	3319	3458	3597	3737	3876	4015
312	4155	4294	4433	4572	4711	4850	4989	5128	5267	5406
313	5544	5683	5822	5960	6099	6238	6376	6515	6653	6791
314	6930	7068	7206	7344	7483	7621	7759	7897	8035	8173
315	8311	8448	8586	8724	8862	8999	9137	9275	9412	9550
316	9687	9824	9962	..99	.236	.374	.511	.648	.785	.922
317	501059	1196	1333	1470	1607	1744	1880	2017	2154	2291
318	2427	2564	2700	2837	2973	3109	3246	3382	3518	3655
319	3791	3927	4063	4199	4335	4471	4607	4743	1878	5014
320	5150	5286	5421	5557	5693	5828	5964	6099	6234	6370
321	6505	6640	6776	6911	7046	7181	7316	7451	7586	7721
322	7856	7991	8126	8260	8395	8530	8664	8799	8934	9008
323	9203	9337	9471	9606	9740	9874	...9	.143	.277	.411
324	510545	0679	0813	0947	1081	1215	1349	1482	1616	1750
325	1883	2017	2151	2284	2418	2551	2684	2818	2951	3084
326	3218	3351	3484	3617	3750	3883	4016	4149	4282	4414
327	4548	4681	4813	4946	5079	5211	5344	5476	5609	5741
328	5874	6006	6139	6271	6403	6535	6668	6800	6932	7064
329	7196	7328	7460	7592	7724	7855	7987	8119	8251	8382
330	8514	8646	8777	8909	9040	9171	9303	9434	9566	9697
331	9828	9959	..90	.221	.353	.484	.615	.745	.876	1007
332	521138	1269	1400	1530	1661	1792	1922	2053	2183	2314
333	2444	2575	2705	2835	2966	3096	3226	3356	3486	3616
334	3746	3876	4006	4136	4266	4396	4526	4656	4785	4915
335	5045	5174	5304	5434	5563	5693	5822	5951	6081	6210
336	6339	6469	6598	6727	6856	6985	7114	7243	7372	7501
337	7630	7759	7888	8016	8145	8274	8402	8531	8660	8788
338	8917	9045	9174	9302	9430	9559	9687	9815	9943	..72
339	530200	0328	0456	0584	0712	0840	0968	1096	1223	1351
340	1479	1607	1734	1862	1960	2117	2245	2372	2500	2627
341	2754	2882	3009	3136	3264	3391	3518	3645	3772	3899
342	4026	4153	4280	4407	4534	4661	4787	4914	5041	5167
343	5294	5421	5547	5674	5800	5927	6053	6180	6306	6432
344	6558	6685	6811	6937	7060	7189	7315	7441	7567	7693
345	7819	7945	8071	8197	8322	8448	8574	8699	8825	8951
346	9076	9202	9327	9452	9578	9703	9829	9954	..79	.204
347	540329	0455	0580	0705	0830	0955	1080	1205	1330	1454
348	1579	1704	1829	1953	2078	2203	2327	2452	2576	2701
349	2825	2950	3074	3199	3323	3447	3571	3696	3820	3944

N.	0	1	2	3	4	5	6	7	8	9
350	544068	4192	4316	4440	4564	4688	4812	4936	5060	5183
351	5307	5431	5555	5678	5805	5925	6049	6172	6296	6419
352	6543	6666	6789	6913	7036	7159	7282	7405	7529	7652
353	7775	7898	8021	8144	8267	8389	8512	8635	8758	8881
354	9003	9126	9249	9371	9494	9616	9739	9861	9984	.196
355	550228	0351	0473	0595	0717	0840	0962	1084	1206	1328
356	1450	1572	1694	1816	1938	2060	2181	2303	2425	2547
357	2668	2790	2911	3033	3155	3276	3393	3519	3640	3762
358	3883	4004	4126	4247	4368	4489	4610	4731	4852	4973
359	5094	5215	5346	5457	5578	5699	5820	5940	6061	6182
360	6303	6423	6544	6664	6785	6905	7026	7146	7267	7387
361	7507	7627	7748	7868	7988	8108	8228	8349	8469	8589
362	8709	8829	8948	9068	9188	9308	9428	9548	9667	9787
363	9907	..26	.146	.265	.385	.504	.624	.743	.863	.982
364	561101	1221	1340	1459	1578	1698	1817	1936	2055	2173
365	2293	2412	2531	2650	2769	2887	3006	3125	3244	3362
366	3481	3600	3718	3837	3955	4074	4192	4311	4429	4548
367	4666	4784	4903	5021	5139	5257	5376	5494	5612	5730
368	5848	5966	6084	6202	6320	6437	6555	6673	6791	6909
369	7026	7144	7262	7379	7497	7614	7732	7849	7967	8084
370	8202	8319	8436	8554	8671	8788	8905	9023	9140	9257
371	9374	9491	9608	9725	9882	9959	..76	.193	.309	.426
372	570543	0660	0776	0893	1010	1126	1243	1359	1476	1592
373	1709	1825	1942	2058	2174	2291	2407	2523	2639	2755
374	2872	2988	3104	3220	3336	3452	3568	3684	3800	3915
375	4031	4147	4263	4379	4494	4610	4726	4841	4957	5072
376	5188	5303	5419	5534	5650	5765	5880	5996	6111	6226
377	6341	6457	6572	6687	6802	6917	7032	7147	7262	7377
378	7492	7607	7722	7836	7951	8066	8181	8295	8410	8525
379	8639	8754	8868	8983	9097	9212	9326	9441	9555	9669
380	9784	9898	..12	.126	.241	.355	.469	.583	.697	.811
381	580925	1039	1153	1267	1381	1495	1608	1722	1836	1950
382	2063	2177	2291	2404	2518	2631	2745	2858	2972	3085
383	3199	3312	3426	3539	3652	3765	3879	3992	4105	4218
384	4331	4444	4557	4670	4783	4896	5009	5122	5235	5348
385	5461	5574	5686	5799	5912	6024	6137	6250	6362	6475
386	6587	6700	6812	6925	7037	7149	7262	7374	7486	7599
387	7711	7823	7935	8047	8160	8272	8384	8496	8608	8720
388	8832	8944	9056	9167	9279	9391	9503	9615	9726	9834
389	9950	..61	.173	.284	.396	.507	.619	.730	.842	.953
390	591065	1176	1287	1399	1510	1621	1732	1843	1955	2066
391	2177	2288	2399	2510	2621	2732	2843	2954	3064	3175
392	3286	3397	3508	3618	3729	3840	3950	4061	4171	4282
393	4393	4503	4614	4724	4834	4945	5055	5165	5276	5386
394	5496	5606	5717	5827	5937	6047	6157	6267	6377	6487
395	6597	6707	6817	6927	7037	7146	7256	7366	7476	7586
396	7695	7805	7914	8024	8134	8243	8353	8462	8572	8681
397	8791	8900	9009	9119	9228	9337	9446	9556	9666	9774
398	9883	9992	.101	.210	.319	.428	.537	.646	.755	.864
399	600973	1082	1191	1299	1408	1517	1625	1734	1843	1951

N.	0	1	2	3	4	5	6	7	8	9
400	602060	2169	2277	2386	2494	2603	2711	2819	2928	3036
401	3144	3253	3361	3469	3573	3686	3794	3902	4010	4118
402	4226	4334	4442	4550	4658	4766	4874	4982	5089	5197
403	5305	5413	5521	5628	5736	5844	5951	6059	6166	6274
404	6381	6489	6596	6704	6811	6919	7026	7133	7241	7348
405	7455	7562	7669	7777	7884	7991	8098	8205	8312	8419
406	8526	8633	8740	8847	8954	9061	9167	9274	9381	9488
407	9594	9701	9808	9914	..21	.128	.234	.341	.447	.554
408	610660	0767	0873	0979	1086	1192	1298	1405	1511	1617
409	1723	1829	1936	2042	2148	2254	2360	2466	2572	2678
410	2784	2890	2996	3102	3207	3313	3419	3525	3630	3736
411	3842	3947	4053	4159	4264	4370	4475	4581	4686	4792
412	4897	5003	5108	5213	5319	5424	5529	5634	5740	5845
413	5950	6055	6160	6265	6370	6476	6581	6686	6790	6895
414	7000	7105	7210	7315	7420	7525	7629	7734	7839	7943
415	8048	8153	8257	8362	8466	8571	8676	8780	8884	8989
416	9293	9198	9302	9406	9511	9615	9719	9824	9928	..32
417	620136	0140	0344	0448	0552	0656	0760	0864	0068	1072
418	1176	1280	1384	1488	1592	1695	1799	1903	2007	2110
419	2214	2318	2421	2525	2628	2732	2835	2939	3042	3146
420	3249	3353	3456	3559	3663	3766	3869	3973	4076	4179
421	4282	4385	4488	4591	4695	4798	4901	5004	5107	5210
422	5312	5415	5518	5621	5724	5827	5929	6032	6135	6238
423	6340	6443	6546	6648	6751	6853	6956	7058	7161	7263
424	7366	7468	7571	7673	7775	7878	7980	8082	8185	8287
425	8389	8491	8593	8695	8797	8900	9002	9104	9206	9308
426	9410	9512	9613	9715	9817	9919	..21	.123	.224	.326
427	630428	0530	0631	0733	0835	0936	1038	1139	1241	1342
428	1444	1545	1647	1748	1849	1951	2052	2153	2255	2356
429	2457	2559	2660	2761	2862	2963	3064	3165	3266	3367
430	3468	3569	3670	3771	3872	3973	4074	4175	4276	4376
431	4477	4578	4679	4779	4880	4981	5081	5182	5283	5383
432	5484	5584	5685	5785	5886	5986	6087	6187	6287	6388
433	6488	6588	6688	6789	6889	6989	7089	7189	7290	7390
434	7490	7590	7690	7790	7890	7990	8090	8190	8290	8389
435	8489	8589	8689	8789	8888	8988	9088	9188	9287	9387
436	9486	9586	9686	9785	9885	9984	..84	.183	.283	.382
437	640481	0581	0680	0779	0879	0978	1077	1177	1276	1375
438	1474	1573	1672	1771	1871	1970	2069	2168	2267	2366
439	2465	2563	2662	2761	2860	2959	3058	3156	3255	3354
440	3453	3551	3650	3749	3847	3946	4044	4143	4242	4340
441	4439	4537	4636	4734	4832	4931	5029	5127	5226	5324
442	5422	5521	5619	5717	5815	5913	6011	6110	6208	6306
443	6404	6502	6600	6698	6796	6894	6992	7089	7187	7285
444	7383	7481	7579	7676	7774	7872	7969	8067	8165	8262
445	8360	8458	8555	8653	8750	8848	8945	9043	9140	9237
446	9335	9432	9530	9627	9724	9821	9919	..16	.113	.210
447	650308	0405	0502	0599	0696	0793	0890	0987	1084	1181
448	1278	1375	1472	1569	1666	1762	1859	1956	2053	2150
449	2246	2343	2440	2530	2633	2730	2826	2923	3019	3116

LOGARITHMS

N.	0	1	2	3	4	5	6	7	8	9
450	653213	3309	3405	3502	3598	3695	3791	3888	3984	4080
451	4177	4273	4369	4465	4562	4658	4754	4850	4946	5042
452	5138	5235	5331	5427	5526	5619	5715	5810	5906	6002
453	6098	6194	6290	6386	6482	6577	6673	6769	6864	6960
454	7056	7152	7247	7343	7438	7534	7629	7725	7820	7916
455	8011	8107	8202	8298	8393	8488	8584	8679	8774	8870
456	8965	9060	9155	9250	9346	9441	9536	9631	9726	9821
457	9916	..11	.106	.201	.296	.391	.486	.581	.676	.771
458	660865	0960	1055	1150	1245	1339	1434	1529	1623	1718
459	1813	1907	2002	2096	2191	2286	2380	2475	2569	2663
460	2758	2852	2947	3041	3135	3230	3324	3418	3512	3607
461	3701	3795	3889	3983	4078	4172	4266	4360	4454	4548
462	4642	4736	4830	4924	5018	5112	5206	5299	5393	5487
463	5581	5675	5769	5862	5956	6050	6143	6237	6331	6424
464	6518	6612	6705	6799	6892	6986	7079	7173	7266	7360
465	7453	7546	7640	7733	7826	7920	8013	8106	8199	8293
466	8386	8479	8572	8665	8759	8852	8945	9038	9131	9324
467	9317	9410	9503	9596	9689	9782	9875	9967	..60	.153
468	670241	0339	0431	0524	0617	0710	0802	0895	0988	1080
469	1173	1265	1358	1451	1543	1636	1728	1821	1913	2005
470	2098	2190	2283	2375	2467	2560	2652	2744	2836	2929
471	3021	3113	3205	3297	3390	3482	3574	3666	3758	3850
472	3942	4034	4126	4218	4310	4402	4494	4586	4677	4769
473	4861	4953	5045	5137	5228	5320	5412	5503	5595	5687
474	5778	5870	5962	6053	6145	6236	6328	6419	6511	6602
475	6694	6785	6876	6968	7059	7151	7242	7333	7424	7516
476	7607	7698	7789	7881	7972	8063	8154	8245	8336	8427
477	8518	8609	8700	8791	8882	8972	9064	9155	9246	9337
478	9428	9519	9610	9700	9791	9882	9973	..63	.154	.245
479	680336	0426	0517	0607	0698	0789	0879	0970	1060	1151
480	1241	1332	1422	1513	1603	1693	1784	1874	1964	2055
481	2145	2235	2326	2416	2506	2596	2686	2777	2867	2957
482	3047	3137	3227	3317	3407	3497	3587	3677	3767	3857
483	3947	4037	4127	4217	4307	4396	4486	4576	4666	4756
484	4854	4935	5025	5114	5204	5294	5383	5473	5563	5652
485	5742	5831	5921	6010	6100	6189	6279	6368	6458	6547
486	6636	6726	6815	6904	6994	7083	7172	7261	7351	7440
487	7529	7618	7707	7796	7886	7975	8064	8153	8242	8331
488	8420	8509	8598	8687	8776	8865	8953	9042	9131	9220
489	9309	9398	9486	9575	9664	9753	9841	9930	..19	.107
490	690196	0285	0373	0362	0550	0639	0728	0816	0905	0993
491	1081	1170	1258	1347	1435	1524	1612	1700	1789	1877
492	1965	2053	2142	2230	2318	2406	2494	2583	2671	2759
493	2847	2935	3023	3111	3199	3287	3375	3463	3551	3639
494	3727	3815	3903	3991	4078	4166	4254	4342	4430	4517
495	4605	4693	4781	4868	4956	5044	5131	5210	5307	5394
496	5482	5569	5657	5744	5832	5919	6007	6094	6182	6269
497	6356	5444	6531	6618	6706	6793	6880	6968	7055	7142
498	7229	7317	7404	7491	7578	7665	7752	7839	7926	8014
499	8101	8188	8275	8362	8449	8535	8622	8709	8796	8883

N.	0	1	2	3	4	5	6	7	8	9
500	698970	9057	9144	9231	9317	9404	9491	9578	9664	9751
501	9838	9924	..11	..98	.184	.271	.358	.444	.531	.617
502	700704	0790	0877	0963	1050	1136	1222	1309	1395	1482
503	1568	1654	1741	1827	1913	1999	2086	2172	2258	2344
504	2431	2517	2603	2689	2775	2861	2947	3033	3119	3205
505	3291	3377	3463	3549	3635	3721	3807	3895	3979	4065
506	4151	4236	4322	4408	4494	4579	4665	4751	4837	4922
507	5008	5094	5179	5265	5350	5436	5522	5607	5693	5778
508	5864	5949	6035	6120	6206	6291	6376	6462	6547	6632
509	6718	6803	6888	6974	7059	7144	7229	7315	7400	7485
510	7570	7655	7740	7826	7910	7996	8081	8166	8251	8336
511	8421	8506	8591	8676	8761	8846	8931	9015	9100	9185
512	9270	9355	9440	9524	9609	9694	9779	9863	9948	..33
513	710117	0202	0287	0371	0456	0540	0625	0710	0794	0879
514	0963	1048	1132	1217	1301	1385	1470	1554	1639	1723
515	1807	1892	1976	2060	2144	2229	2313	2397	2481	2566
516	2650	2734	2818	2902	2986	3070	3154	3238	3326	3407
517	3491	3575	3659	3742	3826	3910	3994	4078	4162	4246
518	4330	4414	4497	4581	4665	4749	4833	4916	5000	5084
519	5167	5251	5335	5418	5502	5586	5669	5753	5836	5920
520	6003	6087	6170	6254	6337	6421	6504	6588	6671	6754
521	6838	6921	7004	7088	7171	7254	7338	7421	7504	7587
522	7671	7754	7837	7920	8003	8086	8169	8253	8336	8419
523	8502	8585	8668	8751	8834	8917	9000	9083	9165	9248
524	9331	9414	9497	9580	9663	9745	9828	9911	9994	..77
525	720159	0242	0325	0407	0490	0573	0655	0738	0821	0903
526	0986	1068	1151	1233	1316	1398	1481	1563	1646	1728
527	1811	1893	.975	2058	2140	2222	2305	2387	2469	2552
528	2634	2716	2798	2881	2963	3045	3127	3209	3291	3374
529	3456	3538	3620	3702	3784	3866	3948	4030	4112	4194
530	4276	4358	4440	4522	4604	4685	4767	4849	4931	5013
531	5095	5176	5258	5340	5422	5503	5585	5667	5748	5830
532	5912	5993	6075	6156	6238	6320	6401	6483	6564	6646
533	6727	6809	6890	6972	7053	7134	7216	7297	7379	7460
534	7541	7623	7704	7785	7866	7948	8029	8110	8191	8273
535	8354	8435	8516	8597	8678	8759	8841	8922	9003	9084
536	9165	9246	9327	9403	9489	9570	9651	9732	9813	9893
537	9974	..55	.136	.217	.298	.378	.459	.440	.621	.702
538	730782	0863	0944	1024	1105	1186	1266	1347	1428	1508
539	1589	1669	1750	1830	1911	1991	2072	2152	2233	2313
540	2394	2474	2555	2635	2715	2796	2876	2956	3037	3117
541	3197	3278	3358	3438	3518	3598	3679	3759	3839	3919
542	3999	4079	4160	4240	4320	4400	4480	4560	4640	4720
543	4800	4880	4960	5040	5120	5200	5279	5359	5439	5519
544	5599	5679	5759	5838	5918	5998	6078	6157	6237	6317
545	6397	6476	6556	6636	6715	6795	6874	6954	7034	7113
546	7193	7272	7352	7431	7511	7590	7670	7749	7829	7908
547	7987	8067	8146	8225	8305	8384	8463	8543	8622	8701
548	8781	8860	8939	9018	9097	9177	9256	9335	9414	9493
549	9572	9651	9731	9810	9889	9968	..47	.126	.205	.284

N.	0	1	2	3	4	5	6	7	8	9
550	740363	0442	0521	0600	0678	0757	0836	0915	0994	1073
551	1152	1230	1309	1388	1467	1546	1624	1703	1782	1860
552	1939	2018	2096	2175	2254	2332	2411	2489	2568	2646
553	2725	2804	2882	2961	3039	3118	3196	3275	3353	3431
554	3510	3588	3667	3745	3823	3902	3980	4058	4136	4215
555	4293	4371	4449	4528	4606	4684	4762	4840	4919	4997
556	5075	5153	5231	5309	5387	5465	5543	5621	5699	5777
557	5855	5933	6011	6089	6167	6245	6323	6401	6479	6556
558	6634	6712	6790	6868	6945	7023	7101	7179	7256	7334
559	7412	7489	7567	7645	7722	7800	7878	7955	8033	8110
560	8188	8266	8343	8421	8498	8576	8653	8731	8808	8885
561	8963	9040	9118	9195	9272	9350	9427	9504	9582	9659
562	9736	9814	9891	9968	..45	.123	.200	.277	.354	.431
563	750508	0586	0663	0740	0817	0894	0971	1048	1125	1202
564	1279	1356	1433	1510	1587	1664	1741	1818	1895	1972
565	2048	2125	2202	2279	2356	2433	2509	2586	2663	2740
566	2816	2893	2970	3047	3123	3200	3277	3353	3430	3506
567	3583	3660	3736	3813	3889	3966	4042	4119	4195	4272
568	4348	4425	4501	4578	4654	4730	4807	4883	4960	5036
569	5112	5189	5265	5341	5417	5494	5570	5646	5722	5799
570	5875	5951	6027	6103	6180	6256	6332	6408	6484	6560
571	6636	6712	6788	6864	6940	7016	7092	7168	7244	7320
572	7396	7472	7548	7624	7700	7775	7851	7927	8003	8079
573	8155	8230	8306	8382	8458	8533	8609	8685	8761	8836
574	8912	8988	9063	9139	9214	9290	9366	9441	9517	9592
575	9668	9743	9819	9894	9970	..45	.121	.196	.272	.347
576	760422	0498	0573	0649	0724	0799	0875	0950	1025	1101
577	1176	1251	1326	1402	1477	1552	1627	1702	1778	1853
578	1928	2003	2078	2153	2228	2303	2378	2453	2529	2604
579	2679	2754	2829	2904	2978	3053	3128	2203	3278	3353
580	3428	3503	3578	3653	3727	3802	3877	3952	4027	4101
581	4176	4251	4326	4400	4475	4550	4624	4699	4774	4848
582	4923	4998	5072	5147	5221	5296	5370	5445	5520	5594
583	5669	5743	5818	5892	5966	6041	6115	6190	6264	6338
584	6413	6487	6562	6636	6710	6785	6859	6933	7007	7082
585	7156	7230	7304	7379	7453	7527	7601	7675	7749	7823
586	7898	7972	8046	8120	8194	8268	8342	8416	8490	8564
587	8638	8712	8786	8860	8934	9008	9082	9156	9230	9303
588	9377	9451	9525	9599	9673	9746	9820	9894	9968	..42
589	770115	0189	0263	0336	0410	0484	0557	0631	0705	0778
590	0852	0926	0999	1073	1146	1220	1293	1367	1440	1514
591	1587	1661	1734	1808	1881	1955	2028	2102	2175	2248
592	2322	2395	2468	2542	2615	2688	2762	2835	2908	2981
593	3055	3128	3201	3274	3348	3421	3494	3567	3640	3713
594	3786	3860	3933	4006	4079	4152	4225	4298	4371	4444
595	4517	4590	4663	4736	4809	4882	4955	5028	5100	5173
596	5246	5319	5392	5465	5538	5610	5683	5756	5829	5902
597	5974	6047	6120	6193	6265	6338	6411	6483	6556	6629
598	6701	6774	6846	6919	6992	7064	7137	7209	7282	7354
599	7427	7499	7572	7644	7717	7789	7862	7934	8006	8079

N.	0	1	2	3	4	5	6	7	8	9
600	778151	8224	8296	8368	8441	8513	8585	8658	8730	8802
601	8874	8947	9019	9091	9163	9236	9308	9380	9452	9524
602	9596	6669	9741	9813	9885	9957	..29	.101	.173	.245
603	780317	0389	0461	0533	0605	0677	0749	0821	0893	0965
604	1037	1109	1181	1253	1324	1396	1468	1540	1612	1684
605	1755	1827	1899	1971	2042	2114	2186	2258	2329	2401
606	2473	2544	2616	2688	2759	2831	2902	2974	3046	3117
607	3189	3260	3332	3403	3475	3546	3618	3689	3761	3832
608	3904	3975	4046	4118	4189	4261	4332	4403	4475	4546
609	4617	4689	4760	4831	4902	4974	5045	5116	5187	5259
610	5330	5401	5472	5543	5615	5686	5757	5828	5899	5970
611	6041	6112	6183	6254	6325	6396	6467	6538	6609	6680
612	6751	6822	6893	6964	7035	7106	7177	7248	7319	7390
613	7460	7531	7602	7673	7744	7815	7885	7956	8027	8098
614	8168	8239	8310	8381	8451	8522	8593	8663	8734	8804
615	8875	8946	9016	9087	9157	9228	9299	9369	9440	9510
616	9581	9651	9722	9792	9863	9933	...4	..74	.144	.215
617	790285	0356	0426	0496	0567	0637	0707	0778	0848	0918
618	0988	1059	1129	1199	1269	1340	1410	1480	1550	1620
619	1691	1761	1831	1901	1971	2041	2111	2181	2252	2322
620	2392	2462	2532	2602	2672	2742	2812	2882	2952	3022
621	3092	3162	3231	3301	3371	3441	3511	3581	3651	3721
622	3790	3860	3930	4000	4070	4139	4209	4279	4349	4418
623	4488	4558	4627	4697	4767	4836	4906	4976	5045	5115
624	5185	5254	5324	5393	5463	5532	5602	5672	5741	5811
625	5880	5949	6019	6088	6158	6227	6297	6366	6436	6505
626	6574	6644	6713	6782	6852	6921	6990	7060	7129	7198
627	7268	7337	7406	7475	7545	7614	7683	7752	7821	7890
628	7960	8029	8098	8167	8236	8305	8374	8443	8513	8582
629	8651	8720	8789	8858	8927	8996	9065	6134	9203	9272
630	9341	9409	9478	9547	9616	9685	9754	9823	9892	9961
631	800026	0098	0167	0236	0305	0373	0442	0511	0580	0648
632	0717	0786	0854	0923	0992	1061	1129	1198	1266	1335
633	1404	1472	1541	1609	1678	1747	1815	1884	1952	2021
634	2089	2158	2226	2295	2363	2432	2500	2568	2637	2705
635	2774	2842	2910	2979	3047	3116	3184	3252	3321	3389
636	3457	3525	3594	3662	3730	3798	3867	3935	4003	4071
637	4139	4208	4276	4354	4412	4480	4548	4616	4685	4753
638	4821	4889	4957	5025	5093	5161	5229	5297	5365	5433
639	5501	5669	5637	5705	5773	5841	5908	5976	6044	6112
640	6180	6248	6316	6384	6451	6519	6587	6655	6723	6790
641	6858	6926	6994	7061	7129	7157	7264	7332	7400	7467
642	7535	7603	7670	7738	7806	7873	7941	8008	8076	8143
643	8211	8279	8346	8414	8481	8549	8616	8684	8751	8818
644	8886	8953	9021	9088	9156	9223	9290	9358	9425	9492
645	9560	9627	9694	9762	9829	9896	9964	..31	..98	.165
646	810233	0300	0367	0434	0501	0596	0636	0703	0770	0837
647	0904	0971	1039	1106	1173	1240	1307	1374	1441	1508
648	1575	1642	1709	1776	1843	1910	1977	2044	2111	2178
649	2245	2312	2379	2445	2512	2579	2646	2713	2780	2847

N.	0	1	2	3	4	5	6	7	8	9
650	812913	2980	3047	3114	3181	3247	3314	3381	3448	3514
651	3581	3648	3714	3781	3848	3914	3981	4048	4114	4181
652	4248	4314	4381	4447	4514	4581	4647	4714	4780	4847
653	4913	4980	5046	5113	5179	5246	5312	5378	5445	5511
654	5578	5644	5711	5777	5843	5910	5976	6042	6109	6175
655	6241	6308	6374	6440	6506	6573	6639	6705	6771	6838
656	6904	6970	7036	7102	7169	7233	7301	7367	7433	7499
657	7565	7631	7698	7764	7830	7896	7962	8028	8094	8160
658	8226	8292	8358	8424	8490	8556	8622	8688	8754	8820
659	8885	8951	9017	9083	9149	9215	9281	9346	9412	9478
660	9544	9610	9676	9741	9807	9873	9939	...4	..70	.136
661	820201	0267	0333	0399	0464	0530	0595	0661	0727	0792
662	0858	0924	0989	1055	1120	1186	1251	1317	1382	1448
663	1514	1579	1645	1710	1775	1841	1906	1972	2037	2103
664	2168	2233	2299	2364	2430	2495	2560	2626	2691	2756
665	2822	2887	2952	3018	3083	3148	3213	3279	3344	3409
666	3474	3539	3605	3670	3735	3800	3865	3930	3996	4061
667	4126	4191	4256	4321	4386	4451	4516	4581	4646	4711
668	4776	4841	4906	4971	5036	5101	5166	5231	5296	5361
669	5426	5491	5556	5621	5686	5751	5815	5880	5945	6010
670	6075	6140	6204	6269	6334	6399	6464	6528	6593	6658
671	6723	6787	6852	6917	6981	7046	7111	7175	7240	7305
672	7369	7434	7499	7563	7628	7692	7757	7821	7886	7951
673	8015	8080	8144	8209	8273	8338	8402	8467	8531	8595
674	8660	8724	8789	8853	8918	8982	9046	9111	9175	9239
675	9304	9368	9432	9497	9561	9625	9690	9754	9818	9882
676	9947	..11	..75	.139	.204	.268	.332	.396	.460	.525
677	830589	0653	0717	0781	0845	0909	0973	1037	1102	1166
678	1230	1294	1358	1422	1486	1550	1614	1678	1742	1806
679	1870	1934	1998	2062	2126	2189	2253	2317	2381	2445
680	2509	2573	2637	2700	2764	2828	2892	2956	3020	3083
681	3147	3211	3275	3338	3402	3466	3530	3593	3657	3721
682	3784	3848	3912	3975	4039	4103	4166	4230	4294	4357
683	4421	4484	4548	4611	4675	4739	4802	4866	4929	4993
684	5056	5120	5183	5247	5310	5373	5437	5500	5564	5627
685	5691	5754	5817	5881	5944	6007	6071	6134	6197	6261
686	6324	6387	6451	6514	6577	6641	6704	6767	6830	6894
687	6957	7020	7083	7146	7210	7273	7336	7399	7462	7525
688	7588	7652	7715	7778	7841	7904	7967	8030	8093	8156
689	8219	8282	8345	8408	8471	8534	8597	8660	8723	8786
690	8849	8912	8975	9038	9109	9164	9227	9289	9352	9415
691	9478	9541	9604	9667	9729	9792	9855	9918	9981	..43
692	840106	0169	0232	0294	0357	0420	0482	0545	0608	0671
693	0733	0796	0859	0921	0984	1046	1109	1172	1234	1297
694	1359	1422	1485	1547	1610	1672	1735	1797	1860	1922
695	1985	2047	2110	2172	2235	2297	2360	2422	2484	2547
696	2609	2672	2734	2796	2859	2921	2983	3046	3108	3170
697	3233	3295	3357	3420	3482	3544	3606	3669	3731	3793
698	3855	3918	3980	4042	4104	4166	4229	4291	4353	4415
699	4477	4539	4601	4664	4726	4788	4850	4912	4974	5036

N.	0	1	2	3	4	5	6	7	8	9
700	845098	5160	5222	5284	5346	5408	5470	5532	5594	5656
701	5718	5780	5842	5904	5966	6028	6090	6151	6213	6275
702	6337	6399	6461	6523	6585	6646	6708	6770	6832	6894
703	6955	7017	7079	7141	7202	7264	7326	7388	7449	7511
704	7573	7634	7676	7758	7819	7831	7943	8004	8066	8128
705	8189	8251	8312	8374	8435	8497	8559	8620	8682	8743
706	8805	8866	8928	8989	9051	9112	9174	9235	9297	9358
707	9419	9481	9542	9604	9665	9726	9788	9849	9911	9972
708	850033	0095	0156	0217	0279	0340	0401	0462	0524	0585
709	0646	0707	0769	0830	0891	0952	1014	1075	1136	1197
710	1258	1320	1381	1442	1503	1564	1625	1686	1747	1809
711	1870	1931	1992	2053	2114	2175	2236	2297	2358	2419
712	2480	2541	2602	2663	2724	2785	2846	2907	2968	3029
713	3090	3150	3211	3272	3333	3394	3455	3516	3577	3637
714	3698	3759	3820	3881	3941	4002	4063	4124	4185	4245
715	4306	4367	4428	4488	4549	4610	4670	4731	4792	4852
716	4913	4974	5034	5095	5156	5216	5277	5337	5398	5459
717	5519	5580	5640	5701	5761	5822	5882	5943	6003	6064
718	6124	6185	6245	6306	6366	6427	6487	6548	6608	6668
719	6729	6789	6850	6910	6970	7031	7091	7152	7212	7272
720	7332	7393	7453	7513	7574	7634	7694	7755	7815	7875
721	7935	7995	8056	8116	8176	8236	8297	8357	8417	8477
722	8537	8597	8657	8718	8778	8838	8898	8958	9018	9078
723	9138	9198	9258	9318	9379	9439	9499	9559	9619	9679
724	9739	9799	9859	9918	9978	..38	..98	.158	.218	.278
725	860338	0398	0458	0518	0578	0637	0697	0757	0817	0877
726	0937	0996	1056	1116	1176	1236	1295	1355	1415	1475
727	1534	1594	1654	1714	1773	1833	1893	1952	2012	2072
728	2131	2191	2251	2310	2370	2430	2489	2549	2608	2668
729	2728	2787	2847	2906	2966	3025	3085	3144	3204	3263
730	3323	3382	3442	3501	3561	3620	3680	3739	3799	3858
731	3917	3977	4036	4096	4155	4214	4274	4333	4392	4452
732	4511	4570	4630	4689	4148	4808	4867	4926	4985	5045
733	5104	5163	5222	5282	5341	5400	5459	5519	5578	5637
734	5696	5755	5814	5874	5933	5992	6051	6110	6169	6228
735	6287	6346	6405	6465	6524	6583	6642	6701	6760	6819
736	6878	6937	6996	7055	7114	7173	7232	7291	7350	7409
737	7467	7526	7585	7644	7703	7762	7821	7880	7939	7998
738	8056	8115	8174	8233	8292	8350	8409	8468	8527	8586
739	8644	8703	8762	8821	8879	8938	8997	9056	9114	9173
740	9232	9290	9349	9408	9466	9525	9584	9642	9701	9760
741	9818	9877	9935	9994	..53	.111	.170	.228	.287	.345
742	870404	0462	0521	0579	0638	0696	0755	0813	0872	0930
743	0989	1047	1106	1164	1223	1281	1339	1398	1456	1515
744	1573	1631	1690	1748	1806	1865	1923	1981	2040	2098
745	2156	2215	2273	2331	2389	2448	2506	2564	2622	2681
746	2739	2797	2855	2913	2972	3030	3088	3146	3204	3262
747	3321	8379	3437	3495	3553	3611	3669	3727	3785	3844
748	3902	3960	4018	4076	4134	4192	4250	4308	4360	4424
749	4482	4540	4598	4656	4714	4772	4830	4888	4945	5003

N.	0	1	2	3	4	5	6	7	8	9
750	875061	5119	5177	5235	5293	5351	5409	5466	5524	5582
751	5640	5698	5756	5813	5871	5929	5987	6045	6102	6160
752	6218	6276	6333	6391	6449	6507	6564	6622	6680	6737
753	6795	6853	6910	6968	7026	7083	7141	7199	7256	7314
754	7371	7429	7487	7544	7602	7659	7717	7774	7832	7889
755	7947	8004	8062	8119	8177	8234	8292	8349	8407	8464
756	8522	8579	8637	8694	8752	8809	8866	8924	8981	9039
757	9096	9153	9211	9268	9325	9383	9440	9497	9555	9612
758	9669	9726	9784	9841	9898	9956	..13	..70	.127	.185
759	880242	0299	0356	0413	0471	0528	0580	0642	0699	0756
760	0814	0871	0928	0985	1042	1099	1156	1213	1271	1328
761	1385	1442	1499	1556	1613	1670	1727	1784	1841	1898
762	1955	2012	2069	2126	2183	2240	2297	2354	2411	2468
763	2525	2581	2638	2695	2752	2809	2866	2923	2980	3037
764	3093	3150	3207	3264	3321	3377	3434	3491	3548	3605
765	3661	3718	3775	3832	3888	3945	4002	4059	4115	4172
766	4229	4285	4342	4399	4455	4512	4569	4625	4682	4739
767	4795	4852	4909	4965	5022	5078	5135	5192	5248	5305
768	5361	5418	5474	5531	5587	5644	5700	5757	5813	5870
769	5926	5983	6039	6096	6152	6209	6265	6321	6378	6434
770	6491	6547	6604	6660	6716	6773	6829	6885	6942	6998
771	7054	7111	7167	7233	7280	7336	7392	7449	7505	7561
772	7617	7674	7730	7786	7842	7898	7955	8011	8067	8123
773	8179	8236	8292	8348	8404	8460	8516	8573	8629	8655
774	8741	8797	8853	8909	8965	9021	9077	9134	9190	9246
775	9302	9358	9414	9470	9526	9582	9638	9694	9750	9806
776	9862	9918	0974	..30	..86	.141	.197	.253	.309	.365
777	890421	0477	0533	0589	0645	0700	0756	0812	0868	0924
778	0980	1035	1091	1147	1203	1259	1314	1370	1426	1482
779	1537	1593	1649	1705	1760	1816	1872	1928	1983	2039
780	2095	2150	2206	2262	2317	2373	2429	2484	2540	2595
781	2651	2707	2762	2818	2873	2929	2985	3040	3096	3151
782	3207	3262	3318	3373	3429	3484	3540	3595	3651	3706
783	3762	3817	3873	3928	3984	4039	4094	4150	4205	4261
784	4316	4371	4427	4482	4538	4593	4648	4704	4759	4814
785	4870	4925	4980	5036	5091	5146	5201	5257	5312	5367
786	5423	5478	5533	5588	5644	5699	5754	5809	5864	5920
787	5975	6030	6085	6140	6195	6251	6306	6361	6416	6471
788	6526	6581	6636	6692	6747	6802	6857	6912	6967	7022
789	7077	7132	7187	7242	7297	7352	7407	7462	7517	7572
790	7627	7683	7737	7792	7847	7902	7957	8012	8067	8122
791	8176	8231	8286	8341	8396	8451	8506	8561	8615	8670
792	8725	8780	8835	8890	8944	8999	9054	9109	9164	9218
793	9273	9328	9383	9437	9492	9547	9602	9656	9711	9766
794	9821	9875	9930	9985	..39	..94	.149	.203	.258	.312
795	900367	0422	0476	0531	0586	0640	0695	0749	0804	0859
796	0913	0968	1022	1077	1131	1186	1240	1295	1349	1404
797	1458	1513	1567	1622	1676	1736	1785	1840	1894	1948
798	2003	2057	2112	2166	2221	2275	2329	2384	2438	2492
799	2547	2601	2655	2710	2764	2818	2873	2927	2981	3036

N.	0	1	2	3	4	5	6	7	8	9
800	903090	3144	3199	3253	3307	3361	3416	3470	3524	3578
801	3633	3687	3741	3795	3849	3904	3958	4012	4066	4120
802	4174	4229	4283	4337	4391	4445	4499	4553	4607	4661
803	4716	4770	4824	4878	4932	4986	5040	5094	5148	5202
804	5256	5310	5364	5418	5472	5526	5580	5634	5688	5742
805	5796	5850	5904	5958	6012	6066	6119	6173	6227	6281
806	6335	6389	6443	6497	6551	6604	6658	6712	6766	6820
807	6874	6927	6981	7035	7089	7143	7196	7250	7304	7358
808	7411	7465	7519	7573	7626	7680	7734	7787	7841	7895
809	7949	8002	8056	8110	8163	8217	8270	8324	8378	8431
810	8485	8539	8592	8646	8699	8753	8807	8860	8914	8967
811	9021	9074	9128	9181	9235	9289	9342	9396	9449	9503
812	9556	9610	9663	9716	9770	9823	9877	9930	9984	..37
813	910091	0144	0197	0251	0304	0358	0411	0464	0518	0571
814	0624	0678	0731	0784	0838	0891	0944	0998	1051	1104
815	1158	1211	1264	1317	1371	1424	1477	1530	1584	1637
816	1690	1743	1797	1850	1903	1956	2009	2063	2115	2169
817	2222	2275	2323	2381	2435	2488	2541	2594	2645	2700
818	2753	2806	2859	2913	2966	3019	3072	3125	3178	3231
819	3284	3337	3390	3443	3496	3549	3602	3655	3708	3761
820	3814	3867	3920	3973	4026	4079	4132	4184	4237	4290
821	4343	4396	4449	4502	4555	4608	4660	4713	4766	4819
822	4872	4925	4977	5030	5083	5136	5189	5241	5594	5347
823	5400	5453	5505	5558	5611	5664	5716	5769	5822	5875
824	5927	5980	6033	6085	6138	6191	6243	6296	6349	6401
825	6454	6507	6559	6612	6664	6717	6770	6822	6875	6927
826	6980	7033	7085	7138	7190	7243	7295	7348	7400	7453
827	7506	7558	7611	7663	7716	7768	7820	7873	7925	7978
828	8030	8083	8185	8188	8240	8293	8345	8397	8450	8502
829	8555	8607	8659	8712	8764	8816	8869	8921	8973	9026
830	9078	9130	9183	9235	9287	9340	9392	9444	9496	9549
831	9601	9653	9706	9758	9810	9862	9914	9967	..19	..71
832	920123	0176	0228	0280	0332	0384	0436	0489	0541	0593
833	0645	0697	0749	0801	0853	0906	0958	1010	1062	1114
834	1166	1218	1270	1322	1374	1426	1478	1530	1582	1634
835	1686	1738	1790	1842	1894	1946	1998	2050	2102	2154
836	2206	2258	2310	2362	2414	2466	2518	2570	2622	2674
837	2725	2777	2829	2881	2933	2985	3037	3089	3140	3192
838	3244	3296	3348	3399	3451	3503	3555	3607	3658	3710
839	3762	3814	3865	3917	3969	4021	4072	4124	4147	4228
840	4279	4331	4383	4434	4486	4538	4589	4641	4693	4744
841	4796	4848	4899	4951	5003	5054	5106	5157	5209	5261
842	5312	5364	5415	5467	5518	5570	5621	5673	5725	5776
843	5828	5874	5931	5982	6034	6085	6137	6188	6240	6291
844	6342	6394	6445	6497	6548	6600	6651	6702	6754	6805
845	6857	6908	6959	7011	7062	7114	7165	7216	7268	7319
846	7370	7422	7473	7524	7576	7627	7678	7730	7783	7832
847	7883	7935	7986	8037	8088	8140	8191	8242	8293	8345
848	8396	8447	8498	8549	8601	8652	8703	8754	8805	8857
849	8908	8959	9010	9061	9112	9163	9216	9266	9317	9368

N.	0	1	2	3	4	5	6	7	8	9
850	929419	9473	9521	9572	9623	9674	9725	9776	9827	9879
851	9930	9981	..32	..83	.134	.185	.236	.287	.338	.389
852	930440	0491	0542	0592	0643	0694	0745	0796	0847	0898
853	0949	1000	1051	1102	1153	1204	1254	1305	1356	1407
854	1458	1509	1560	1610	1661	1712	1763	1814	1865	1915
855	1966	2017	2068	2118	2169	2220	2271	2322	2372	2423
856	2474	2524	2575	2626	2677	2727	2778	2829	2879	2930
857	2981	3031	3082	3133	3183	3234	3285	3335	3386	3437
858	3487	3538	3589	3639	3690	3740	3791	3841	3892	3943
859	3993	4044	4094	4145	4195	4246	4269	4347	4397	4448
860	4498	4549	4599	4650	4700	4751	4801	4852	4902	4953
861	5003	5054	5104	5154	5205	5255	5306	5356	5406	5457
862	5507	5558	5608	5658	5709	5759	5809	5860	5910	5960
863	6011	6061	6111	6162	6212	6262	6313	6363	6413	6463
864	6514	6564	6614	6665	6715	6765	6815	6865	6916	6966
865	7016	7066	7117	7167	7217	7267	7317	7367	7418	7468
866	7518	7568	7618	7668	7718	7769	7819	7869	7919	7969
867	8019	8069	8119	8169	8219	8269	8320	8370	8420	8470
868	8520	8570	8620	8670	8720	8770	8820	8870	8919	8970
869	9020	9070	9120	9170	9220	9270	9320	9369	9419	9469
870	9519	9569	9616	9669	9719	9769	9819	9869	9918	9968
871	940018	0068	0118	0168	0218	0267	0317	0367	0417	0467
872	0516	0566	0616	0666	0716	0765	0815	0865	0915	0964
873	1014	1064	1114	1163	1213	1263	1313	1362	1412	1462
874	1511	1561	1611	1660	1710	1760	1809	1859	1909	1958
875	2008	2058	2107	2157	2207	2256	2306	2355	2405	2455
876	2504	2554	2603	2653	2702	2752	2801	2851	2901	2950
877	3000	3049	3099	3148	3198	3247	3297	3346	3396	3445
878	3495	3544	3593	3643	3692	3742	3791	3841	3890	3939
879	3989	4038	4088	4137	4186	4236	4285	4335	4384	4433
880	4483	4532	4581	4631	4680	4729	4779	4828	4877	4927
881	4976	5025	5074	5124	5173	5222	5272	5321	5370	5419
882	5469	5518	5567	5616	5665	5715	5764	5813	5862	5912
883	5961	6010	6059	6108	6157	6207	6256	6305	6354	6403
884	6452	6501	6551	6600	6649	6698	6747	6796	6845	6894
885	6943	6992	7041	7090	7140	7189	7238	7287	7336	7385
886	7434	7483	7532	7581	7630	7679	7728	7777	7826	7875
887	7924	7973	8022	8070	8119	8168	8217	8266	8315	8365
888	8413	8462	8511	8560	8609	8657	8706	8755	8804	8853
889	8902	8951	8999	9048	9097	9146	9195	9244	9292	9341
890	9390	9439	9488	9536	9585	9634	9683	9731	9780	9829
891	9878	9926	9975	..24	..73	.121	.170	.219	.267	.316
892	950365	0414	0462	0511	0560	0608	0657	0706	0754	0803
893	0851	0900	0949	0997	1046	1095	1143	1192	1240	1289
894	1338	1386	1435	1483	1532	1580	1629	1677	1726	1775
895	1823	1872	1920	1969	2017	2066	2114	2163	2211	2260
896	2308	2356	2405	2453	2502	2550	2599	2647	5696	2744
897	2792	2841	2889	2938	2986	3034	3083	3131	3180	3228
898	3276	3325	3373	3421	3470	3518	3566	3615	3663	3711
899	3760	3808	3856	3905	3953	4001	4049	4098	4146	4194

N.	0	1	2	3	4	5	6	7	8	9
900	954243	4291	4339	4387	4435	4484	4532	4580	4628	4677
901	4725	4773	4821	4869	4918	4966	5014	5062	5110	5158
902	5207	5255	5303	5351	5399	5447	5495	5543	5592	5640
903	5688	5736	5784	5832	5880	5928	5976	6024	6072	6120
904	6168	6216	6265	6313	6361	6409	6457	6505	6553	6601
905	6649	6697	6745	6793	6840	6888	6936	6984	7032	7080
906	7128	7176	7224	7272	7320	7368	7416	7464	7512	7559
907	7607	7655	7703	7751	7799	7847	7894	7942	7990	8038
908	8086	8134	8181	8229	8277	8325	8373	8421	8468	8516
909	8564	8612	8659	8707	8755	8803	8850	8898	8946	8994
910	9041	9089	9137	9185	9232	9280	9328	9375	9423	9471
911	9518	9566	9614	9661	9709	9757	9804	9852	9900	9947
912	9995	..42	..90	.138	.185	.233	.280	.328	.376	.423
913	960471	0518	0566	0613	0661	0709	0756	0804	0851	0899
914	0946	0994	1041	1089	1136	1184	1231	1279	1326	1374
915	1421	1469	1516	1563	1611	1658	1706	1753	1801	1848
916	1895	1943	1990	2038	2085	2132	2180	2227	2275	2322
917	2369	2417	2464	2511	2559	2606	2653	2701	2748	2795
918	2843	2890	2937	2985	3032	3079	3126	3174	3221	3268
919	3316	3363	3410	3457	3504	3552	3599	3646	3693	3741
920	3788	3835	3882	3929	3977	4024	4071	4118	4165	4212
921	4260	4307	4354	4401	4448	4495	4542	4590	4637	4684
922	4731	4778	4825	4872	4919	4966	5013	5061	5108	5155
923	5202	5249	5296	5343	5390	5437	5484	5531	5578	5625
924	5672	5719	5766	5813	5860	5907	5954	6001	6048	6095
925	6142	6189	6236	6283	6329	6376	6423	6470	6517	6564
926	6611	6658	6705	6752	6799	6845	6892	6939	6986	7033
927	7080	7127	7173	7220	7267	7314	7361	7408	7454	7501
928	7548	7595	7642	7688	7735	7782	7829	7875	7922	7969
929	8016	8062	8109	8156	8203	8249	8296	8343	8390	8436
930	8483	8530	8576	8623	8670	8716	8763	8810	8856	8903
931	8950	8996	9043	9090	9136	9183	9229	9276	9323	9369
932	9416	9463	9509	9556	9602	9649	9695	9742	9789	9835
933	9882	9928	9975	..21	..68	.114	.161	.207	.254	.300
934	970347	0393	0440	0486	0533	0579	0626	0672	0719	0765
935	0812	0858	0904	0951	0997	1044	1090	1137	1183	1229
936	1276	1322	1369	1415	1461	1508	1554	1601	1647	1693
937	1740	1786	1832	1879	1925	1971	2018	2064	2110	2157
938	2203	2249	2295	2342	2388	2434	2481	2527	2573	2619
939	2666	2712	2758	2804	2851	2897	2943	2989	3035	3082
940	3128	3174	3220	3266	3313	3359	3405	3451	3497	3543
941	3590	3636	3682	3728	3774	3820	3866	3913	3959	4005
942	4051	4097	4143	4189	4235	4281	4327	4374	4420	4466
943	4512	4558	4604	4650	4696	4742	4788	4834	4880	4926
944	4972	5018	5064	5110	5156	5202	5248	5294	5340	5386
945	5432	5478	5524	5570	5616	5662	5707	5753	5799	5845
946	5891	5937	5983	6029	6075	6121	6167	6212	6258	6304
947	6350	6396	6442	6488	6533	6579	6925	6671	6717	6763
948	6808	6854	6900	6946	6992	7037	7083	7129	7175	7220
949	7266	7312	7358	7403	7449	7495	7541	7586	7632	7678

N.	0	1	2	3	4	5	6	7	8	9
950	977724	7769	7815	7861	7906	7952	7998	8043	8089	8135
951	8181	8226	8272	8317	8363	8409	8454	8500	8546	8591
952	8637	8683	8728	8774	8819	8865	8911	8956	9002	9047
953	9093	9138	9184	9230	9275	9321	9366	9412	9457	9503
954	9548	9594	9639	9685	9730	9776	9821	9867	9912	9958
955	980003	0049	0094	0140	0185	0231	0276	0322	0367	0412
956	0458	0503	0549	0594	0640	0685	0730	0776	0821	0867
957	0912	0957	1003	1048	1093	1139	1184	1229	1275	1320
958	1366	1411	1456	1501	1547	1592	1637	1683	1728	1773
959	1819	1864	1909	1954	2000	2045	2090	2135	2181	2226
960	2271	2316	2362	2407	2452	2497	2543	2588	2633	2678
961	2723	2769	2814	2859	2904	2949	2994	3040	3085	3130
962	3175	3220	3265	3310	3356	3401	3446	3491	3536	3581
963	3626	3671	3716	3762	3807	3852	3897	3942	3987	4032
964	4077	4122	4167	4212	4257	4302	4347	4392	4437	4482
965	4527	4572	4617	4662	4707	4752	4797	4842	4887	4932
966	4977	5022	5067	5112	5157	5202	5247	5292	5337	5382
967	5426	5471	5516	5561	5606	5651	5699	5741	5786	5830
968	5875	5920	5965	6010	6055	6100	6144	6189	6234	6279
969	6324	6369	6413	6458	6503	6548	6593	6637	6682	6727
970	6772	6817	6861	6906	6951	6996	7040	7085	7130	7175
971	7219	7264	7309	7353	7398	7443	7488	7532	7577	7622
972	7666	7711	7756	7800	7845	7890	7934	7979	8024	8068
973	8113	8157	8202	8247	8291	8336	8381	8425	8470	8514
974	8559	8604	8648	8693	8737	8782	8826	8871	8916	8960
975	9005	9049	9093	9138	9183	9227	9272	9316	9361	9405
976	9450	9494	9539	9583	9628	9672	9717	9761	9806	9850
977	9895	9939	9983	..28	..72	.117	.161	.206	.250	.294
978	990339	0383	0428	0472	0516	0561	0605	0650	0694	0738
979	0783	0827	0871	0916	0960	1004	1049	1093	1137	1182
980	1226	1270	1315	1359	1403	1448	1492	1536	1580	1625
981	1669	1713	1758	1802	1846	1890	1935	1979	2023	2067
982	2111	2156	2200	2244	2288	2333	2377	2421	2465	2509
983	2554	2598	2642	2686	2730	2774	2819	2863	2907	2951
984	2995	3039	3083	3127	3172	3216	3260	3304	3348	3392
985	3436	3480	3524	3568	3613	3657	3701	3745	3789	3833
986	3877	3921	3965	4009	4053	4097	4141	4185	4229	4273
987	4317	4361	4405	4449	4493	4537	4581	4625	4669	4713
988	4757	4801	4845	4886	4933	4977	5021	5065	5108	5152
989	5196	5240	5284	5328	5372	5416	5460	5504	5547	5591
990	5635	5679	5723	5767	5811	5854	5898	5942	5986	6030
991	6074	6117	6161	6205	6249	6293	6337	6380	6424	6468
992	6512	6555	6599	6643	6687	6731	6774	6818	6862	6906
993	6949	6993	7037	7080	7124	7168	7212	7255	7299	7343
994	7386	7430	7474	7517	7561	7605	7648	7692	7736	7779
995	7823	7867	7910	7954	7998	8041	8085	8129	8172	8216
996	8259	8303	8347	8390	8434	8477	8521	8564	8608	8652
997	8695	8739	8782	8826	8869	8913	8956	9000	9043	9087
998	9131	9174	9218	9261	9305	9348	9392	9435	9479	9522
999	9565	9609	9652	9696	9739	9783	9826	9870	9913	9957

TABLE II. Log. Sines and Tangents. (0°) Natural Sines.

′	Sine.	D.10″	Cosine.	D.10″	Tang.	D.10″	Cotang.	N.sine.	N. cos.	
0	0.000000		10.000000		0.000000		Infinite.	00000	100000	60
1	6.463726		000000		6.463726		13.536274	00029	100000	59
2	764756		000000		764756		235244	00058	100000	58
3	940847		000000		940847		059153	00087	100000	57
4	7.065786		000000		7.065786		12.934214	00116	100000	56
5	162696		000000		162696		837304	00145	100000	55
6	241877		9.999999		241878		758122	00175	100000	54
7	308824		999999		308825		691175	00204	100000	53
8	366816		999999		366817		633183	00233	100000	52
9	417968		999999		417970		582030	00262	100000	51
10	463725		999998		463727		536273	00291	100000	50
11	7.505118		9.999998		7.505120		12.494880	00320	99999	49
12	542906		999997		542909		457091	00349	99999	48
13	577668		999997		577672		422328	00378	99999	47
14	609853		999996		609857		390143	00407	99999	46
15	639816		999996		639820		360180	00436	99999	45
16	667845		999995		667849		332151	00465	99999	44
17	694173		999995		694179		305821	00495	99999	43
18	718997		999994		719003		280997	00524	99999	42
19	742477		999993		742484		257516	00553	99998	41
20	764754		999993		764761		235239	00582	99998	40
21	7.785943		9.999992		7.785951		12.214049	00611	99998	39
22	806146		999991		806155		193845	00640	99998	38
23	825451		999990		825460		174540	00669	99998	37
24	843934		999989		843944		156056	00698	99998	36
25	861663		999988		861674		138326	00727	99997	35
26	878695		999988		878708		121292	00756	99997	34
27	895085		999987		895099		104901	00785	99997	33
28	910879		999986		910894		089106	00814	99997	32
29	926119		999985		926134		073866	00844	99996	31
30	940842		999983		940858		059142	00873	99996	30
31	7.955082	2298	9.999982	0.2	7.955100	2298	12.044900	00902	99996	29
32	968870	2227	999981	0.2	968889	2227	031111	00931	99996	28
33	982233	2161	999980	0.2	982253	2161	017747	00960	99995	27
34	995198	2098	999979	0.2	995219	2098	004781	00989	99995	26
35	8.007787	2039	999977	0.2	8.007809	2039	11.992191	01018	99995	25
36	020021	1983	999976	0.2	020045	1983	979955	01047	99995	24
37	031919	1930	999975	0.2	031945	1930	968055	01076	99994	23
38	043501	1880	999973	0.2	043527	1880	956473	01105	99994	22
39	054781	1832	999972	0.2	054809	1833	945191	01134	99994	21
40	065776	1787	999971	0.2	065806	1787	934194	01164	99993	20
41	8.076500	1744	9.999969	0.2	8.076531	1744	11.923469	01193	99993	19
42	086965	1703	999968	0.2	086997	1703	913003	01222	99993	18
43	097183	1664	999966	0.2	097217	1664	902783	01251	99992	17
44	107167	1626	999964	0.3	107202	1627	892797	01280	99992	16
45	116926	1591	999963	0.3	116963	1591	883037	01309	99991	15
46	126471	1557	999961	0.3	126510	1557	873490	01338	99991	14
47	135810	1524	999959	0.3	135851	1524	864149	01367	99991	13
48	144953	1492	999958	0.3	144996	1493	855004	01396	99990	12
49	153907	1462	999956	0.3	153952	1463	846048	01425	99990	11
50	162681	1433	999954	0.3	162727	1434	837273	01454	99989	10
51	8.171280	1405	9.999952	0.3	8.171328	1406	11.828672	01483	99989	9
52	179713	1379	999950	0.3	179763	1379	820237	01513	99989	8
53	187985	1353	999948	0.3	188036	1353	811964	01542	99988	7
54	196102	1328	999946	0.3	196156	1328	803844	01571	99988	6
55	204070	1304	999944	0.3	204126	1304	795874	01600	99987	5
56	211895	1281	999942	0.4	211953	1281	788047	01629	99987	4
57	219581	1259	999940	0.4	219641	1259	780359	01658	99986	3
58	227134	1237	999938	0.4	227195	1238	772805	01687	99986	2
59	234557	1216	999936	0.4	234621	1217	765379	01716	99985	1
60	241855		999934		241921		758079	01745	99985	0
	Cosine.		Sine.		Cotang.		Tang.	N. cos.	N. sine.	′

89 Degrees.

′	Sine.	D.10″	Cosine.	D.10″	Tang.	D.10″	Cotang.	N. sine.	N. cos.	
0	8.241855	1196	9.999934	0.4	8.241921	1197	11.758079	01742	99985	60
1	249033	1177	999932	0.4	249102	1177	750898.	01774	99984	59
2	256094	1158	999929	0.4	256165	1158	743835	01803	99984	58
3	263042	1140	999927	0.4	263115	1140	736885	01832	99983	57
4	269881	1122	999925	0.4	269956	1122	730044	01862	99983	56
5	276614	1105	999922	0.4	276691	1105	723309	01891	99982	55
6	283243	1088	999920	0.4	283323	1089	716677	01920	99982	54
7	289773	1072	999918	0.4	289856	1073	710144	01949	99981	53
8	296207	1056	999915	0.4	296292	1057	703708	01978	99980	52
9	302546	1041	999913	0.4	302634	1042	697366	02007	99980	51
10	308794	1027	999910	0.4	308884	1027	691116	02036	99979	50
11	8.314954	1012	9.999907	0.4	8.315046	1013	11.684954	02065	99979	49
12	321027	998	999905	0.4	321122	999	678878	02094	99978	48
13	327016	985	999902	0.4	327114	985	672886	02123	99977	47
14	332924	971	999899	0.5	333025	972	666975	02152	99977	46
15	338753	959	999897	0.5	333856	959	661144	02181	99976	45
16	344504	946	999894	0.5	344610	946	655390	02211	99976	44
17	350181	934	999891	0.5	350289	934	649711	02240	99975	43
18	355783	922	999888	0.5	355895	922	644105	02269	99974	42
19	361315	910	999885	0.5	361430	911	638570	02298	99974	41
20	366777	899	999882	0.5	366895	899	633105	02327	99973	40
21	8.372171	888	9.999879	0.5	8.372292	888	11.627708	02356	99972	39
22	377499	877	999876	0.5	377622	879	622378	02385	99972	38
23	382762	867	999873	0.5	382889	867	617111	02414	99971	37
24	387962	856	999870	0.5	388092	857	611908	02443	99970	36
25	393101	846	999867	0.5	393234	847	606766	02472	99969	35
26	398179	837	999864	0.5	398315	837	601685	02501	99969	34
27	403199	827	999861	0.5	403338	828	596662	02530	99968	33
28	408161	818	999858	0.5	408304	818	591696	02560	99967	32
29	413068	809	999854	0.5	413213	809	586787	02589	99966	31
30	417919	800	999851	0.6	418068	800	581932	02618	99966	30
31	8.422717	791	9.999848	0.6	8.422869	791	11.577131	02647	99965	29
32	427462	782	999844	0.6	427618	783	572382	02676	99964	28
33	432156	774	999841	0.6	432315	774	567685	02705	99963	27
34	436800	766	999838	0.6	436962	766	563038	02734	99963	26
35	441394	758	999834	0.6	441560	758	558440	02763	99962	25
36	445941	750	999831	0.6	446110	750	553890	02792	99961	24
37	450440	742	999827	0.6	450613	743	549387	02821	99960	23
38	454893	735	999823	0.6	455070	735	544930	02850	99959	22
39	459301	727	999820	0.6	459481	728	540519	02879	99959	21
40	463665	720	999816	0.6	463849	720	536151	02908	99958	20
41	8.467985	712	9.999812	0.6	8.468172	713	11.531828	02938	99957	19
42	472263	706	999809	0.6	472454	707	527546	02967	99956	18
43	476498	699	999805	0.6	476693	700	523307	02996	99955	17
44	480693	692	999801	0.6	480892	693	519108	03025	99954	16
45	484848	686	999797	0.7	485050	686	514950	03054	99953	15
46	488963	679	999793	0.7	489170	680	510830	03083	99952	14
47	493040	673	999790	0.7	493250	674	506750	03112	99952	13
48	497078	667	999786	0.7	497293	668	502707	03141	99951	12
49	501080	661	999782	0.7	501298	661	498702	03170	99950	11
50	505045	655	999778	0.7	505267	655	494733	03199	99949	10
51	8.508974	649	9.999774	0.7	8.509200	650	11.490800	03228	99948	9
52	512867	643	999769	0.7	513098	644	486902	03257	99947	8
53	516726	637	999765	0.7	516961	638	483039	03286	99946	7
54	520551	632	999761	0.7	520790	633	479210	03316	99945	6
55	524343	626	999757	0.7	524586	627	475414	03345	99944	5
56	528102	621	999753	0.7	528349	622	471651	03374	99943	4
57	531828	616	999748	0.7	532080	616	467920	03403	99942	3
58	535523	611	999744	0.7	535779	611	464221	03432	99941	2
59	539186	605	999740	0.7	539447	606	460553	03461	99940	1
60	542819		999735		543084		456916	03490	99939	0
	Cosine.		Sine.		Cotang.		Tang.	N. cos.	N.sine.	′

′	Sine.	D. 10″	Cosine.	D. 10″	Tang.	D. 10″	Cotang.	N. sine.	N. cos.	
0	8.542819	600	9.999735	0.7	8.543084	602	11.456916	03490	99939	60
1	546422	595	999731	0.7	546691	596	453309	03519	99938	59
2	549995	591	999726	0.7	550268	591	449732	03548	99937	58
3	553539	586	999722	0.8	553817	587	446183	03577	99936	57
4	557054	581	999717	0.8	557336	582	442664	03606	99935	56
5	560540	576	999713	0.8	560828	577	439172	03635	99934	55
6	563999	572	999708	0.8	564291	573	435709	03664	99933	54
7	567431	567	999704	0.8	567727	568	432273	03693	99932	53
8	570836	563	999699	0.8	571137	564	428863	03723	99931	52
9	574214	559	999694	0.8	574520	559	425480	03752	99930	51
10	577566	554	999689	0.8	577877	555	422123	03781	99929	50
11	8.580892	550	9.999685	0.8	8.581208	551	11.418792	03810	99927	49
12	584193	546	999680	0.8	584514	547	415486	03839	99926	48
13	587469	542	999675	0.8	587795	543	412205	03868	99925	47
14	590721	538	999670	0.8	591051	539	408949	03897	99924	46
15	593948	534	999665	0.8	594283	535	405717	03926	99923	45
16	597152	530	999660	0.8	597492	531	402508	03955	99922	44
17	600332	526	999655	0.8	600677	527	399323	03984	99921	43
18	603489	522	999650	0.8	603839	523	396161	04013	99919	42
19	606623	519	999645	0.8	606978	519	393022	04042	99918	41
20	609734	515	999640	0.9	610094	516	389906	04071	99917	40
21	8.612823	511	9.999635	0.9	8.613189	512	11.386811	04100	99916	39
22	615891	508	999629	0.9	616262	508	383738	03129	99915	38
23	618937	504	999324	0.9	619313	505	380687	04159	99913	37
24	621962	501	999619	0.9	622343	501	377657	04188	99912	36
25	624965	497	999614	0.9	625352	498	374648	04217	99911	35
26	627948	494	999608	0.9	628340	495	371660	04246	99910	34
27	630911	490	999603	0.9	631308	491	368692	04275	99909	33
28	633854	487	999597	0.9	634256	488	365744	04304	99907	32
29	636776	484	999592	0.9	637184	485	362816	04333	99906	31
30	639680	481	999586	0.9	640093	482	359907	04362	99905	30
31	8.642563	477	9.999581	0.9	8.642982	478	11.357018	04391	99904	29
32	645428	474	999575	0.9	645853	475	354147	04420	99902	28
33	648274	471	999570	0.9	648704	472	351296	04449	99901	27
34	651102	468	999564	0.9	651537	469	348463	04478	99900	26
35	653911	465	999558	1.0	654352	466	345648	04507	99898	25
36	656702	462	999553	1.0	657149	463	342851	04536	99897	24
37	659475	459	999547	1.0	659928	460	340072	04565	99896	23
38	662230	456	999541	1.0	662689	457	337311	04594	99894	22
39	664968	453	999535	1.0	665433	454	334567	04623	99893	21
40	667689	451	999529	1.0	668160	453	331840	04653	99892	20
41	8.670393	448	9.999524	1.0	8.670870	449	11.329130	04682	99890	19
42	673080	445	999518	1.0	673563	446	326437	04711	99889	18
43	675751	442	999512	1.0	676239	443	323761	04740	99888	17
44	678405	440	999506	1.0	678900	442	321100	04769	99886	16
45	681043	437	999500	1.0	681544	438	318456	04798	99885	15
46	683665	434	999493	1.0	684172	435	315828	04827	99883	14
47	686272	432	999487	1.0	686784	433	313216	04856	99882	13
48	688863	429	999481	1.0	689381	430	310619	04885	99881	12
49	691438	427	999475	1.0	691963	428	308037	04914	99879	11
50	693998	424	999469	1.0	694529	425	305471	04943	99878	10
51	8.696543	422	9.999463	1.1	8.697081	423	11.302919	04972	99876	9
52	699073	419	999456	1.1	699617	420	300383	05001	99875	8
53	701589	417	999450	1.1	702139	418	297861	05030	99873	7
54	704090	414	999443	1.1	704246	415	295354	05059	99872	6
55	706577	412	999437	1.1	707140	413	292860	05088	99870	5
56	709049	410	999431	1.1	709618	411	290382	05117	99869	4
57	711507	407	999424	1.1	702083	408	287917	05146	99867	3
58	713952	405	999418	1.1	714534	406	285465	05175	99866	2
59	716383	403	999411	1.1	716972	404	283028	05205	99864	1
60	718800		999404		719396		280604	05234	99863	0
	Cosine.		Sine.		Cotang.		Tang.	N. cos.	N.sine.	′

87 Degrees.

′	Sine.	D. 10″	Cosine.	D. 10″	Tang.	D. 10″	Cotang.	N. sine.	N. cos.	
0	8.718800	401	9.999404	1.1	8.719396	402	11.280604	05234	99863	60
1	721204	398	999398	1.1	721806	399	278194	05263	99861	59
2	723595	396	999391	1.1	724204	397	275796	05292	99860	58
3	725972	394	999384	1.1	726588	395	273412	05321	99858	57
4	728337	392	999378	1.1	728959	393	271041	05350	99857	56
5	730688	390	999371	1.1	731317	391	268683	05379	99855	55
6	733027	388	999364	1.2	733663	389	266337	05408	99854	54
7	735354	386	999357	1.2	735996	387	264004	05437	99852	53
8	737667	384	999350	1.2	738317	385	261683	05466	99851	52
9	739969	382	999343	1.2	740626	383	259374	05495	99849	51
10	742259	380	999336	1.2	742922	381	257078	05524	99847	50
11	8.744536	378	9.999329	1.2	8.745207	379	11.254793	05553	99846	49
12	746802	376	999322	1.2	747479	377	252521	05582	99844	48
13	749055	374	999315	1.2	749740	375	250260	05611	99842	47
14	751297	372	999308	1.2	751989	373	248011	05640	99841	46
15	753528	370	999301	1.2	754227	371	245773	05669	99839	45
16	755747	368	999294	1.2	756453	369	243547	05698	99838	44
17	757955	366	999286	1.2	758668	367	241332	05727	99836	43
18	760151	364	999279	1.2	760872	365	239128	05756	99834	42
19	762337	362	999272	1.2	763065	364	236935	05785	99833	41
20	764511	361	999265	1.2	765246	362	234754	05814	99831	40
21	8.766675	359	9.999257	1.2	8.767417	360	11.232583	05844	99829	39
22	768828	357	999250	1.3	769578	358	230422	05873	99827	38
23	770970	355	999242	1.3	771727	356	228273	05902	99826	37
24	773101	353	999235	1.3	773866	355	226134	05931	99824	36
25	775223	352	999227	1.3	775995	353	224005	05960	99822	35
26	777333	350	999220	1.3	778114	351	221886	05989	99821	34
27	779434	348	999212	1.3	780222	350	219778	06018	99819	33
28	781524	347	999205	1.3	782320	348	217680	06047	99817	32
29	783605	345	999197	1.3	784408	346	215592	06076	99815	31
30	785675	343	999189	1.3	786486	345	213514	06105	99813	30
31	8.787736	342	9.999181	1.3	8.788554	343	11.211446	06134	99812	29
32	789787	340	999174	1.3	790613	341	209387	06163	99810	28
33	791828	339	999166	1.3	792662	340	207338	06192	99808	27
34	793859	337	999158	1.3	794701	338	205299	06221	99806	26
35	795881	335	999150	1.3	796731	337	203269	06250	99804	25
36	797894	334	999142	1.3	798752	335	201248	06279	99803	24
37	799897	332	999134	1.3	800763	334	199237	06308	99801	23
38	801892	331	999126	1.3	802765	332	197235	06337	99799	22
39	803876	329	999118	1.3	804758	331	195242	06366	99797	21
40	805852	328	999110	1.3	806742	329	193258	06395	99795	20
41	8.807819	326	9.999102	1.3	8.808717	328	11.191283	06424	99793	19
42	809777	325	999094	1.4	810683	326	189317	06453	99792	18
43	811726	323	999086	1.4	812641	325	187359	06482	99790	17
44	813667	322	999077	1.4	814589	323	185411	06511	99788	16
45	815599	320	999069	1.4	816529	322	183471	06540	99786	15
46	817522	319	999061	1.4	818461	320	181539	06569	99784	14
47	819436	318	999053	1.4	820384	319	179616	06598	99782	13
48	821343	316	999044	1.4	822298	318	177702	06627	99780	12
49	823240	315	999036	1.4	824205	316	175795	06656	99778	11
50	825130	313	999027	1.4	826103	315	173897	06685	99776	10
51	8.827011	312	9.999019	1.4	8.827992	314	11.172008	06714	99774	9
52	828884	311	999010	1.4	829874	312	170126	06743	99772	8
53	830749	309	999002	1.4	831748	311	168252	06773	99770	7
54	832607	308	998993	1.4	833613	310	166387	06802	99768	6
55	834456	307	998984	1.4	835471	308	164529	06831	99766	5
56	836297	306	998976	1.4	837321	307	162679	06860	99764	4
57	838130	304	998967	1.5	839163	306	160837	06889	99762	3
58	839956	303	998958	1.5	840998	304	159002	06918	99760	2
59	841774	302	998950	1.5	842825	303	157175	06947	99758	1
60	843585		998941		844644		155356	06976	99756	0
	Cosine.		Sine.		Cotang.		Tang.	N. cos.	N. sine.	′

86 Degrees.

′	Sine.	D. 10″	Cosine.	D. 10″	Tang.	D. 10″	Cotang.	N. sine.	N. cos.	
0	8.843585	300	9.998941	1.5	8.844644	302	11.155356	06976	99756	60
1	845387	299	998932	1.5	846455	301	153545	07005	99754	59
2	847183	298	998923	1.5	848260	299	151740	07034	99752	58
3	848971	297	998914	1.5	850057	298	149943	07063	99750	57
4	850751	295	998905	1.5	851846	297	148154	07092	99748	56
5	852525	294	998896	1.5	853628	296	146372	07121	99746	55
6	854291	293	998887	1.5	855403	295	144597	07150	99744	54
7	856049	292	998878	1.5	857171	293	142829	07179	99742	53
8	857801	291	998869	1.5	858932	292	141068	07208	99740	52
9	859546	290	998860	1.5	860686	291	139314	07237	99738	51
10	861283	288	998851	1.5	862433	290	137567	07266	99736	50
11	8.863014	287	9.998841	1.5	8.864173	289	11.135827	07295	99734	49
12	864738	286	998832	1.5	865906	288	134094	07324	99731	48
13	866455	285	998823	1.6	867632	287	132368	07353	99729	47
14	868165	284	998813	1.6	869351	285	130649	07382	99727	46
15	869868	283	998804	1.6	871064	284	128936	07411	99725	45
16	871565	282	998795	1.6	872770	283	127230	07440	99723	44
17	873255	281	998785	1.6	874469	282	125531	07469	99721	43
18	874938	279	998776	1.6	876162	281	123838	07498	99719	42
19	876615	279	998766	1.6	877849	280	122151	07527	99716	41
20	878285	277	998757	1.6	879529	279	120471	07556	99714	40
21	8.879949	276	9.998747	1.6	8.881202	278	11.118798	07585	99712	39
22	881607	275	998738	1.6	882869	277	117131	07614	99710	38
23	883258	274	998728	1.6	884530	276	115470	07643	99708	37
24	884903	273	998718	1.6	886185	275	113815	07672	99705	36
25	886542	272	998708	1.6	887833	274	112167	07701	99703	35
26	888174	271	998699	1.6	889476	273	110524	07730	99701	34
27	889801	270	998689	1.6	891112	272	108888	07759	99699	33
28	891421	269	998679	1.6	892742	271	107258	07788	99696	32
29	893035	268	998669	1.7	894366	270	105634	07817	99694	31
30	894643	267	998659	1.7	895984	269	104016	07846	99692	30
31	8.896246	266	9.998649	1.7	8.897596	268	11.102404	07875	99689	29
32	897842	265	998639	1.7	899203	267	100797	07904	99687	28
33	899432	264	998629	1.7	900803	266	099197	07933	99685	27
34	901017	263	998619	1.7	902398	265	097602	07962	99683	26
35	902596	262	998609	1.7	903987	264	096013	07991	99680	25
36	904169	261	998599	1.7	905570	263	094430	08020	99678	24
37	905736	260	998589	1.7	907147	262	092853	08049	99676	23
38	907297	259	998578	1.7	908719	261	091281	08078	99673	22
39	908853	258	998568	1.7	910285	260	089715	08107	99671	21
40	910404	257	998558	1.7	911846	259	088154	08136	99668	20
41	8.911949	257	9.998548	1.7	8.913401	258	11.086599	08165	99666	19
42	913488	256	998537	1.7	914951	257	085049	08194	99664	18
43	915022	255	998527	1.7	916495	256	083505	08223	99661	17
44	916550	254	998516	1.8	918034	256	081966	08252	99659	16
45	918073	253	998506	1.8	919568	255	080432	08281	99657	15
46	919591	252	998495	1.8	921096	254	078904	08310	99654	14
47	921103	251	998485	1.8	922619	253	077381	08339	99652	13
48	922610	250	998474	1.8	924136	252	075864	08368	99649	12
49	924112	249	998464	1.8	925649	251	074351	08397	99647	11
50	925609	249	998453	1.8	927156	250	072844	08426	99644	10
51	8.927100	248	9.998442	1.8	8.928658	249	11.071342	08455	99642	9
52	928587	247	998431	1.8	930155	249	069845	08484	99639	8
53	930068	246	998421	1.8	931647	248	068353	08513	99637	7
54	931544	245	998410	1.8	933134	247	066866	08542	99635	6
55	933015	244	998399	1.8	934616	246	065384	08571	99632	5
56	934481	243	998388	1.8	936093	245	063907	08600	99630	4
57	935942	243	998377	1.8	937565	244	062435	08629	99627	3
58	937398	242	998366	1.8	939032	244	060968	08658	99625	2
59	938850	241	998355	1.8	940494	243	059506	08687	99622	1
60	940296		998344		941952		058048	08716	99619	0
	Cosine.		Sine.		Cotang.		Tang.	N. cos.	N.sine.	′

′	Sine.	D. 10″	Cosine.	D. 10″	Tang.	D. 10″	Cotang.	N. sine.	N. cos.	
0	8.940296	240	9.998344	1.9	8.941952	242	11.058048	08716	99619	60
1	941738	239	998333	1.9	943404	241	056596	08745	99617	59
2	943174	239	998322	1.9	944852	240	055148	08774	99614	58
3	944606	238	998311	1.9	946295	240	053705	08803	99612	57
4	946034	237	998300	1.9	947734	239	052266	08831	99609	56
5	947456	236	998289	1.9	949168	238	050832	08860	99607	55
6	948874	235	998277	1.9	950597	237	049403	08889	99604	54
7	950287	235	998266	1.9	952021	237	047979	08918	99602	53
8	951696	234	998255	1.9	953441	236	046559	08947	99599	52
9	953100	233	998243	1.9	954856	235	045144	08976	99596	51
10	954499	232	998232	1.9	956267	234	043733	09005	99594	50
11	8.955894	232	9.998220	1.9	8.957674	234	11.042326	09034	99591	49
12	957284	231	998209	1.9	959075	233	040925	09063	99588	48
13	958670	230	998197	1.9	960473	232	039527	09092	99586	47
14	960052	229	998186	1.9	961866	231	038134	09121	99583	46
15	961429	229	998174	1.9	963255	231	036745	09150	99580	45
16	962801	228	998163	1.9	964639	230	035361	09179	99578	44
17	964170	227	998151	1.9	966019	229	033981	09208	99575	43
18	965534	227	998139	2.0	967394	229	032606	09237	99572	42
19	966893	226	998128	2.0	968766	228	031234	09266	99570	41
20	968249	225	998116	2.0	970133	227	029867	09295	99567	40
21	8.969600	224	9.998104	2.0	8.971496	226	11.028504	09324	99564	39
22	970947	224	998092	2.0	972855	226	027145	09353	99562	38
23	972289	223	998080	2.0	974209	225	025791	09382	99559	37
24	973628	222	998068	2.0	975560	224	024440	09411	99556	36
25	974962	222	998056	2.0	976906	224	023094	09440	99553	35
26	976293	221	998044	2.0	978248	223	021752	09469	99551	34
27	977619	220	998032	2.0	979586	222	020414	09498	99548	33
28	978941	220	998020	2.0	980921	222	019079	09527	99545	32
29	980259	219	998008	2.0	982251	221	017749	09556	99542	31
30	981573	218	997996	2.0	983577	220	016423	09585	99540	30
31	8.982883	218	9.997984	2.0	8.984899	220	11.015101	09614	99537	29
32	984189	217	997972	2.0	986217	219	013783	09642	99534	28
33	985491	216	997959	2.0	987532	218	012468	09671	99531	27
34	986789	216	997947	2.0	988842	218	011158	09700	99528	26
35	988083	215	997935	2.1	990149	217	009851	09729	99526	25
36	989374	214	997922	2.1	991451	216	008549	06758	99523	24
37	990660	214	997910	2.1	992750	216	007250	09787	99520	23
38	991943	213	997897	2.1	994045	215	005955	09816	99517	22
39	993222	212	997885	2.1	995337	215	004663	09845	99514	21
40	994497	212	997872	2.1	996624	214	003376	09874	99511	20
41	8.995768	211	9.997860	2.1	8.997908	213	11.002092	09903	99508	19
42	997036	211	997847	2.1	999188	213	000812	09932	99506	18
43	998299	210	997835	2.1	9.000465	212	10.999535	09961	99503	17
44	999560	209	997822	2.1	001738	211	998262	09990	99500	16
45	9.000816	209	997809	2.1	003007	211	996993	10019	99497	15
46	002069	208	997797	2.1	004272	210	995728	10048	99494	14
47	003318	208	997784	2.1	005534	210	994466	10077	99491	13
48	004563	207	997771	2.1	006792	209	993208	10106	99488	12
49	005805	206	997758	2.1	008047	208	991953	10135	99485	11
50	007044	206	997745	2.1	009298	208	990702	10164	99482	10
51	9.008278	205	9.997732	2.1	9.010546	207	10.989454	10192	99479	9
52	009510	205	997719	2.1	011790	207	988210	10221	99476	8
53	010737	204	997706	2.1	013031	206	686969	10250	99473	7
54	011962	203	997693	2.2	014268	206	985732	10279	99470	6
55	013182	203	997680	2.2	015502	205	984498	10308	99467	5
56	014400	202	997667	2.2	016732	204	983268	10337	99464	4
57	015613	202	997654	2.2	017959	204	983041	10366	99461	3
58	016824	201	997641	2.2	019183	203	980817	10395	99458	2
59	018031	201	997628	2.2	020403	203	979597	10424	99455	1
60	019235		997614		021620		978380	10453	99452	0
	Cosine.		Sine.		Cotang.		Tang.	N. cos.	N.sine.	′

′	Sine.	D. 10″	Cosine.	D. 10″	Tang.	D. 10″	Cotang.	N. sine.	N. cos.	
0	9.019235	200	9.997614	2.2	9.021620	202	10.978380	10453	99452	60
1	020435	199	997601	2.2	022834	202	977166	10482	99449	59
2	021632	199	997588	2.2	024044	201	975956	10511	99446	58
3	022825	198	997574	2.2	025251	201	974749	10540	99443	57
4	024016	198	997561	2.2	026455	200	973545	10569	99440	56
5	025203	197	997547	2.2	027655	199	972345	10597	99437	55
6	026386	197	997534	2.3	028852	199	971148	10626	99434	54
7	027567	196	997520	2.3	030046	198	969954	10655	99431	53
8	028744	196	997507	2.3	031237	198	968763	10684	99428	52
9	029918	195	997493	2.3	032425	197	967575	10713	99424	51
10	031089	195	997480	2.3	033609	197	966391	10742	99421	50
11	9.032257	194	9.997466	2.3	9.034791	196	10.965209	10771	99418	49
12	033421	194	997452	2.3	035969	196	964031	10800	99415	48
13	034582	193	997439	2.3	037144	195	962856	10829	99412	47
14	035741	192	997425	2.3	038316	195	961684	10858	99409	46
15	036896	192	997411	2.3	039485	194	960515	10887	99406	45
16	038048	191	997397	2.3	040651	194	959349	10916	99402	44
17	039197	191	997383	2.3.	041813	193	958187	10945	99399	43
18	040342	190	997369	2.3	042973	193	957027	10973	99396	42
19	041485	190	997355	2.3	044130	192	955870	11002	99393	41
20	042625	189	997341	2.3	045284	192	954716	11031	99390	40
21	9.043762	189	9.997327	2.4	9.046434	191	10.953566	11060	99386	39
22	044895	180	997313	2.4	047582	191	952418	11089	99383	38
23	046026	188	997299	2.4	048727	190	951273	11118	99380	37
24	047154	187	997285	2.4	049869	190	950131	11147	99377	36
25	048279	187	997271	2.4	051008	189	948992	11176	99374	35
26	049400	186	997257	2.4	052144	189	947856	11205	99370	34
27	050519	186	997242	2.4	053277	188	946723	11234	99367	33
28	051635	185	997228	2.4	054407	188	945593	11263	99364	32
29	052749	185	997214	2.4	055535	187	944465	11291	99360	31
30	053859	184	997199	2.4	056659	187	943341	11320	99357	30
31	9.054966	184	9.997185	2.4	9.057781	186	10.942219	11349	99354	29
32	056071	184	997170	2.4	058900	186	941100	11378	99351	28
33	057172	183	997156	2.4	060016	185	939984	11407	99347	27
34	058271	183	997141	2.4	061130	185	938870	11436	99344	26
35	059367	182	997127	2.4	062240	185	937760	11465	99341	25
36	060460	182	997112	2.4	063348	184	936652	11494	99337	24
37	061551	181	997098	2.4	064453	184	935547	11523	99334	23
38	062639	181	997083	2.5	065556	183	934444	11552	99331	22
39	063724	180	997068	2.5	066655	183	933345	11580	99327	21
40	064806	180	997053	2.5	067752	182	932248	11609	99324	20
41	9.065885	179	9.997039	2.5	9.068846	182	10.931154	11638	99320	19
42	066962	179	997024	2.5	069038	181	930062	11667	99317	18
43	068036	179	997009	2.5	071027	181	928973	11696	99314	17
44	069107	178	996994	2.5	072113	181	927887	11725	99310	16
45	070176	178	996979	2.5	073197	180	926803	11754	99307	15
46	071242	177	996964	2.5	074278	180	925722	11783	99303	14
47	072306	177	996949	2.5	075356	179	924644	11812	99300	13
48	073366	176	996934	2.5	076432	179	923568	11840	99297	12
49	074424	176	996919	2.5	077505	178	922495	11869	99293	11
50	075480	175	996904	2.5	078576	178	921424	11898	99290	10
51	9.076533	175	9.996889	2.5	9.079644	178	10.920356	11927	99286	9
52	077583	175	996874	2.5	080710	177	919290	11956	99283	8
53	078631	174	996858	2.5	081773	177	918227	11985	99279	7
54	079676	174	996843	2.5	082833	176	917167	12014	99276	6
55	080719	173	996828	2.5	083891	176	916109	12043	99272	5
56	081759	173	996812	2.6	084947	175	915053	12071	99269	4
57	082797	172	996797	2.6	086000	175	914000	12100	99265	3
58	083832	172	996782	2.6	087050	175	912950	12129	99262	2
59	084864	172	996766	2.6	088098	174	911902	12158	99258	1
60	085894		996751		089144		910856	12187	99255	0
	Cosine.		Sine.		Cotang.		Tang.	N. cos.	N.sine.	′

83 Degrees.

′	Sine.	D. 10″	Cosine.	D. 10″	Tang.	D. 10″	Cotang.	N. sine.	N. cos.	
0	9.085894	171	9.996751	2.6	9.089144	174	10.910856	12187	99255	60
1	086922	171	996735	2.6	090187	173	909813	12216	99251	59
2	087947	170	996720	2.6	091228	173	908772	12245	99248	58
3	088970	170	996704	2.6	092266	173	907734	12274	99244	57
4	089990	170	996688	2.6	093302	172	906698	12302	99240	56
5	091008	169	996673	2.6	094336	172	905664	12331	99237	55
6	092024	169	996657	2.6	095367	171	904633	12360	99233	54
7	093037	168	996641	2.6	096395	171	903605	12389	99230	53
8	094047	168	996625	2.6	097422	171	902578	12418	99226	52
9	095056	168	996610	2.6	098446	170	901554	12447	99222	51
10	096062	167	996594	2.6	099468	170	900532	12476	99219	50
11	9.097065	167	9.996578	2.7	9.100487	169	10.899513	12504	99215	49
12	098066	166	996562	2.7	101504	169	898496	12533	99211	48
13	099065	166	996546	2.7	102519	169	897481	12562	99208	47
14	100062	166	996530	2.7	103532	168	896468	12591	99204	46
15	101056	165	996514	2.7	104542	168	895458	12620	99200	45
16	102048	165	996498	2.7	105550	168	894450	12649	99197	44
17	103037	164	996482	2.7	106556	167	893444	12678	99193	43
18	104025	164	996465	2.7	107559	167	892441	12706	99189	42
19	105010	164	996449	2.7	108560	166	891440	12735	99186	41
20	105992	163	996433	2.7	109559	166	890441	12764	99182	40
21	9.106973	163	9.996417	2.7	9.110556	166	10.889444	12793	99178	39
22	107951	163	996400	2.7	111551	165	888449	12822	99175	38
23	108927	162	996384	2.7	112543	165	887457	12851	99171	37
24	109901	162	996368	2.7	113533	165	886467	12880	99167	36
25	110873	162	996351	2.7	114521	164	885479	12908	99163	35
26	111842	161	996335	2.7	115507	164	884493	12937	99160	34
27	112809	161	996318	2.7	116491	164	883509	12966	99156	33
28	113774	160	996302	2.8	117472	163	882528	12995	99152	32
29	114737	160	996285	2.8	118452	163	881548	13024	99148	31
30	115698	160	996269	2.8	119429	162	880571	13053	99144	30
31	9.116656	159	9.996252	2.8	9.120404	162	10.879596	13081	99141	29
32	117613	159	996235	2.8	121377	162	878623	13110	99137	28
33	118567	159	996219	2.8	122348	161	877652	13139	99133	27
34	119519	158	996202	2.8	123317	161	876683	13168	99129	26
35	120469	158	996185	2.8	124284	161	875716	13197	99125	25
36	121417	158	996168	2.8	125249	160	874751	13226	99122	24
37	122362	157	996151	2.8	126211	160	873789	13254	99118	23
38	123306	157	996134	2.8	127172	160	872828	13283	99114	22
39	124248	157	996117	2.8	128130	159	871870	13312	99110	21
40	125187	156	996100	2.8	129087	159	870913	13341	99106	20
41	9.126125	156	9.996083	2.9	9.130041	159	10.869959	13370	99102	19
42	127060	156	996066	2.9	130994	158	869006	13399	99098	18
43	127993	155	996049	2.9	131944	158	868056	13427	99094	17
44	128925	155	996032	2.9	132893	158	867107	13456	99091	16
45	129854	154	996015	2.9	133839	157	866161	13485	99087	15
46	130781	154	995998	2.9	134784	157	865216	13514	99083	14
47	131706	154	995980	2.9	135726	157	864274	13543	99079	13
48	132630	153	995963	2.9	136667	156	863333	13572	99075	12
49	133551	153	995946	2.9	137605	156	862395	13600	99071	11
50	134470	153	995928	2.9	138542	156	861458	13629	99067	10
51	9.135387	152	9.995911	2.9	9.139476	155	10.860524	13658	99063	9
52	136303	152	995894	2.9	140409	155	859591	13687	99059	8
53	137216	152	995876	2.9	141340	155	858660	13716	99055	7
54	138128	152	995859	2.9	142269	154	857731	13744	99051	6
55	139037	151	995841	2.9	143196	154	856804	13773	99047	5
56	139944	151	995823	2.9	144121	154	855879	13802	99043	4
57	140850	151	995806	2.9	145044	153	854956	13831	99039	3
58	141754	150	995788	2.9	145966	153	854034	13860	99035	2
59	142655	150	995771	2.9	146885	153	853115	13889	99031	1
60	143555		995753		147803		852197	13917	99027	0
	Cosine.		Sine.		Cotang.		Tang.	N. cos.	N. sine.	′

82 Degrees.

′	Sine.	D. 10″	Cosine.	D. 10″	Tang.	D. 10″	Cotang.	N. sine.	N. cos.	
0	9.143555	150	9.995753	3.0	9.147803	153	10.852197	13917	99027	60
1	144453	149	995735	3.0	148718	152	851282	13946	99023	59
2	145349	149	995717	3.0	149632	152	850368	13975	99019	58
3	146243	149	995699	3.0	150544	152	849456	14004	99015	57
4	147136	148	995681	3.0	151454	151	848546	14033	99011	56
5	148026	148	995664	3.0	152363	151	847637	14061	99006	55
6	148915	148	995646	3.0	153269	151	846731	14090	99002	54
7	149802	147	995628	3.0	154174	150	845826	14119	98998	53
8	150686	147	995610	3.0	155077	150	844923	14148	98994	52
9	151569	147	995591	3.0	155978	150	844022	14177	98990	51
10	152451	147	995573	3.0	156877	150	843123	14205	98986	50
11	9.153330	146	9.995555	3.0	9.157775	149	10.842225	14234	98982	49
12	154208	146	995537	3.0	158671	149	841329	14263	98978	48
13	155083	146	995519	3.0	159565	149	840435	14292	98973	47
14	155957	145	995501	3.1	160457	148	839543	14320	98969	46
15	156830	145	995482	3.1	161347	148	838653	14349	98965	45
16	157700	145	995464	3.1	162236	148	837764	14378	98961	44
17	158569	144	995446	3.1	163123	148	836877	14407	98957	43
18	159435	144	995427	3.1	164008	147	835992	14436	98953	42
19	160301	144	995409	3.1	164892	147	835108	14464	98948	41
20	161164	144	995390	3.1	165774	147	834226	14493	98944	40
21	9.162025	143	9.995372	3.1	9.166654	146	10.833346	14522	98940	39
22	162885	143	995353	3.1	167532	146	832468	14551	98936	38
23	163743	143	995334	3.1	168409	146	831591	14580	98931	37
24	164600	142	995316	3.1	169284	145	830716	14608	98927	36
25	165454	142	995297	3.1	170157	145	829843	14637	98923	35
26	166307	142	995278	3.1	171029	145	828971	14666	98919	34
27	167159	142	995260	3.1	171899	145	828101	14695	98914	33
28	168008	141	995241	3.2	172767	144	827233	14723	98910	32
29	168856	141	995222	3.2	173634	144	826366	14752	98906	31
30	169702	141	995203	3.2	174499	144	825501	14781	98902	30
31	9.170547	140	9.995184	3.2	9.175362	144	10.824638	14810	98897	29
32	171389	140	995165	3.2	176224	143	823776	14838	98893	28
33	172230	140	995146	3.2	177084	143	822916	14867	98889	27
34	173070	140	995127	3.2	177942	143	822058	14896	98884	26
35	173908	139	995108	3.2	178799	142	821201	14925	98880	25
36	174744	139	995089	3.2	179655	142	820345	14954	98876	24
37	175578	139	995070	3.2	180508	142	819492	14982	98871	23
38	176411	139	995051	3.2	181360	142	818640	15011	98867	22
39	177242	138	995032	3.2	182211	141	817789	15040	98863	21
40	178072	138	995013	3.2	183059	141	816941	15069	98858	20
41	9.178900	138	9.994993	3.2	9.183907	141	10.816093	15097	98854	19
42	179726	137	994974	3.2	184752	141	815248	15126	98849	18
43	180551	137	994955	3.2	185597	140	814403	15155	98845	17
44	181374	137	994935	3.2	186439	140	813561	15184	98841	16
45	182196	137	994916	3.3	187280	140	812720	15212	98836	15
46	183016	136	994896	3.3	188120	140	811880	15241	98832	14
47	183834	136	994877	3.3	188958	139	811042	15270	98827	13
48	184651	136	994857	3.3	189794	139	810206	15299	98823	12
49	185466	136	994838	3.3	190629	139	809371	15327	98818	11
50	186280	135	994818	3.3	191462	139	808538	15356	98814	10
51	9.187092	135	9.994798	3.3	9.192294	138	10.807706	15385	98809	9
52	187903	135	994779	3.3	193124	138	806876	15414	98805	8
53	188712	135	994759	3.3	193953	138	806047	15442	98800	7
54	189519	134	994739	3.3	194780	138	805220	15471	98796	6
55	190325	134	994719	3.3	195606	137	804394	15500	98791	5
56	191130	134	994700	3.3	196430	137	803570	15529	98787	4
57	191933	134	994680	3.3	197253	137	802747	15557	98782	3
58	192734	133	994660	3.3	198074	137	801926	15586	98778	2
59	193534	133	994640	3.3	198894	136	801106	15615	98773	1
60	194332		994620		199713		800287	15643	98769	0
	Cosine.		Sine.		Cotang.		Tang.	N. cos.	N. sine	′

81 Degrees.

′	Sine.	D. 10″	Cosine.	D. 10″	Tang.	D. 10″	Cotang.	N. sine.	N. cos.	
0	9.194332	133	9.994620	3.3	9.199713	136	10.800287	15643	98769	60
1	195129	133	994600	3.3	200529	136	799471	15672	98764	59
2	195925	132	994580	3.3	201345	136	798655	15701	98760	58
3	196719	182	994560	3.4	202159	135	797841	15730	98755	57
4	197511	132	994540	3.4	202971	135	797029	15758	98751	56
5	198302	132	994519	3.4	203782	135	796218	15787	98746	55
6	199091	131	994499	3.4	204592	135	795408	15816	98741	54
7	199879	131	994479	3.4	205400	134	794600	15845	98737	53
8	200666	131	994459	3.4	206207	134	793793	15873	98732	52
9	201451	131	994438	3.4	207013	134	792987	15902	98728	51
10	202234	130	994418	3.4	207817	134	792183	15931	98723	50
11	9.203017	130	9.994397	3.4	9.208619	133	10.791381	15959	98718	49
12	203797	130	994377	3.4	209420	133	790580	15988	98714	48
13	204577	130	994357	3.4	210220	133	789780	16017	98709	47
14	205354	129	994336	3.4	211018	133	788982	16046	98704	46
15	206131	129	994316	3.4	211815	133	788185	16074	98700	45
16	206906	129	994295	3.4	212611	132	787389	16103	98695	44
17	207679	129	994274	3.5	213405	132	786595	16132	98690	43
18	208452	128	994254	3.5	214198	132	785802	16160	98686	42
19	209222	128	994233	3.5	214989	132	735011	16189	98681	41
20	209992	128	994212	3.5	215780	131	784220	16218	98676	40
21	9.210760	128	9.994191	3.5	9.216568	131	10.783432	16246	98671	39
22	211526	127	994171	3.5	217356	131	782644	16275	98667	38
23	212291	127	994150	3.5	218142	131	781858	16304	98662	37
24	213055	127	994129	3.5	218926	130	781074	16333	98657	36
25	213818	127	994108	3.5	219710	130	780290	16361	98652	35
26	214579	127	994087	3.5	220492	130	779508	16390	98648	34
27	215338	126	994066	3.5	221272	130	778728	16419	98643	33
28	216097	126	994045	3.5	222052	130	777948	16447	98638	32
29	216854	126	994024	3.5	222830	129	777170	16476	98633	31
30	217609	126	994003	3.5	223606	129	776394	16505	98629	30
31	9.218363	125	9.993981	3.5	9.224382	129	10.775618	16533	98624	29
32	219116	125	993960	3.5	225156	129	774844	16562	98619	28
33	219868	125	993939	3.5	225929	129	774071	16591	98614	27
34	220618	125	993918	3.5	226700	128	773300	16620	98609	26
35	221367	125	993896	3.6	227471	128	772529	16648	98604	25
36	222115	124	993875	3.6	228239	128	771761	16677	98600	24
37	222861	124	993854	3.6	229007	128	770993	16706	98595	23
38	223606	124	993832	3.6	229773	127	770227	16734	98590	22
39	224349	124	993811	3.6	230539	127	769461	16763	98585	21
40	225092	123	993789	3.6	231302	127	768698	16792	98580	20
41	9.225833	123	9.993768	3.6	9.232065	127	10.767935	16820	98575	19
42	226573	123	993746	3.6	232826	127	767174	16849	98570	18
43	227311	123	993725	3.6	233586	126	766414	16878	98565	17
44	228048	123	993703	3.6	234345	126	765655	16906	98561	16
45	228784	122	993681	3.6	235103	126	764897	16935	98556	15
46	229518	122	993660	3.6	235859	126	764141	16964	98551	14
47	230252	122	993638	3.6	236614	126	763386	16992	98546	13
48	230984	122	993616	3.6	237368	125	762632	17021	98541	12
49	231714	122	993594	3.7	238120	125	761880	17050	98536	11
50	232444	121	993572	3.7	238872	125	761128	17078	98531	10
51	9.233172	121	9.993550	3.7	9.239622	125	10.760378	17107	98526	9
52	233899	121	994528	3.7	240371	125	759629	17136	98521	8
53	234625	121	993506	3.7	241118	124	758882	17164	98516	7
54	235349	120	993484	3.7	241865	124	758135	17193	98511	6
55	236073	120	993462	3.7	242610	124	757390	17222	98506	5
56	236795	120	993440	3.7	243354	124	756646	17250	98501	4
57	237515	120	993418	3.7	244097	124	755903	17279	98496	3
58	238235	120	993396	3.7	244839	123	755161	17308	98491	2
59	238953	119	993374	3.7	245579	123	754421	17336	98486	1
60	239670		993351		246319		753681	17365	98481	0
	Cosine.		Sine.		Cotang.		Tang.	N. cos.	N.sine.	′

′	Sine.	D. 10″	Cosine.	D. 10″	Tang.	D. 10″	Cotang.	N.sine.	N. cos.	
0	9.239670	119	9.993351	3.7	9.246319	123	10.753681	17365	98481	60
1	240386	119	993329	3.7	247057	123	752943	17393	98476	59
2	241101	119	993307	3.7	247794	123	752206	17422	98471	58
3	241814	119	993285	3.7	248530	122	751470	17451	98466	57
4	242526	118	993262	3.7	249264	122	750736	17479	98461	56
5	243237	118	993240	3.7	249998	122	750002	17508	98455	55
6	243947	118	993217	3.8	250730	122	749270	17537	98450	54
7	244656	118	993195	3.8	251461	122	748539	17565	98445	53
8	245363	118	993172	3.8	252191	121	747809	17594	98440	52
9	246069	117	993149	3.8	252920	121	747080	17623	98435	51
10	246775	117	993127	3.8	253648	121	746352	17651	98430	50
11	9.247478	117	9.993104	3.8	9.254374	121	10.745626	17680	98425	49
12	248181	117	993081	3.8	255100	121	744900	17708	98420	48
13	248883	117	993059	3.8	255824	120	744176	17737	98414	47
14	249583	116	993036	3.8	256547	120	743453	17766	98409	46
15	250282	116	993013	3.8	257269	120	742731	17794	98404	45
16	250980	116	992990	3.8	257990	120	742010	17823	98399	44
17	251677	116	992967	3.8	258710	120	741290	17852	98394	43
18	252373	116	992944	3.8	259429	120	740571	17880	98389	42
19	253067	116	992921	3.8	260146	119	739854	17909	98383	41
20	253761	115	992898	3.8	260863	119	739137	17937	98378	40
21	9.254453	115	9.992875	3.8	9.261578	119	10.738422	17966	98373	39
22	255144	115	992852	3.8	262292	119	737708	17995	98368	38
23	255834	115	992829	3.9	263005	119	736995	18023	98362	37
24	256523	115	992806	3.9	263717	118	736283	18052	98357	36
25	257211	114	992783	3.9	264428	118	735572	18081	98352	35
26	257898	114	992759	3.9	265138	118	734862	18109	98347	34
27	258583	114	992736	3.9	265847	118	734153	18138	98341	33
28	259268	114	992713	3.9	266555	118	733445	18166	98336	32
29	259951	114	992690	3.9	267261	118	732739	18195	98331	31
30	260633	113	992666	3.9	267967	117	732033	18224	98325	30
31	9.261314	113	9.992643	3.9	9.268671	117	10.731329	18252	98320	29
32	261994	113	992619	3.9	269375	117	730625	18281	98315	28
33	262673	113	992596	3.9	270077	117	729923	18309	98310	27
34	263351	113	992572	3.9	270779	117	729221	18338	98304	26
35	264027	113	992549	3.9	271479	116	728521	18367	98299	25
36	264703	112	992525	3.9	272178	116	727822	18395	98294	24
37	265377	112	992501	3.9	272876	116	727124	18424	98288	23
38	266051	112	992478	4.0	273573	116	726427	18452	98283	22
39	266723	112	992454	4.0	274269	116	725731	18481	98277	21
40	267395	112	992430	4.0	274964	116	725036	18509	98272	20
41	9.268065	111	9.992406	4.0	9.275658	115	10.724342	18538	98267	19
42	268734	111	992382	4.0	276351	115	723649	18567	98261	18
43	269402	111	992359	4.0	277043	115	722957	18595	98256	17
44	270069	111	992335	4.0	277734	115	722266	18624	98250	16
45	270735	111	992311	4.0	278424	115	721576	18652	98245	15
46	271400	111	992287	4.0	279113	115	720887	18681	98240	14
47	272064	110	992263	4.0	279801	114	720199	18710	98234	13
48	272726	110	992239	4.0	280488	114	719512	18738	98229	12
49	273388	110	992214	4.0	281174	114	718826	18767	98223	11
50	274049	110	992190	4.0	281858	114	718142	18795	98218	10
51	9.274708	110	9.992166	4.0	9.282542	114	10.717458	18824	98212	9
52	275367	110	992142	4.0	283225	114	716775	18852	98207	8
53	276024	109	992117	4.1	283907	113	716093	18881	98201	7
54	276681	109	992093	4.1	284588	113	715412	18910	98196	6
55	277337	109	992069	4.1	285268	113	714732	18938	98190	5
56	277991	109	992044	4.1	285947	113	714053	18967	98185	4
57	278644	109	992020	4.1	286624	113	713376	18995	98179	3
58	279297	109	991996	4.1	287301	113	712699	19024	98174	2
59	279948	108	991971	4.1	287977	112	712023	19052	98168	1
60	280599		991947		288652		711348	19081	98163	0
	Cosine.		Sine.		Cotang.		Tang.	N. cos.	N.sine.	′

′	Sine.	D. 10″	Cosine.	D. 10″	Tang.	D. 10″	Cotang.	N. sine.	N. cos.	
0	9.280599	108	9.991947	4.1	9.288652	112	10.711348	19081	98163	60
1	281248	108	991922	4.1	289326	112	710674	19109	98157	59
2	281897	108	991897	4.1	289999	112	710001	19138	98152	58
3	282544	108	991873	4.1	290671	112	709329	19167	98146	57
4	283190	108	991848	4.1	291342	112	708658	19195	98140	56
5	283836	107	991823	4.1	292013	111	707987	19224	98135	55
6	284480	107	991799	4.1	292682	111	707318	19252	98129	54
7	285124	107	991774	4.2	293350	111	706650	19281	98124	53
8	285766	107	991749	4.2	294017	111	705983	19309	98118	52
9	286408	107	991724	4.2	294684	111	705316	19338	98112	51
10	287048	107	991699	4.2	295349	111	704651	19366	98107	50
11	9.287687	106	9.991674	4.2	9.296013	111	10.703987	19395	98101	49
12	288326	106	991649	4.2	296677	110	703323	19423	98096	48
13	288964	106	991624	4.2	297339	110	702661	19452	98090	47
14	289600	106	991599	4.2	298001	110	701999	19481	98084	46
15	290236	106	991574	4.2	298662	110	701338	19509	98079	45
16	290870	106	991549	4.2	299322	110	700678	19538	98073	44
17	291504	105	991524	4.2	299980	110	700020	19566	98067	43
18	292137	105	991498	4.2	300638	109	699362	19595	98061	42
19	292768	105	991473	4.2	301295	109	698705	19623	98056	41
20	293399	105	991448	4.2	301951	109	698049	19652	98050	40
21	9.294029	105	9.991422	4.2	9.302607	109	10.697393	19680	98044	39
22	294658	105	991397	4.2	303261	109	696739	19709	98039	38
23	295286	104	991372	4.3	303914	109	696086	19737	98033	37
24	295913	104	991346	4.3	304567	109	695433	19766	98027	36
25	296539	104	991321	4.3	305218	108	694782	19794	98021	35
26	297164	104	991295	4.3	305869	108	694131	19823	98016	34
27	297788	104	991270	4.3	306519	108	693481	19851	98010	33
28	298412	104	991244	4.3	307168	108	692832	19880	98004	32
29	299034	104	991218	4.3	307815	108	692185	19908	97998	31
30	299655	103	991193	4.3	308463	108	691537	19937	97992	30
31	9.300276	103	9.991167	4.3	9.309109	107	10.690891	19965	97987	29
32	300895	103	991141	4.3	309754	107	690246	19994	97981	28
33	301514	103	991115	4.3	310398	107	689602	20022	97975	27
34	302132	103	991090	4.3	311042	107	688958	20051	97969	26
35	302748	103	991064	4.3	311685	107	688315	20079	97963	25
36	303364	102	991038	4.3	312327	107	687673	20108	97958	24
37	303979	102	991012	4.3	312967	107	687033	20136	97952	23
38	304593	102	990986	4.3	313608	106	686392	20165	97946	22
39	305207	102	990960	4.3	314247	106	685753	20193	97940	21
40	305819	102	990934	4.4	314885	106	685115	20222	97934	20
41	9.306430	102	9.990908	4.4	9.315523	106	10.684477	20250	97928	19
42	307041	102	990882	4.4	316159	106	683841	20279	97922	18
43	307650	101	990855	4.4	316795	106	683205	20307	97916	17
44	308259	101	990829	4.4	317430	106	682570	20336	97910	16
45	308867	101	990803	4.4	318064	105	681936	20364	97905	15
46	309474	101	990777	4.4	318697	105	681303	20393	97899	14
47	310080	101	990750	4.4	319329	105	680671	20421	97893	13
48	310685	101	990724	4.4	319961	105	680039	20450	97887	12
49	311289	100	990697	4.4	320592	105	679408	20478	97881	11
50	311893	100	990671	4.4	321222	105	678778	20507	97875	10
51	9.312495	100	9.990644	4.4	9.321851	105	10.678149	20535	97869	9
52	313097	100	990618	4.4	322479	104	677521	20563	97863	8
53	313698	100	990591	4.4	323106	104	676894	20592	97857	7
54	314297	100	990565	4.4	323733	104	676267	20620	97851	6
55	314897	100	990538	4.4	324358	104	675642	20649	97845	5
56	315495	100	990511	4.5	324983	104	675017	20677	97839	4
57	316092	99	990485	4.5	325607	104	674393	20706	97833	3
58	316689	99	990458	4.5	326231	104	673769	20734	97827	2
59	317284	99	990431	4.5	326853	104	673147	20763	97821	1
60	317879		990404		327475		672525	20791	97815	0
	Cosine.		Sine.		Cotang.		Tang.	N. cos.	N. sine.	′

′	Sine.	D. 10″	Cosine.	D. 10″	Tang.	D. 10″	Cotang.	N. sine.	N. cos.	
0	9.317879	99.0	9.990404	4.5	9.327474	103	10.672526	20791	97815	60
1	318473	98.8	990378	4.5	328095	103	671905	20820	97809	59
2	319066	98.7	990351	4.5	328715	103	671285	20848	97803	58
3	319658	98.6	990324	4.5	329334	103	670666	20877	97797	57
4	320249	98.4	990297	4.5	329953	103	670047	20905	97791	56
5	320840	98.3	990270	4.5	330570	103	669430	20933	97784	55
6	321430	98.2	990243	4.5	331187	103	668813	20962	97778	54
7	322019	98.0	990215	4.5	331803	102	668197	20990	97772	53
8	322607	97.9	990188	4.5	332418	102	667582	21019	97766	52
9	323194	97.7	990161	4.5	333033	102	666967	21047	97760	51
10	323780	97.6	990134	4.5	333646	102	666354	21076	97754	50
11	9.324366	97.5	9.990107	4.6	9.334259	102	10.665741	21104	97748	49
12	324950	97.3	990079	4.6	334871	102	665129	21132	97742	48
13	325534	97.2	990052	4.6	335482	102	664518	21161	97735	47
14	326117	97.0	990025	4.6	336093	102	663907	21189	97729	46
15	326700	96.9	989997	4.6	336702	101	663298	21218	97723	45
16	327281	96.8	989970	4.6	337311	101	662689	21246	97717	44
17	327862	96.6	989942	4.6	337919	101	662081	21275	97711	43
18	328442	96.5	989915	4.6	338527	101	661473	21303	97705	42
19	329021	96.4	989887	4.6	339133	101	660867	21331	97698	41
20	329599	96.2	989860	4.6	339739	101	660261	21360	97692	40
21	9.330176	96.1	9.989832	4.6	9.340344	101	10.659656	21388	97686	39
22	330753	96.0	989804	4.6	340948	101	659052	21417	97680	38
23	331329	95.8	989777	4.6	341552	100	658448	21445	97673	37
24	331903	95.7	989749	4.7	342155	100	657845	21474	97667	36
25	332478	95.6	989721	4.7	342757	100	657243	21502	97661	35
26	333051	95.4	989693	4.7	343358	100	656642	21530	97655	34
27	333624	95.3	989665	4.7	343958	100	656042	21559	97648	33
28	334195	95.2	989637	4.7	344558	100	655442	21587	97642	32
29	334766	95.0	989609	4.7	345157	100	654843	21616	97636	31
30	335337	94.9	989582	4.7	345755	100	654245	21644	97630	30
31	9.335906	94.8	9.989553	4.7	9.346353	99.4	10.653647	21672	97623	29
32	336475	94.6	989525	4.7	346949	99.3	653051	21701	97617	28
33	337043	94.5	989497	4.7	347545	99.2	652455	21729	97611	27
34	337610	94.4	989469	4.7	348141	99.1	651859	21758	97604	26
35	338176	94.3	989441	4.7	348735	99.0	651265	21786	97598	25
36	338742	94.1	989413	4.7	349329	98.8	650671	21814	97592	24
37	339306	94.0	989384	4.7	349922	98.7	650078	21843	97585	23
38	339871	93.9	989356	4.7	350514	98.6	649486	21871	97579	22
39	340434	93.7	989328	4.7	351106	98.5	648894	21899	97573	21
40	340996	93.6	989300	4.7	351697	98.3	648303	21928	97566	20
41	9.341558	93.5	9.989271	4.7	9.352287	98.2	10.647713	21956	97560	19
42	342119	93.4	989243	4.7	352876	98.1	647124	21985	97553	18
43	342679	93.2	989214	4.7	353465	98.0	646535	22013	97547	17
44	343239	93.1	989186	4.7	354053	97.9	645947	22041	97541	16
45	343797	93.0	989157	4.7	354640	97.7	645360	22070	97534	15
46	344355	92.9	989128	4.8	355227	97.6	644773	22098	97528	14
47	344912	92.7	989100	4.8	355813	97.5	644187	22126	97521	13
48	345469	92.6	989071	4.8	356398	97.4	643602	22155	97515	12
49	346024	92.5	989042	4.8	356982	97.3	643018	22183	97508	11
50	346579	92.4	989014	4.8	357566	97.1	642434	22212	97502	10
51	9.347134	92.2	9.988985	4.8	9.358149	97.0	10.641851	22240	97496	9
52	347687	92.1	988956	4.8	358731	96.9	641269	22268	97489	8
53	348240	92.0	988927	4.8	359313	96.8	640687	22297	97483	7
54	348792	91.9	988898	4.8	359893	96.7	640107	22325	97476	6
55	349343	91.7	988869	4.8	360474	96.6	639526	22353	97470	5
56	349893	91.6	988840	4.8	361053	96.5	638947	22382	97463	4
57	350443	91.5	988811	4.9	361632	96.3	638368	22410	97457	3
58	350992	91.4	988782	4.9	362210	96.2	637790	22438	97450	2
59	351540	91.3	988753	4.9	362787	96.1	637213	22467	97444	1
60	352088		988724		363364		636636	22495	97437	0
	Cosine.		Sine.		Cotang.		Tang.	N. cos.	N.sine.	′

′	Sine.	D. 10″	Cosine.	D. 10″	Tang.	D. 10″	Cotang.	N.sine.	N. cos.	
0	9.352088	91.1	9.988724	4.9	9.363364	96.0	10.636636	22495	97437	60
1	352635	91.0	988695	4.9	363940	95.9	636060	22523	97430	59
2	353181	90.9	988666	4.9	364515	95.8	635485	22552	97424	58
3	353726	90.8	988636	4.9	365090	95.7	634910	22580	97417	57
4	354271	90.7	988607	4.9	365664	95.5	634336	22608	97411	56
5	354815	90.5	988578	4.9	366237	95.4	633763	22637	97404	55
6	355358	90.4	988548	4.9	366810	95.3	633190	22665	97398	54
7	355901	90.3	988519	4.9	367382	95.2	632618	22693	97391	53
8	356443	90.2	988489	4.9	367953	95.1	632047	22722	97384	52
9	356984	90.1	988460	4.9	368524	95.0	631476	22750	97378	51
10	357524	89.9	988430	4.9	369094	94.9	630906	22778	97371	50
11	9.358064	89.8	9.988401	4.9	9.369663	94.8	10.630337	22807	97365	49
12	358603	89.7	988371	4.9	370232	94.6	629768	22835	97358	48
13	359141	89.6	988342	4.9	370799	94.5	629201	22863	97351	47
14	359678	89.5	988312	5.0	371367	94.4	628633	22892	97345	46
15	360215	89.3	988282	5.0	371933	94.3	628067	22920	97338	45
16	360752	89.2	988252	5.0	372499	94.2	627501	22948	97331	44
17	361287	89.1	988223	5.0	373064	94.1	626936	22977	97325	43
18	361822	89.0	988193	5.0	373629	94.0	626371	23005	97318	42
19	362356	88.9	988163	5.0	374193	93.9	625807	23033	97311	41
20	362889	88.8	988133	5.0	374756	93.8	625244	23062	97304	40
21	9.363422	88.7	9.988103	5.0	9.375319	93.7	10.624681	23090	97298	39
22	363954	88.5	988073	5.0	375881	93.5	624119	23118	97291	38
23	364485	88.4	988043	5.0	376442	93.4	623558	23146	97284	37
24	365016	88.3	988013	5.0	377003	93.3	622997	23175	97278	36
25	365546	88.2	987983	5.0	377563	93.2	622437	23203	97271	35
26	366075	88.1	987953	5.0	378122	93.1	621878	23231	97264	34
27	366604	88.0	987922	5.0	378681	93.0	621319	23260	97257	33
28	367131	87.9	987892	5.0	379239	92.9	620761	23288	97251	32
29	367659	87.7	987862	5.0	379797	92.8	620203	23316	97244	31
30	368185	87.6	987832	5.1	380354	92.7	619646	23345	97237	30
31	9.368711	87.5	9.987801	5.1	9.380910	92.6	10.619090	23373	97230	29
32	369236	87.4	987771	5.1	381466	92.5	618534	23401	97223	28
33	369761	87.3	987740	5.1	382020	92.4	617980	23429	97217	27
34	370285	87.2	987710	5.1	382575	92.3	617425	23458	97210	26
35	370808	87.1	987679	5.1	383129	92.2	616871	23486	97203	25
36	371330	87.0	987649	5.1	383682	93.1	616318	23514	97196	24
37	371852	86.9	987618	5.1	384234	92.0	615766	23542	97189	23
38	372373	86.7	987588	5.1	384786	91.9	615214	23571	97182	22
39	372894	86.6	987557	5.1	385337	91.8	614663	23599	97176	21
40	373414	86.5	987526	5.1	385888	91.7	614112	23627	97169	20
41	9.373933	86.4	9.987496	5.1	9.386438	91.5	10.613562	23656	97162	19
42	3744[illegible]	86.3	987465	5.1	386987	91.4	613013	23684	97155	18
43	374970	86.2	987434	5.1	387536	91.3	612464	23712	97148	17
44	375487	86.1	987403	5.2	388084	91.2	611916	23740	97141	16
45	376003	86.0	987372	5.2	388631	91.1	611369	23769	97134	15
46	376519	85.9	987341	5.2	389178	91.0	610822	23797	97127	14
47	377035	85.8	987310	5.2	389724	90.9	610276	23825	97120	13
48	377549	85.7	987279	5.2	390270	90.8	609730	23853	97113	12
49	378063	85.6	987248	5.2	390815	90.7	609185	23882	97106	11
50	378577	85.4	987217	5.2	391360	90.6	608640	23910	97100	10
51	9.379089	85.3	9.987186	5.2	9.391903	90.5	10.608097	23938	97093	9
52	379601	85.2	987155	5.2	392447	90.4	607553	23966	97086	8
53	380113	85.1	987124	5.2	392989	90.3	607011	23995	97079	7
54	380624	85.0	987092	5.2	393531	90.2	606469	24023	97072	6
55	381134	84.9	987061	5.2	394073	90.1	605927	24051	97065	5
56	381643	84.8	987030	5.2	394614	90.0	605386	24079	97058	4
57	382152	84.7	986998	5.2	395154	89.9	604846	24108	97051	3
58	382661	84.6	986967	5.2	395694	89.8	604306	24136	97044	2
59	383168	84.5	986936	5.2	396233	89.7	603767	24164	97037	1
60	383675		986904		396771		603229	24192	97030	0
	Cosine.		Sine.		Cotang.		Tang.	N. cos.	N.sine.	′

′	Sine.	D. 10″	Cosine.	D. 10″	Tang.	D. 10″	Cotang.	N. sine.	N. cos.	
0	9.383675	84.4	9.986904	5.2	9.396771	89.6	10.603229	24192	97030	60
1	384182	84.3	986873	5.3	397309	89.6	602691	24220	97023	59
2	384687	84.2	986841	5.3	397846	89.5	602154	24249	97015	58
3	385192	84.1	986809	5.3	398383	89.4	601617	24277	97008	57
4	385697	84.0	986778	5.3	398919	89.3	601081	24305	97001	56
5	386201	83.9	986746	5.3	399455	89.2	600545	24333	96994	55
6	386704	83.8	986714	5.3	399990	89.1	600010	24362	96987	54
7	387207	83.7	986683	5.3	400524	89.0	599476	24390	96980	53
8	387709	83.6	986651	5.3	401058	88.9	598942	24418	96973	52
9	388210	83.5	986619	5.3	401591	88.8	598409	24446	96966	51
10	388711	83.4	986587	5.3	402124	88.7	597876	24474	96959	50
11	9.389211	83.3	9.986555	5.3	9.402656	88.6	10.597344	24503	96952	49
12	389711	83.2	986523	5.3	403187	88.5	596813	24531	96945	48
13	390210	83.1	986491	5.3	403718	88.4	596282	24559	96937	47
14	390708	83.0	986459	5.3	404249	88.3	595751	24587	96930	46
15	391206	82.8	986427	5.3	404778	88.2	595222	24615	96923	45
16	391703	82.7	986395	5.3	405308	88.1	594692	24644	96916	44
17	392199	82.6	986363	5.4	405836	88.0	594164	24672	96909	43
18	392695	82.5	986331	5.4	406364	87.9	593636	24700	96902	42
19	393191	82.4	986299	5.4	406892	87.8	593108	24728	96894	41
20	393685	82.3	986266	5.4	407419	87.7	592581	24756	96887	40
21	9.394179	82.2	9.986234	5.4	9.407945	87.6	10.592055	24784	96880	39
22	394673	82.1	986202	5.4	408471	87.5	591529	24813	96873	38
23	395166	82.0	986169	5.4	408997	87.4	591003	24841	96866	37
24	395658	81.9	986137	5.4	409521	87.4	590479	24869	96858	36
25	396150	81.8	986104	5.4	410045	87.3	589955	24897	96851	35
26	396641	81.7	986072	5.4	410569	87.2	589431	24925	96844	34
27	397132	81.7	986039	5.4	411092	87.1	588908	24954	96837	33
28	397621	81.6	986007	5.4	411615	87.0	588385	24982	96829	32
29	398111	81.5	985974	5.4	412137	86.9	587863	25010	96822	31
30	398600	81.4	985942	5.4	412658	86.8	587342	25038	96815	30
31	9.399088	81.3	9.985909	5.5	9.413179	86.7	10.586821	25066	96807	29
32	399575	81.2	985876	5.5	413699	86.6	586301	25094	96800	28
33	400062	81.1	985843	5.5	414219	86.5	585781	25122	96793	27
34	400549	81.0	985811	5.5	414738	86.4	585262	25151	96786	26
35	401035	80.9	985778	5.5	415257	86.4	584743	25179	96778	25
36	401520	80.8	985745	5.5	415775	86.3	584225	25207	96771	24
37	402005	80.7	985712	5.5	416293	86.2	583707	25235	96764	23
38	402489	80.6	985679	5.5	416810	86.1	583190	25263	96756	22
39	402972	80.5	985646	5.5	417326	86.0	582674	25291	96749	21
40	403455	80.4	985613	5.5	417842	85.9	582158	25320	96742	20
41	9.403938	80.3	9.985580	5.5	9.418358	85.8	10.581642	25348	96734	19
42	404420	80.2	985547	5.5	418873	85.7	581127	25376	96727	18
43	404901	80.1	985514	5.5	419387	85.6	580613	25404	96719	17
44	405382	80.0	985480	5.5	419901	85.5	580099	25432	96712	16
45	405862	79.9	985447	5.5	420415	85.5	579585	25460	96705	15
46	406341	79.8	985414	5.6	420927	85.4	579073	25488	96697	14
47	406820	79.7	985380	5.6	421440	85.3	578560	25516	96690	13
48	407299	79.6	985347	5.6	421952	85.2	578048	25545	96682	12
49	407777	79.5	985314	5.6	422463	85.1	577537	25573	96675	11
50	408254	79.4	985280	5.6	422974	85.0	577026	25601	96667	10
51	9.408731	79.4	9.985247	5.6	9.423484	84.9	10.576516	25629	96660	9
52	409207	79.3	985213	5.6	423993	84.8	576007	25657	96653	8
53	409682	79.2	985180	5.6	424503	84.8	575497	25685	96645	7
54	410157	79.1	985146	5.6	425011	84.7	574989	25713	96638	6
55	410632	79.0	985113	5.6	425519	84.6	574481	25741	96630	5
56	411106	78.9	985079	5.6	426027	84.5	573973	25766	96623	4
57	411579	78.8	985045	5.6	426534	84.4	573466	25798	96615	3
58	412052	78.7	985011	5.6	427041	84.3	572959	25826	96608	2
59	412524	78.6	984978	5.6	427547	84.3	572453	25854	96600	1
60	412996		984944		428052		571948	25882	96593	0
	Cosine.		Sine.		Cotang.		Tang.	N. cos.	N. sine.	′

75 Degrees.

′	Sine.	D. 10″	Cosine.	D. 10″	Tang.	D. 10″	Cotang.	N. sine.	N. cos.	
0	9.412996	78.5	9.984944	5.7	9.428052	84.2	10.571948	25882	96593	60
1	413467	78.4	984910	5.7	428557	84.1	571443	25910	96585	59
2	413938	78.3	984876	5.7	429062	84.0	570938	25938	96578	58
3	414408	78.3	984842	5.7	429566	83.9	570434	25966	96570	57
4	414878	78.2	984808	5.7	430070	83.8	569930	25994	96562	56
5	415347	78.1	984774	5.7	430573	83.8	569427	26022	96555	55
6	415815	78.0	984740	5.7	431075	83.7	568925	26050	96547	54
7	416283	77.9	984706	5.7	431577	83.6	568423	26079	96540	53
8	416751	77.8	984672	5.7	432079	83.5	567921	26107	96532	52
9	417217	77.7	984637	5.7	432580	83.4	567420	26135	96524	51
10	417684	77.6	984603	5.7	433080	83.3	566920	26163	96517	50
11	9.418150	77.5	9.984569	5.7	9.433580	83.2	10.566420	26191	96509	49
12	418615	77.4	984535	5.7	434080	83.2	565920	26219	96502	48
13	419079	77.3	984500	5.7	434579	83.1	565421	26247	96494	47
14	419544	77.3	984466	5.7	435078	83.0	564922	26275	96486	46
15	420007	77.2	984432	5.8	435576	82.9	564424	26303	96479	45
16	420470	77.1	984397	5.8	436073	82.8	563927	26331	96471	44
17	420933	77.0	984363	5.8	436570	82.8	563430	26359	96463	43
18	421395	76.9	984328	5.8	437067	82.7	562933	26387	96456	42
19	421857	76.8	984294	5.8	437563	82.6	562437	26415	96448	41
20	422318	76.7	984259	5.8	438059	82.5	561941	26443	96440	40
21	9.422778	76.7	9.984224	5.8	9.438554	82.4	10.561446	26471	96433	39
22	423238	76.6	984190	5.8	439048	82.3	560952	26500	96425	38
23	423697	76.5	984155	5.8	439543	82.3	560457	26528	96417	37
24	424156	76.4	984120	5.8	440036	82.2	559964	26556	96410	36
25	424615	76.3	984085	5.8	440529	82.1	559471	26584	96402	35
26	425073	76.2	984050	5.8	441022	82.0	558978	26612	96394	34
27	425530	76.1	984015	5.8	441514	81.9	558486	26640	96386	33
28	425987	76.0	983981	5.8	442006	81.9	557994	26668	96379	32
29	426443	76.0	983946	5.8	442497	81.8	557503	26696	96371	31
30	426899	75.9	983911	5.8	442988	81.7	557012	26724	96363	30
31	9.427354	75.8	9.983875	5.8	9.443479	81.6	10.556521	26752	96355	29
32	427809	75.7	983840	5.9	443968	81.6	556032	26780	96347	28
33	428263	75.6	983805	5.9	444458	81.5	555542	26808	96340	27
34	428717	75.5	983770	5.9	444947	81.4	555053	26836	96332	26
35	429170	75.4	983735	5.9	445435	81.3	554565	26864	96324	25
36	429623	75.3	983700	5.9	445923	81.2	554077	26892	96316	24
37	430075	75.2	983664	5.9	446411	81.2	553589	26920	96308	23
38	430527	75.2	983629	5.9	446898	81.1	553102	26948	96301	22
39	430978	75.1	983594	5.9	447384	81.0	552616	26976	96293	21
40	431429	75.0	983558	5.9	447870	80.9	552130	27004	96285	20
41	9.431879	74.9	9.983523	5.9	9.448356	80.9	10.551644	27032	96277	19
42	432329	74.9	983487	5.9	448841	80.8	551159	27060	96269	18
43	432778	74.8	983452	5.9	449326	80.7	550674	27088	96261	17
44	433226	74.7	983416	5.9	449810	80.6	550190	27116	96253	16
45	433675	74.6	983381	5.9	450294	80.6	549706	27144	96246	15
46	434122	74.5	983345	5.9	450777	80.5	549223	27172	96238	14
47	434569	74.4	983309	5.9	451260	80.4	548740	27200	96230	13
48	435016	74.4	983273	6.0	451743	80.3	548257	27228	96222	12
49	435462	74.3	983238	6.0	452225	80.2	547775	27256	96214	11
50	435908	74.2	983202	6.0	452706	80.2	547294	27284	96206	10
51	9.436353	74.1	9.983166	6.0	9.453187	80.1	10.546813	27312	96198	9
52	436798	74.0	983130	6.0	453668	80.0	546332	27340	96190	8
53	437242	74.0	983094	6.0	454148	79.9	545852	27368	96182	7
54	437686	73.9	983058	6.0	454628	79.9	545372	27396	96174	6
55	438129	73.8	983022	6.0	455107	79.8	544893	27424	96166	5
56	438572	73.7	982986	6.0	455586	79.7	544414	27452	96158	4
57	439014	73.6	982950	6.0	456064	79.6	543936	27480	96150	3
58	439456	73.6	982914	6.0	456542	79.6	543458	27508	96142	2
59	439897	73.5	982878	6.0	457019	79.5	542981	27536	96134	1
60	440338		982842		457496		542504	27564	96126	0
	Cosine.		Sine.		Cotang.		Tang.	N. cos.	N.sine.	′

′	Sine.	D. 10″	Cosine.	D. 10″	Tang.	D. 10″	Cotang.	N. sine.	N. cos.	
0	9.440338	73.4	9.982842	6.0	9.457496	79.4	10.542504	27564	96126	60
1	440778	73.3	982805	6.0	457973	79.3	542027	27592	96118	59
2	441218	73.2	982769	6.1	458449	79.3	541551	27620	96110	58
3	441658	73.1	982733	6.1	458925	79.2	541075	27648	96102	57
4	442096	73.1	982696	6.1	459400	79.1	540600	27676	96094	56
5	442535	73.0	982660	6.1	459875	79.0	540125	27704	96086	55
6	442973	72.9	982624	6.1	460349	79.0	539651	27731	96078	54
7	443410	72.8	982587	6.1	460823	78.9	539177	27759	96070	53
8	443847	72.7	982551	6.1	461297	78.8	538703	27787	96062	52
9	444284	72.7	982514	6.1	461770	78.9	538230	27815	96054	51
10	444720	72.6	982477	6.1	462242	78.7	537758	27843	96046	50
11	9.445155	72.5	9.982441	6.1	9.462714	78.6	10.537286	27871	96037	49
12	445590	72.4	982404	6.1	463186	78.5	536814	27899	96029	48
13	446025	72.3	982367	6.1	463658	78.5	536342	27927	96021	47
14	446459	72.3	982331	6.1	464129	78.4	535871	27955	96013	46
15	446893	72.2	982294	6.1	464599	78.3	535401	27983	96005	45
16	447326	72.1	982257	6.1	465069	78.3	534931	28011	95997	44
17	447759	72.0	982220	6.2	465539	78.2	534461	28039	95989	43
18	448191	72.0	982183	6.2	466008	78.1	533992	28067	95981	42
19	448623	71.9	982146	6.2	466476	78.0	533524	28095	95972	41
20	449054	71.8	982109	6.2	466945	78.0	533055	28123	95964	40
21	9.449485	71.7	9.982072	6.2	9.467413	77.9	10.532587	28150	95956	39
22	449915	71.6	982035	6.2	467880	77.8	532120	28178	95948	38
23	450345	71.6	981998	6.2	468347	77.8	531653	28206	95940	37
24	450775	71.5	981961	6.2	468814	77.7	531186	28234	95931	36
25	451204	71.4	981924	6.2	469280	77.6	530720	28262	95923	35
26	451632	71.3	981886	6.2	469746	77.5	530254	28290	95915	34
27	452060	71.3	981849	6.2	470211	77.5	529789	28318	95907	33
28	452488	71.2	981812	6.2	470676	77.4	529324	28346	95898	32
29	452915	71.1	981774	6.2	471141	77.3	528859	28374	95890	31
30	453342	71.0	981737	6.2	471605	77.3	528395	28402	95882	30
31	9.453768	71.0	9.981699	6.2	9.472068	77.2	10.527932	28429	95874	29
32	454194	70.9	981662	6.3	472532	77.1	527468	28457	95865	28
33	454619	70.8	981625	6.3	472995	77.1	527005	28485	95857	27
34	455044	70.7	981587	6.3	473457	77.0	526543	28513	95849	26
35	455469	70.7	981549	6.3	473919	76.9	526081	28541	95841	25
36	455893	70.6	981512	6.3	474381	76.9	525619	28569	95832	24
37	456316	70.5	981474	6.3	474842	76.8	525158	28597	95824	23
38	456739	70.4	981436	6.3	475303	76.7	524697	28625	95816	22
39	457162	70.4	981399	6.3	475763	76.7	524237	28652	95807	21
40	457584	70.3	981361	6.3	476223	76.6	523777	28680	95799	20
41	9.458006	70.2	9.981323	6.3	9.476683	76.5	10.523317	28708	95791	19
42	458427	70.1	981285	6.3	477142	76.5	522858	28736	95782	18
43	458848	70.1	981247	6.3	477601	76.4	522399	28764	95774	17
44	459268	70.0	981209	6.3	478059	76.3	521941	28792	95766	16
45	459688	69.9	981171	6.3	478517	76.3	521483	28820	95757	15
46	460108	69.8	981133	6.3	478975	76.2	521025	28847	95749	14
47	460527	69.8	981095	6.4	479432	76.1	520568	28875	95740	13
48	460946	69.7	981057	6.4	479889	76.1	520111	28903	95732	12
49	461364	69.6	981019	6.4	480345	76.0	519655	28931	95724	11
50	461782	69.5	980981	6.4	480801	75.9	519199	28959	95715	10
51	9.462199	69.5	9.980942	6.4	9.481257	75.9	10.518743	28987	95707	9
52	462616	69.4	980904	6.4	481712	75.8	518288	29015	95698	8
53	463032	69.3	980866	6.4	482167	75.7	517833	29042	95690	7
54	463448	69.3	980827	6.4	482621	75.7	517379	29070	95681	6
55	463864	69.2	980789	6.4	483075	75.6	516925	29098	95673	5
56	464279	69.1	980750	6.4	483529	75.5	516471	29126	95664	4
57	464694	69.0	980712	6.4	483982	75.5	516018	29154	95656	3
58	465108	69.0	980673	6.4	484435	75.4	515565	29182	95647	2
59	465522	68.9	980635	6.4	484887	75.3	515113	29209	95639	1
60	465935		980596		485339		514661	29247	95630	0
	Cosine.		Sine.		Cotang.		Tang.	N. cos.	N. sine.	′

′	Sine.	D. 10″	Cosine.	D. 10″	Tang.	D. 10″	Cotang.	N. sine.	N. cos.	
0	9.465935	68.8	9.980596	6.4	9.485339	75.3	10.514661	29237	95630	60
1	466348	68.8	980558	6.4	485791	75.2	514209	29265	95622	59
2	466761	68.7	980519	6.5	486242	75.1	513758	29293	95613	58
3	467173	68.6	980480	6.5	486693	75.1	513307	29321	95605	57
4	467585	68.5	980442	6.5	487143	75.0	512857	29348	95596	56
5	467996	68.5	980403	6.5	487593	74.9	512407	29376	95588	55
6	468407	68.4	980364	6.5	488043	74.9	511957	29404	95579	54
7	468817	68.3	980325	6.5	488492	74.8	511508	29432	95571	53
8	469227	68.3	980286	6.5	488941	74.7	511059	29460	95562	52
9	469637	68.2	980247	6.5	489390	74.7	510610	29487	95554	51
10	470046	68.1	980208	6.5	489838	74.6	510162	29515	95545	50
11	9.470455	68.0	9.980169	6.5	9.490286	74.6	10.509714	29543	95536	49
12	470863	68.0	980130	6.5	490733	74.5	509267	29571	95528	48
13	471271	67.9	980091	6.5	491180	74.4	508820	29599	95519	47
14	471679	67.8	980052	6.5	491627	74.4	508373	29626	95511	46
15	472086	67.8	980012	6.5	492073	74.3	507927	29654	95502	45
16	472492	67.7	979973	6.5	492519	74.3	507481	29682	95493	44
17	472898	67.6	979934	6.6	492965	74.2	507035	29710	95485	43
18	473304	67.6	979895	6.6	493410	74.1	506590	29737	95476	42
19	473710	67.5	979855	6.6	493854	74.0	506146	29765	95467	41
20	474115	67.4	979816	6.6	494299	74.0	505701	29793	95459	40
21	9.474519	67.4	9.979776	6.6	9.494743	74.0	10.505257	29821	95450	39
22	474923	67.3	979737	6.6	495186	73.9	504814	29849	95441	38
23	475327	67.2	979697	6.6	495630	73.8	504370	29876	95433	37
24	475730	67.2	979658	6.6	496073	73.7	503927	29904	95424	36
25	476133	67.1	979618	6.6	496515	73.7	503485	29932	95415	35
26	476536	67.0	979579	6.6	496957	73.6	503043	29960	95407	34
27	476938	66.9	979539	6.6	497399	73.6	502601	29987	95398	33
28	477340	66.9	979499	6.6	497841	73.5	502159	30015	95389	32
29	477741	66.8	979459	6.6	468282	73.4	501718	30043	95380	31
30	478142	66.7	979420	6.6	498722	73.4	501278	30071	95372	30
31	9.478542	66.7	9.979380	6.6	9.499163	73.3	10.500837	30098	95363	29
32	478942	66.6	979340	6.6	499603	73.3	500397	30126	95354	28
33	479342	66.5	979300	6.7	500042	73.2	499958	30154	95345	27
34	479741	66.5	979260	6.7	500481	73.1	499519	30182	95337	26
35	480140	66.4	979220	6.7	500920	73.1	499080	30209	95328	25
36	480539	66.3	979180	6.7	501359	73.0	498641	30237	95319	24
37	480937	66.3	979140	6.7	501797	73.0	498203	30265	95310	23
38	481334	66.2	979100	6.7	502235	72.9	497765	30292	95301	22
39	481731	66.1	979059	6.7	502672	72.8	497328	30320	95293	21
40	482128	66.1	979019	6.7	503109	72.8	496891	30348	95284	20
41	9.482525	66.0	9.978979	6.7	9.503546	72.7	10.496454	30376	95275	19
42	482921	65.9	978939	6.7	503982	72.7	496018	30403	95266	18
43	483316	65.9	978898	6.7	504418	72.6	495582	30431	95257	17
44	483712	65.8	978858	6.7	504854	72.5	495146	30459	95248	16
45	484107	65.7	978817	6.7	505289	72.5	494711	30486	95240	15
46	484501	65.7	978777	6.7	505724	72.4	494276	30514	95231	14
47	484895	65.6	978736	6.7	506159	72.4	493841	30542	95222	13
48	485289	65.5	978696	6.8	506593	72.3	493407	30570	95213	12
49	485682	65.5	978655	6.8	507027	72.2	492973	30597	95204	11
50	486075	65.4	978615	6.8	507460	72.2	492540	30625	95195	10
51	9.486467	66.3	9.978574	6.8	9.507893	72.1	10.492107	30653	95186	9
52	486860	65.3	978533	6.8	508326	72.1	491674	30680	95177	8
53	487251	65.2	978493	6.8	508759	72.0	491241	30708	95168	7
54	487643	65.1	978452	6.8	509191	71.9	490809	30736	95159	6
55	488034	65.1	978411	6.8	509622	71.9	490378	30763	95150	5
56	488424	65.0	978370	6.8	510054	71.8	489946	30791	95142	4
57	488814	65.0	978329	6.8	510485	71.8	489515	30819	95133	3
58	489204	64.9	978288	6.8	510916	71.7	489084	30846	95124	2
59	489593	64.8	978247	6.8	511346	71.6	488654	30874	95115	1
60	489982		978206		511776		488224	30902	95106	0
	Cosine.		Sine.		Cotang.		Tang.	N. cos.	N. sine.	′

′	Sine.	D. 10″	Cosine.	D. 10″	Tang.	D. 10″	Cotang.	N. sine.	N. cos.	
0	9.489982	64.8	9.978206	6.8	9.511776	71.6	10.488224	30902	95106	60
1	490371	64.8	978165	6.8	512206	71.6	487794	30929	95097	59
2	490759	64.7	978124	6.8	512635	71.5	487365	30957	95088	58
3	491147	64.6	978083	6.9	513064	71.4	486936	30985	95079	57
4	491535	64.6	978042	6.9	513493	71.4	486507	31012	95070	56
5	491922	64.5	978001	6.9	513921	71.3	486079	31040	95061	55
6	492308	64.4	977959	6.9	514349	71.3	485651	31068	95052	54
7	492695	64.4	977918	6.9	514777	71.2	485223	31095	95043	53
8	493081	64.3	977877	6.9	515204	71.2	484796	31123	95033	52
9	493466	64.2	977835	6.9	515631	71.1	484369	31151	95024	51
10	493851	64.2	977794	6.9	516057	71.0	483943	31178	95015	50
11	9.494236	64.1	9.977752	6.9	9.516484	71.0	10.483516	31206	95006	49
12	494621	64.1	977711	6.9	516910	70.9	483090	31233	94997	48
13	495005	64.0	977669	6.9	517335	70.9	482665	31261	94988	47
14	495388	63.9	977628	6.9	517761	70.8	482239	31289	94979	46
15	495772	63.9	977586	6.9	518185	70.8	481815	31316	94970	45
16	496154	63.8	977544	7.0	518610	70.7	481390	31344	94961	44
17	496537	63.7	977503	7.0	519034	70.6	480966	31372	94952	43
18	496919	63.7	977461	7.0	519458	70.6	480542	31399	94943	42
19	497301	63.6	977419	7.0	519882	70.5	480118	31427	94933	41
20	497682	63.6	977377	7.0	520305	70.5	479695	31454	94924	40
21	9.498064	63.5	9.977335	7.0	9.520728	70.4	10.479272	31482	94915	39
22	498444	63.4	977293	7.0	521151	70.3	478849	31510	94906	38
23	498825	63.4	977251	7.0	521573	70.3	478427	31537	94897	37
24	499204	63.3	977209	7.0	521995	70.3	478005	31565	94888	36
25	499584	63.2	977167	7.0	522417	70.2	477583	31593	94878	35
26	499963	63.2	977125	7.0	522838	70.2	477162	31620	94869	34
27	500342	63.1	977083	7.0	523259	70.1	476741	31648	94860	33
28	500721	63.1	977041	7.0	523680	70.1	476320	31675	94851	32
29	501099	63.0	976999	7.0	524100	70.0	475900	31703	94842	31
30	501476	62.9	976957	7.0	524520	69.9	475480	31730	94832	30
31	9.501854	62.9	9.976914	7.0	9.524939	69.9	10.475061	31758	94823	29
32	502231	62.8	976872	7.1	525359	69.8	474641	31786	94814	28
33	502607	62.8	976830	7.1	525778	69.8	474222	31813	94805	27
34	502984	62.7	976787	7.1	526197	69.7	473803	31841	94795	26
35	503360	62.6	976745	7.1	526615	69.7	473385	31868	94786	25
36	503735	62.6	976702	7.1	527033	69.6	472967	31896	94777	24
37	504110	62.5	976660	7.1	527451	69.6	472549	31923	94768	23
38	504485	62.5	976617	7.1	527868	69.5	472132	31951	94758	22
39	504860	62.4	976574	7.1	528285	69.5	471715	31979	94749	21
40	505234	62.3	976532	7.1	528702	69.4	471298	32006	94740	20
41	9.505608	62.3	9.976489	7.1	9.529119	69.3	10.470881	32034	94730	19
42	505981	62.2	976446	7.1	529535	69.3	470465	32061	94721	18
43	506354	62.2	976404	7.1	529950	69.3	470050	32089	94712	17
44	506727	62.1	976361	7.1	530366	69.2	469634	32116	94702	16
45	507099	62.0	976318	7.1	530781	69.1	469219	32144	94693	15
46	507471	62.0	976275	7.1	531196	69.1	468804	32171	94684	14
47	507843	61.9	976232	7.2	531611	69.0	468389	32199	94674	13
48	508214	61.9	976189	7.2	532025	69.0	467975	32227	94665	12
49	508585	61.8	976146	7.2	532439	68.9	467561	32250	94656	11
50	508956	61.8	976103	7.2	532853	68.9	467147	32282	94646	10
51	9.509326	61.7	9.976060	7.2	9.533266	68.8	10.466734	32309	94637	9
52	509696	61.6	976017	7.2	533679	68.8	466321	32337	94627	8
53	510065	61.6	975974	7.2	534092	68.7	465908	32364	94618	7
54	510434	61.5	975930	7.2	534504	68.7	465496	32392	94609	6
55	510803	61.5	975887	7.2	534916	68.6	465084	32419	94599	5
56	511172	61.4	975844	7.2	535328	68.6	464672	32447	94590	4
57	511540	61.3	975800	7.2	535739	68.5	464261	32474	94580	3
58	511907	61.3	975757	7.2	536150	68.5	463850	32502	94571	2
59	512275	61.2	975714	7.2	536561	68.4	463439	32529	94561	1
60	512642		975670		536972		463028	32557	94552	0
	Cosine.		Sine.		Cotang.		Tang.	N. cos.	N.sine.	′

71 Degrees.

′	Sine.	D. 10″	Cosine.	D. 10″	Tang	D. 10″	Cotang.	N. sine.	N. cos.	
0	9.512642	61.2	9.975670	7.3	9.536972	68.4	10.463028	32557	94552	60
1	513009	61.1	975627	7.3	537382	68.3	462618	32584	94542	59
2	513375	61.1	975583	7.3	537792	68.3	462208	32612	94533	58
3	513741	61.0	975539	7.3	538202	68.2	461798	32639	94523	57
4	514107	60.9	975496	7.3	538611	68.2	461389	32667	94514	56
5	514472	60.9	975452	7.3	539020	68.1	460980	32694	94504	55
6	514837	60.8	975408	7.3	539429	68.1	460571	32722	94495	54
7	515202	60.8	975365	7.3	539837	68.0	460163	32749	94485	53
8	515566	60.7	975321	7.3	540245	68.0	459755	32777	94476	52
9	515930	60.7	975277	7.3	540653	67.9	459347	32804	94466	51
10	516294	60.6	975233	7.3	541061	67.9	458939	32832	94457	50
11	9.516657	60.5	9.975189	7.3	9.541468	67.8	10.458532	32859	94447	49
12	517020	60.5	975145	7.3	541875	67.8	458125	32887	94438	48
13	517382	60.4	975101	7.3	542281	67.7	457719	32914	94428	47
14	517745	60.4	975057	7.3	542688	67.7	457312	32942	94418	46
15	518107	60.3	975013	7.3	543094	67.6	456906	32969	94409	45
16	518468	60.3	974969	7.4	543499	67.6	456501	32997	94399	44
17	518829	60.2	974925	7.4	543905	67.5	456095	33024	94390	43
18	519190	60.1	974880	7.4	544310	67.5	455690	33051	94380	42
19	519551	60.1	974836	7.4	544715	67.4	455285	33079	94370	41
20	519911	60.0	974792	7.4	545119	67.4	454881	33106	94361	40
21	9.520271	60.0	9.974748	7.4	9.545524	67.3	10.454476	33134	94351	39
22	520631	59.9	974703	7.4	545928	67.3	454072	33161	94342	38
23	520990	59.9	974659	7.4	546331	67.2	453669	33189	94332	37
24	521349	59.8	974614	7.4	546735	67.2	453265	33216	94322	36
25	521707	59.8	974570	7.4	547138	67.1	452862	33244	94313	35
26	522066	59.7	974525	7.4	547540	67.1	452460	33271	94303	34
27	522424	59.6	974481	7.4	547943	67.0	452057	33298	94293	33
28	522781	59.6	974436	7.4	548345	67.0	451655	33326	94284	32
29	523138	59.5	974391	7.4	548747	66.9	451253	33353	94274	31
30	523495	59.5	974347	7.5	549149	66.9	450851	33381	94264	30
31	9.523852	59.4	9.974302	7.5	9.549550	66.8	10.450450	33408	94254	29
32	524208	59.4	974257	7.5	549951	66.8	450049	33436	94245	28
33	524564	59.3	974212	7.5	550352	66.7	449648	33463	94235	27
34	524920	59.3	974167	7.5	550752	66.7	449248	33490	94225	26
35	525275	59.2	974122	7.5	551152	66.6	448848	33518	94215	25
36	525630	59.1	974077	7.5	551552	66.6	448448	33545	94206	24
37	525984	59.1	974032	7.5	551952	66.5	448048	33573	94196	23
38	526339	59.0	973987	7.5	552351	66.5	447649	33600	94186	22
39	526693	59.0	973942	7.5	552750	66.5	447250	33627	94176	21
40	527046	58.9	973897	7.5	553149	66.4	446851	33655	94167	20
41	9.527400	58.9	9.973852	7.5	9.553548	66.4	10.446452	33682	94157	19
42	527753	58.8	973807	7.5	553946	66.3	446054	33710	94147	18
43	528105	58.8	973761	7.5	554344	66.3	445656	33737	94137	17
44	528458	58.7	973716	7.6	554741	66.2	445259	33764	94127	16
45	528810	58.7	973671	7.6	555139	66.2	444861	33792	94118	15
46	529161	58.6	973625	7.6	555536	66.1	444464	33819	94108	14
47	529513	58.6	973580	7.6	555933	66.1	444067	33846	94098	13
48	529864	58.5	973535	7.6	556329	66.0	443671	33874	94088	12
49	530215	58.5	973489	7.6	556725	66.0	443275	33901	94078	11
50	530565	58.4	973444	7.6	557121	65.9	442879	33929	94068	10
51	9.530915	58.4	9.973398	7.6	9.557517	65.9	10.442483	33956	94058	9
52	531265	58.3	973352	7.6	557913	65.9	442087	33983	94049	8
53	531614	58.2	973307	7.6	558308	65.8	441692	34011	94039	7
54	531963	58.2	973261	7.6	558702	65.8	441298	34038	94029	6
55	532312	58.1	973215	7.6	559097	65.7	440903	34065	94019	5
56	532661	58.1	973169	7.6	559491	65.7	440509	34093	94009	4
57	533009	58.0	973124	7.6	559885	65.6	440115	34120	93999	3
58	533357	58.0	973078	7.6	560279	65.6	439721	34147	93989	2
59	533704	57.9	973032	7.7	560673	65.5	439327	34175	93979	1
60	534052		972986		561066		438934	34202	93969	0
	Cosine.		Sine.		Cotang.		Tang.	N. cos.	N.sine.	′

70 Degrees.

′	Sine.	D. 10″	Cosine.	D. 10″	Tang.	D. 10″	Cotang.	N. sine.	N. cos.	
0	9.534052	57.8	9.972986	7.7	9.561066	65.5	10.438934	34202	93969	60
1	534399	57.7	972940	7.7	561459	65.4	438541	34229	93959	59
2	534745	57.7	972894	7.7	561851	65.4	438149	34257	93949	58
3	535092	57.7	972848	7.7	562244	65.3	437756	34284	93939	57
4	535438	57.6	972802	7.7	562636	65.3	437364	34311	93929	56
5	535783	57.6	972755	7.7	563028	65.3	436972	34339	93919	55
6	536129	57.5	972709	7.7	563419	65.2	436581	34366	93909	54
7	536474	57.4	972663	7.7	563811	65.2	436189	34393	93899	53
8	536818	57.4	972617	7.7	564202	65.1	435798	34421	93889	52
9	537163	57.3	972570	7.7	564592	65.1	435408	34448	93879	51
10	537507	57.3	972524	7.7	564983	65.0	435017	34475	93869	50
11	9.537851	57.2	9.972478	7.7	9.565373	65.0	10.434627	34503	93859	49
12	538194	57.2	972431	7.8	565763	64.9	434237	34530	93849	48
13	538538	57.1	972385	7.8	566153	64.9	433847	34557	93839	47
14	538880	57.1	972338	7.8	566542	64.9	433458	34584	93829	46
15	539223	57.0	972291	7.8	566932	64.8	433068	34612	93819	45
16	539565	57.0	972245	7.8	567320	64.8	432680	34639	93809	44
17	539907	56.9	972198	7.8	567709	64.7	432291	34666	93799	43
18	540249	56.9	972151	7.8	568098	64.7	431902	34694	93789	42
19	540590	56.8	972105	7.8	568486	64.6	431514	34721	93779	41
20	540931	56.8	972058	7.8	568873	64.6	431127	34748	93769	40
21	9.541272	56.7	9.972011	7.8	9.569261	64.5	10.430739	34775	93759	39
22	541613	56.7	971964	7.8	569648	64.5	430352	34803	93748	38
23	541953	56.6	971917	7.8	570035	64.5	429965	34830	93738	37
24	542293	56.6	971870	7.8	570422	64.4	429578	34857	93728	36
25	542632	56.5	971823	7.8	570809	64.4	429191	34884	93718	35
26	542971	56.5	971776	7.8	571195	64.3	428805	34912	93708	34
27	543310	56.4	971729	7.9	571581	64.3	428419	34939	93698	33
28	543649	56.4	971682	7.9	571967	64.2	428033	34966	93688	32
29	543987	56.3	971635	7.9	572352	64.2	427648	34993	93677	31
30	544325	56.3	971588	7.9	572738	64.2	427262	35021	93667	30
31	9.544663	56.2	9.971540	7.9	9.573123	64.1	10.426877	35048	93657	29
32	545000	56.2	971493	7.9	573507	64.1	426493	35075	93647	28
33	545338	56.1	971446	7.9	573892	64.0	426108	35102	93637	27
34	545674	56.1	971398	7.9	574276	64.0	425724	35130	93626	26
35	546011	56.0	971351	7.9	574660	63.9	425340	35157	93616	25
36	546347	56.0	971303	7.9	575044	63.9	424956	35184	93606	24
37	546683	55.9	971256	7.9	575427	63.9	424573	35211	93596	23
38	547019	55.9	971208	7.9	575810	63.8	424190	35239	93585	22
39	547354	55.8	971161	7.9	576193	63.8	423807	35266	93575	21
40	547689	55.8	971113	7.9	576576	63.7	423424	35293	93565	20
41	9.548024	55.7	9.971066	8.0	9.576958	63.7	10.423041	35320	93555	19
42	548359	55.7	971018	8.0	577341	63.6	422659	35347	93544	18
43	548693	55.6	970970	8.0	577723	63.6	422277	35375	93534	17
44	549027	55.6	970922	8.0	578104	63.6	421896	35402	93524	16
45	549360	55.5	970874	8.0	578486	63.5	421514	35429	93514	15
46	549693	55.5	970827	8.0	578867	63.5	421133	35456	93503	14
47	550026	55.4	970779	8.0	579248	63.4	420752	35484	93493	13
48	550359	55.4	970731	8.0	579629	63.4	420371	35511	93483	12
49	550692	55.3	970683	8.0	580009	63.4	419991	35538	93472	11
50	551024	55.3	970635	8.0	580389	63.3	419611	35565	93462	10
51	9.551356	55.2	9.970586	8.0	9.580769	63.3	10.419231	35592	93452	9
52	551687	55.2	970538	8.0	581149	63.2	418851	35619	93441	8
53	552018	55.2	970490	8.0	581528	63.2	418472	35647	93431	7
54	552349	55.1	970442	8.0	581907	63.2	418093	35674	93420	6
55	552680	55.1	970394	8.0	582286	63.1	417714	35701	93410	5
56	553010	55.0	970345	8.1	582665	63.1	417335	35728	93400	4
57	553341	55.0	970297	8.1	583043	63.0	416957	35755	93389	3
58	553670	54.9	970249	8.1	583422	63.0	416578	35782	93379	2
59	554000	54.9	970200	8.1	583800	62.9	416200	35810	93368	1
60	554329		970152		584177		415823	35837	93358	0
	Cosine.		Sine.		Cotang.		Tang.	N. cos.	N.sine.	′

69 Degrees.

′	Sine.	D. 10″	Cosine.	D. 10″	Tang.	D. 10″	Cotang.	N.sine.	N. cos.	
0	9.554329	54.8	9.970152	8.1	9.584177	62.9	10.415823	35837	93358	60
1	554658	54.8	970103	8.1	584555	62.9	415445	35864	93348	59
2	554987	54.7	970055	8.1	584932	62.8	415068	35891	93337	58
3	555315	54.7	970006	8.1	585309	62.8	414691	35918	93327	57
4	555643	54.6	969957	8.1	585686	62.7	414314	35945	93316	56
5	555971	54.6	969909	8.1	586062	62.7	413938	35973	93306	55
6	556299	54.5	969860	8.1	586439	62.7	413561	36000	93295	54
7	556626	54.5	969811	8.1	586815	62.6	413185	36027	93285	53
8	556953	54.4	969762	8.1	587190	62.6	412810	36054	93274	52
9	557280	54.4	969714	8.1	587566	62.5	412434	36081	93264	51
10	557606	54.3	969665	8.1	587941	62.5	412059	36108	93253	50
11	9.557932	54.3	9.969616	8.2	9.588316	62.5	10.411684	36135	93243	49
12	558258	54.3	969567	8.2	588691	62.4	411309	36162	93232	48
13	558583	54.2	969518	8.2	589066	62.4	410934	36190	93222	47
14	558909	54.2	969469	8.2	589440	62.3	410560	36217	93211	46
15	559234	54.1	969420	8.2	589814	62.3	410186	36244	93201	45
16	559558	54.1	969370	8.2	590188	62.3	409812	36271	93190	44
17	559883	54.0	969321	8.2	590562	62.2	409438	36298	93180	43
18	560207	54.0	969272	8.2	590935	62.2	409065	36325	93169	42
19	560531	53.9	969223	8.2	591308	62.2	408692	36352	93159	41
20	560855	53.9	969173	8.2	591681	62.1	408319	36379	93148	40
21	9.561178	53.8	9.969124	8.2	9.592054	62.1	10.407946	36406	93137	39
22	561501	53.8	969075	8.2	592426	62.0	407574	36434	93127	38
23	561824	53.7	969025	8.2	592798	62.0	407202	36461	93116	37
24	562146	53.7	968976	8.2	593170	61.9	406829	36488	93106	36
25	562468	53.6	968926	8.3	593542	61.9	406458	36515	93095	35
26	562790	53.6	968877	8.3	593914	61.8	406086	36542	93084	34
27	563112	53.6	968827	8.3	594285	61.8	405715	36569	93074	33
28	563433	53.5	968777	8.3	594656	61.8	405344	36596	93063	32
29	563755	53.5	968728	8.3	595027	61.7	404973	36623	93052	31
30	564075	53.4	968678	8.3	595398	61.7	404602	36650	93042	30
31	9.564396	53.4	9.968628	8.3	9.595768	61.7	10.404232	36677	93031	29
32	564716	53.3	968578	8.3	596138	61.6	403862	36704	93020	28
33	565036	53.3	968528	8.3	596508	61.6	403492	36731	93010	27
34	565356	53.2	968479	8.3	596878	61.6	403122	36758	92999	26
35	565676	53.2	968429	8.3	597247	61.5	402753	36785	92988	25
36	565995	53.1	968379	8.3	597616	61.5	402384	36812	92978	24
37	566314	53.1	968329	8.3	597985	61.5	402015	36839	92967	23
38	566632	53.1	968278	8.3	598354	61.4	401646	36867	92956	22
39	566951	53.0	968228	8.4	598722	61.4	401278	36894	92945	21
40	567269	53.0	968178	8.4	599091	61.3	400909	36921	92935	20
41	9.567587	52.9	9.968128	8.4	9.599459	61.3	10.400541	36948	92926	19
42	567904	52.9	968078	8.4	599827	61.3	400173	36975	92913	18
43	568222	52.8	968027	8.4	600194	61.2	399806	37002	92902	17
44	568539	52.8	967977	8.4	600562	61.2	399438	37029	92892	16
45	568856	52.8	967927	8.4	600929	61.1	399071	37056	92881	15
46	569172	52.7	967876	8.4	601296	61.1	398704	37083	92870	14
47	569488	52.7	967826	8.4	601662	61.1	398338	37110	92859	13
48	569804	52.6	967775	8.4	602029	61.0	397971	37137	92849	12
49	570120	52.6	967725	8.4	602395	61.0	397605	37164	92838	11
50	570435	52.5	967674	8.4	602761	61.0	397239	37191	92827	10
51	9.570751	52.5	9.967624	8.4	9.603127	60.9	10.396873	37218	92816	9
52	571066	52.4	967573	8.4	603493	60.9	396507	37245	92805	8
53	571380	52.4	967522	8.5	603858	60.9	396142	37272	92794	7
54	571695	52.3	967471	8.5	604223	60.8	395777	37299	92784	6
55	572009	52.3	967421	8.5	604588	60.8	395412	37326	92773	5
56	572323	52.3	967370	8.5	604953	60.7	395047	37353	92762	4
57	572636	52.2	967319	8.5	605317	60.7	394683	37380	92751	3
58	572950	52.2	967268	8.5	605682	60.7	394318	37407	92740	2
59	573263	52.1	967217	8.5	606046	60.6	393954	37434	92729	1
60	573575		967166		606410		393590	37461	92718	0
	Cosine.		Sine.		Cotang.		Tang.	N. cos.	N.Sine.	′

68 Degrees.

′	Sine.	D. 10″	Cosine.	D. 10″	Tang.	D. 10″	Cotang.	N. sine.	N. cos.	
0	9.573575	52.1	9.967166	8.5	9.606410	60.6	10.393590	37461	92718	60
1	573888	52.0	967115	8.5	606773	60.6	393227	37488	92707	59
2	574200	52.0	967064	8.5	607137	60.5	392863	37515	92697	58
3	574512	51.9	967013	8.5	607500	60.5	392500	37542	92686	57
4	574824	51.9	966961	8.5	607863	60.4	392137	37569	92675	56
5	575136	51.9	966910	8.5	608225	60.4	391775	37595	92664	55
6	575447	51.8	966859	8.5	608588	60.4	391412	37622	92653	54
7	575758	51.8	966808	8.5	608950	60.3	391050	37649	92642	53
8	576069	51.7	966756	8.6	609312	60.3	390688	37676	92631	52
9	576379	51.7	966705	8.6	609674	60.3	390326	37703	92620	51
10	576689	51.6	966653	8.6	610036	60.2	389964	37730	92609	50
11	9.576999	51.6	9.966602	8.6	9.610397	60.2	10.389603	37757	92598	49
12	577309	51.6	966550	8.6	610759	60.2	389241	37784	92587	48
13	577618	51.5	966499	8.6	611120	60.1	388880	37811	92576	47
14	577927	51.5	966447	8.6	611480	60.1	388520	37838	92565	46
15	578236	51.4	966395	8.6	611841	60.1	388159	37865	92554	45
16	578545	51.4	966344	8.6	612201	60.0	387799	37892	92543	44
17	578853	51.3	966292	8.6	612561	60.0	387439	37919	92532	43
18	579162	51.3	966240	8.6	612921	60.0	387079	37946	92521	42
19	579470	51.3	966188	8.6	613281	59.9	386719	37973	92510	41
20	579777	51.2	966136	8.6	613641	59.9	386359	37999	92499	40
21	9.580085	51.2	9.966085	8.7	9.614000	59.8	10.386000	38026	92488	39
22	580392	51.1	966033	8.7	614359	59.8	385641	38053	92477	38
23	580699	51.1	965981	8.7	614718	59.8	385282	38080	92466	37
24	581005	51.1	965928	8,7	615077	59.7	384923	38107	92455	36
25	581312	51.0	965876	8.7	615435	59.7	384565	38134	92444	35
26	581618	51.0	965824	8.7	615793	59.7	384207	38161	92432	34
27	581924	50.9	965772	8.7	616151	59.6	383849	38188	92421	33
28	582229	50.9	965720	8.7	616509	59.6	383491	38215	92410	32
29	582535	50.9	965668	8.7	616867	59.6	383133	38241	92399	31
30	582840	50.8	965615	8.7	617224	59.5	382776	38268	92388	30
31	9.583145	50.8	9.965563	8.7	9.617582	59.5	10.382418	38295	92377	29
32	583449	50.7	965511	8.7	617939	59.5	382061	38322	92366	28
33	583754	50.7	965458	8.7	618295	59.4	381705	38349	92355	27
34	584058	50.6	965406	8.7	618652	59.4	381348	38376	92343	26
35	584361	50.6	965353	8.8	619008	59.4	380992	38403	92332	25
36	584665	50.6	965301	8.8	619364	59.3	380636	38430	92321	24
37	584968	50.5	965248	8.8	619721	59.3	380279	38456	92310	23
38	585272	50.5	965195	8.8	620076	59.3	379924	38483	92299	22
39	585574	50.4	965143	8.8	620432	59.2	379568	38510	92287	21
40	585877	50.4	965090	8.8	620787	59.2	379213	38537	92276	20
41	9.586179	50.3	9.965037	8.8	9.621142	59.2	10.378858	38564	92265	19
42	586482	50.3	964984	8.8	621497	59.1	378503	38591	92254	18
43	586783	50.3	964931	8.8	621852	59.1	378148	38617	92243	17
44	587085	50.2	964879	8.8	622207	59.0	377793	38644	92231	16
45	587386	50.2	964826	8.8	622561	59.0	377439	38671	92220	15
46	587688	50.1	964773	8.8	622915	59.0	377085	38698	92209	14
47	587989	50.1	964719	8.8	623269	58.9	376731	38725	92198	13
48	588289	50.1	964666	8.9	623623	58.9	376377	38752	92186	12
49	588590	50.0	964613	8.9	623976	58.9	376024	38778	92175	11
50	588890	50.0	964560	8.9	624330	58.8	375670	38805	92164	10
51	9.589190	49.9	9.964507	8.9	9.624683	58.8	10.375317	38832	92152	9
52	589489	49.9	964454	8.9	625036	58.8	374964	38859	92141	8
53	589789	49.9	964400	8.9	625388	58.7	374612	38886	92130	7
54	590088	49.8	964347	8.9	625741	58.7	374259	38912	92119	6
55	590387	49.8	964294	8.9	626093	58.7	373907	38939	92107	5
56	590686	49.7	964240	8.9	626445	58.6	373555	38966	92096	4
57	590984	49.7	964187	8.9	626797	58.6	373203	38993	92085	3
58	591282	49.7	964133	8.9	627149	58.6	372851	39020	92073	2
59	591580	49.6	964080	8.9	627501	58.5	372499	39046	92062	1
60	591878		964026		627852		372148	39073	92050	0
	Cosine.		Sine.		Cotang.		Tang.	N. cos.	N.sine.	′

′	Sine.	D. 10″	Cosine.	D. 10″	Tang.	D. 10″	Cotang.	N. sine.	N. cos.	
0	9.591878	49.6	9.964026	8.9	9.627852	58.5	10.372148	39073	92050	60
1	592176	49.5	963972	8.9	628203	58.5	371797	39100	92039	59
2	592473	49.5	963919	8.9	628554	58.5	371446	39127	92028	58
3	592770	49.5	963865	9.0	628905	58.4	371095	39153	92016	57
4	593067	49.4	963811	9.0	629255	58.4	370745	39180	92005	56
5	593363	49.4	963757	9.0	629606	58.3	370394	39207	91994	55
6	593659	49.3	963704	9.0	629956	58.3	370044	39234	91982	54
7	593955	49.3	963650	9.0	630306	58.3	369694	39260	91971	53
8	594251	49.3	963596	9.0	630656	58.3	369344	39287	91959	52
9	594547	49.2	963542	9.0	631005	58.2	368995	39314	91948	51
10	594842	49.2	963488	9.0	631355	58.2	368645	39341	91936	50
11	9.595137	49.1	9.963434	9.0	9.631704	58.2	10.368296	39367	91925	49
12	595432	49.1	963379	9.0	632053	58.1	367947	39394	91914	48
13	595727	49.1	963325	9.0	632401	58.1	367599	39421	91902	47
14	596021	49.0	963271	9.0	632750	58.1	367250	39448	91891	46
15	596315	49.0	963217	9.0	633098	58.0	366902	39474	91879	45
16	596609	48.9	963163	9.0	633447	58.0	366553	39501	91868	44
17	596903	48.9	963108	9.1	633795	58.0	366205	39528	91856	43
18	597196	48.9	963054	9.1	634143	57.9	365857	39555	91845	42
19	597490	48.8	962999	9.1	634490	57.9	365510	39581	91833	41
20	597783	48.8	962945	9.1	634838	57.9	365162	39608	91822	40
21	9.598075	48.7	9.962890	9.1	9.635185	57.8	10.364815	39635	91810	39
22	598368	48.7	962836	9.1	635532	57.8	364468	39661	91799	38
23	598660	48.7	962781	9.1	635879	57.8	364121	39688	91787	37
24	598952	48.6	962727	9.1	636226	57.7	363774	39715	91775	36
25	599244	48.6	962672	9.1	636572	57.7	363428	39741	91764	35
26	599536	48.5	962617	9.1	636919	57.7	363081	39768	91752	34
27	599827	48.5	962562	9.1	637265	57.7	362735	39795	91741	33
28	600118	48.5	962508	9.1	637611	57.6	362389	39822	91729	32
29	600409	48.4	962453	9.1	637956	57.6	362044	39848	91718	31
30	600700	48.4	962398	9.2	638302	57.6	361698	39875	91706	30
31	9.600990	48.4	9.962343	9.2	9.638647	57.5	10.361353	39902	91694	29
32	601280	48.3	962288	9.2	638992	57.5	361008	39928	91683	28
33	601570	48.3	962233	9.2	639337	57.5	360663	39955	91671	27
34	601860	48.2	962178	9.2	639682	57.4	360318	39982	91660	26
35	602150	48.2	962123	9.2	640027	57.4	359973	40008	91648	25
36	602439	48.2	962067	9.2	640371	57.4	359629	40035	91636	24
37	602728	48.1	962012	9.2	640716	57.3	359284	40062	91625	23
38	603017	48.1	961957	9.2	641060	57.3	358940	40088	91613	22
39	603305	48.1	961902	9.2	641404	57.3	358596	40115	91601	21
40	603594	48.0	961846	9.2	641747	57.2	358253	40141	91590	20
41	9.603882	48.0	9.961791	9.2	9.642091	57.2	10.357909	40168	91578	19
42	604170	47.9	961735	9.2	642434	57.2	357566	40195	91566	18
43	604457	47.9	961680	9.2	642777	57.2	357223	40221	91555	17
44	604745	47.9	961624	9.3	643120	57.1	356880	40248	91543	16
45	605032	47.8	961569	9.3	643463	57.1	356537	40275	91531	15
46	605319	47.8	961513	9.3	643806	57.1	356194	40301	91519	14
47	605606	47.8	961458	9.3	644148	57.0	355852	40328	91508	13
48	605892	47.7	961402	9.3	644490	57.0	355510	40355	91496	12
49	606179	47.7	961346	9.3	644832	57.0	355168	40381	91484	11
50	606465	47.6	961290	9.3	645174	56.9	354826	40408	91472	10
51	9.606751	47.6	9.961235	9.3	9.645516	56.9	10.354484	40434	91461	9
52	607036	47.6	961179	9.3	645857	56.9	354143	40461	91449	8
53	607322	47.5	961123	9.3	646199	56.9	353801	40488	91437	7
54	607607	47.5	961067	9.3	646540	56.8	353460	40514	91425	6
55	607892	47.4	961011	9.3	646881	56.8	353119	40541	91414	5
56	608177	47.4	960955	9.3	647222	56.8	352778	40567	91402	4
57	608461	47.4	960899	9.3	647562	56.7	352438	40594	91390	3
58	608745	47.3	960843	9.4	647903	56.7	352097	40621	91378	2
59	609029	47.3	960786	9.4	648243	56.7	351757	40647	91366	1
60	609313		960730		648583		351417	40674	91355	0
	Cosine.		Sine.		Cotang.		Tang.	N. cos.	N.sine.	′

66 Degrees.

′	Sine.	D. 10″	Cosine.	D. 10″	Tang.	D. 10″	Cotang.	N. sine.	N. cos.	
0	9.609313	47.3	9.960730	9.4	9.648583	56.6	10.351417	40674	91355	60
1	609597	47.2	960674	9.4	648923	56.6	351077	40700	91343	59
2	609880	47.2	960618	9.4	649263	56.6	350737	40727	91331	58
3	610164	47.2	960561	9.4	649602	56.6	350398	40753	91319	57
4	610447	47.1	960505	9.4	649942	56.5	350058	40780	91307	56
5	610729	47.1	960448	9.4	650281	56.5	349719	40806	91295	55
6	611012	47.0	960392	9.4	650620	59.5	349380	40833	91283	54
7	611294	47.0	960335	9.4	650959	56.4	349041	40860	91272	53
8	611576	47.0	960279	9.4	651297	56.4	348703	40886	91260	52
9	611858	46.9	960222	9.4	651636	56.4	348364	40913	91248	51
10	612140	46.9	960165	9.4	651974	56.3	348026	40939	91236	50
11	9.612421	46.9	9.960109	9.5	9.652312	56.3	10.347688	40966	91224	49
12	612702	46.8	960052	9.5	652650	56.3	347350	40992	91212	48
13	612983	46.8	959995	9.5	652988	56.3	347012	41019	91200	47
14	613264	46.7	959938	9.5	653326	56.2	346674	41045	91188	46
15	613545	46.7	959882	9.5	653663	56.2	346337	41072	91176	45
16	613825	46.7	959825	9.5	654000	56.2	346000	41098	91164	44
17	614105	46.6	959768	9.5	654337	56.1	345663	41125	91152	43
18	614385	46.6	959711	9.5	654174	56.1	345326	41151	91140	42
19	614665	46.6	959654	9.5	655011	56.1	344989	41178	91128	41
20	614944	46.5	959596	9.5	655348	56.1	344652	41204	91116	40
21	9.615223	46.5	9.959539	9.5	9.655684	56.0	10.344316	41231	91104	39
22	615502	46.5	959482	9.5	656020	56.0	343980	41257	91092	38
23	615781	46.4	959425	9.5	656356	56.0	343644	41284	91080	37
24	616060	46.4	959368	9.5	656692	55.9	343308	41310	91068	36
25	616338	46.4	959310	9.6	657028	55.9	342972	41337	91056	35
26	616616	46.3	959253	9.6	657364	55.9	342636	41363	91044	34
27	616894	46.3	959195	9.6	657699	55.9	342301	41390	91032	33
28	617172	46.2	959138	9.6	658034	55.8	341966	41416	91020	32
29	617450	46.2	959081	9.6	658369	55.8	341631	41443	91008	31
30	617727	46.2	959023	9.6	658704	55.8	341296	41469	90996	30
31	9.618004	46.1	9.958965	9.6	9.659039	55.8	10.340961	41496	90984	29
32	618281	46.1	958908	9.6	659373	55.7	340627	41522	90972	28
33	618558	46.1	958850	9.6	659708	55.7	340292	41549	90960	27
34	618834	46.0	958792	9.6	660042	55.7	339958	41575	90948	26
35	619110	46.0	958734	9.6	660376	55.7	339624	41602	90936	25
36	619386	46.0	958577	9.6	660710	55.6	339290	41628	90924	24
37	619662	45.9	958619	9.6	661043	55.6	338957	41655	90911	23
38	619938	45.9	958561	9.6	661377	55.6	338623	41681	90899	22
39	620213	45.9	958503	9.7	661710	55.5	338290	41707	90887	21
40	620488	45.8	958445	9.7	662043	55.5	337957	41734	90875	20
41	9.620763	45.8	9.958387	9.7	9.662376	55.5	10.337624	41760	90863	19
42	621038	45.7	958329	9.7	662709	55.4	337291	41787	90851	18
43	621313	45.7	958271	9.7	663042	55.4	336958	41813	90839	17
44	621587	45.7	958213	9.7	663375	55.4	336625	41840	90826	16
45	621861	45.6	958154	9.7	663707	55.4	336293	41866	90814	15
46	622135	45.6	958096	9.7	664039	55.3	335961	41892	90802	14
47	622409	45.6	958038	9.7	664371	55.3	335629	41919	90790	13
48	622682	45.5	957979	9.7	664703	55.3	335297	41945	90778	12
49	622956	45.5	957921	9.7	665035	55.3	334965	41972	90766	11
50	623229	45.5	957863	9.7	665366	55.2	334634	41998	90753	10
51	9.623512	45.4	9.957804	9.7	9.665697	55.2	10.334303	42024	90741	9
52	623774	45.4	957746	9.8	666029	55.2	333971	42051	90729	8
53	624047	45.4	957687	9.8	666360	55.1	333620	42077	90717	7
54	624319	45.3	957628	9.8	666691	55.1	333309	42104	90704	6
55	624591	45.3	957570	9.8	667021	55.1	332979	42130	90692	5
56	624863	46.3	957511	9.8	667352	55.1	332648	42156	90680	4
57	625135	45.2	957452	9.8	667682	55.0	332318	42183	90668	3
58	625406	45.2	957393	9.8	668013	55.0	331987	42209	90655	2
59	625677	45.2	957335	9.8	668343	55.0	331657	42235	90643	1
60	625948		957276		668672		331328	42262	90631	0
	Cosine.		Sine.		Cotang.		Tang.	N. cos.	N.sine.	′

′	Sine.	D. 10″	Cosine.	D. 10″	Tang.	D. 10″	Cotang.	N. sine.	N. cos.	
0	9.625948	45.1	9.957276	9.8	9.668673	55.0	10.331327	42262	90631	60
1	626219	45.1	957217	9.8	669002	54.9	330998	42288	90613	59
2	626490	45.1	957158	9.8	669332	54.9	330668	42315	90606	58
3	626760	45.0	957099	9.8	669661	54.9	330339	42341	90594	57
4	627030	45.0	957040	9.8	669991	54.8	330009	42367	90582	56
5	627300	45.0	956981	9.8	670320	54.8	329680	42394	90569	55
6	627570	44.9	956921	9.9	670649	54.8	329351	42420	90557	54
7	627840	44.9	956862	9.9	670977	54.8	329023	42446	90545	53
8	628109	44.9	956803	9.9	671306	54.7	328694	42473	90532	52
9	628378	44.8	956744	9.9	671634	54.7	328366	42499	90520	51
10	628647	44.8	956684	9.9	671963	54.7	328037	42525	90507	50
11	9.628916	44.7	9.956625	9.9	9.672291	54.7	10.327709	42552	90495	49
12	629185	44.7	956566	9.9	672619	54.6	327381	42578	90483	48
13	629453	44.7	956506	9.9	672947	54.6	327053	42604	90470	47
14	629721	44.6	956447	9.9	673274	54.6	326726	42631	90458	46
15	629989	44.6	956387	9.9	673602	54.6	326398	42657	90446	45
16	630257	44.6	956327	9.9	673929	54.5	326071	42683	90433	44
17	630524	44.6	956268	9.9	674257	54.5	325743	42709	90421	43
18	630792	44.5	956208	10.0	674584	54.5	325416	42736	90408	42
19	631059	44.5	956148	10.0	674910	54.4	325090	42762	90396	41
20	631326	44.5	956089	10.0	675237	54.4	324763	42788	90383	40
21	9.631593	44.4	9.956029	10.0	9.675564	54.4	10.324436	42815	90371	39
22	631859	44.4	955969	10.0	675890	54.4	324110	42841	90358	38
23	632125	44.4	955909	10.0	676216	54.3	323784	42867	90346	37
24	632392	44.3	955849	10.0	676543	54.3	323457	42894	90334	36
25	632658	44.3	955789	10.0	676869	54.3	323131	42920	90321	35
26	632923	44.3	955729	10.0	677194	54.3	322806	42946	90309	34
27	633189	44.2	955669	10.0	677520	54.2	322480	42972	90296	33
28	633454	44.2	955609	10.0	677846	54.2	322154	42999	90284	32
29	633719	44.2	955548	10.0	678171	54.2	321829	43025	90271	31
30	633984	44.1	955488	10.0	678496	54.2	321504	43051	90259	30
31	9.634249	44.1	9.955428	10.1	9.678821	54.1	10.321179	43077	90246	29
32	634514	44.0	955368	10.1	679146	54.1	320854	43104	90233	28
33	634778	44.0	955307	10.1	679471	54.1	320529	43130	90221	27
34	635042	44.0	955247	10.1	679795	54.1	320205	43156	90208	26
35	635306	43.9	955186	10.1	680120	54.0	319880	43182	90196	25
36	635570	43.9	955126	10.1	680444	54.0	319556	43209	90183	24
37	635834	43.9	955065	10.1	680768	54.0	319232	43235	90171	23
38	636097	43.8	955005	10.1	681092	54.0	318908	43261	90158	22
39	636360	43.8	954944	10.1	681416	53.9	318584	43287	90146	21
40	636623	43.8	954883	10.1	681740	53.9	318260	43313	90133	20
41	9.636886	43.7	9.954823	10.1	9.682063	53.9	10.317937	43340	90120	19
42	637148	43.7	954762	10.1	682387	53.9	317613	43366	90108	18
43	637411	43.7	954701	10.1	682710	53.8	317290	43392	90095	17
44	637673	43.7	954640	10.1	683033	53.8	316967	43418	90082	16
45	637935	43.6	954579	10.1	683356	53.8	316644	43445	90070	15
46	638197	43.6	954518	10.2	683679	53.8	316321	43471	90057	14
47	638458	43.6	954457	10.2	684001	53.7	315999	43497	90045	13
48	638720	43.5	954396	10.2	684324	53.7	315676	43523	90032	12
49	638981	43.5	954335	10.2	684646	53.7	315354	43549	90019	11
50	639242	43.5	954274	10.2	684968	53.7	315032	43575	90007	10
51	9.639503	43.4	9.954213	10.2	9.685290	53.6	10.314710	43602	89994	9
52	639764	43.4	954152	10.2	685612	53.6	314388	43628	89981	8
53	640024	43.4	954090	10.2	685934	53.6	314066	43654	89968	7
54	640284	43.3	954029	10.2	686255	53.6	313745	43680	89956	6
55	640544	43.3	953968	10.2	686577	53.5	313423	43706	89943	5
56	640804	43.3	953906	10.2	686898	53.5	313102	43733	89930	4
57	641064	43.2	953845	10.2	687219	53.5	312781	43759	89918	3
58	641324	43.2	953783	10.2	687540	53.5	312460	43785	89905	2
59	641584	43.2	953722	10.3	687861	53.4	312139	43811	89892	1
60	641842		953660		688182		311818	43837	89879	0
	Cosine.		Sine.		Cotang.		Tang.	N. cos.	N. sine.	′

64 Degrees.

′	Sine.	D. 10″	Cosine.	D. 10″	Tang.	D. 10″	Cotang.	N. sine.	N. cos.	
0	9.641842	43.1	9.953660	10.3	9.688182	53.4	10.311818	43837	89879	60
1	642101	43.1	953599	10.3	688502	53.4	311498	43863	89867	59
2	642360	43.1	953537	10.3	688823	53.4	311177	43889	89854	58
3	642618	43.0	953475	10.3	689143	53.3	310857	43916	89841	57
4	642877	43.0	953413	10.3	689463	53.3	310537	43942	89828	56
5	643135	43.0	953352	10.3	689783	53.3	310217	43968	89816	55
6	643393	43.0	953290	10.3	690103	53.3	309897	43994	89803	54
7	643650	42.9	953228	10.3	690423	53.3	309577	44020	89790	53
8	643908	42.9	953166	10.3	690742	53.2	309258	44046	89777	52
9	644165	42.9	953104	10.3	691062	53.2	308938	44072	89764	51
10	644423	42.8	953042	10.3	691381	53.2	308619	44098	89752	50
11	9.644680	42.8	9.952980	10.4	9.691700	53.1	10.308300	44124	89739	49
12	644936	42.8	952918	10.4	692019	53.1	307981	44151	89726	48
13	645193	42.7	952855	10.4	692338	53.1	307662	44177	89713	47
14	645450	42.7	952793	10.4	692656	53.1	307344	44203	89700	46
15	645706	42.7	952731	10.4	692975	53.1	307025	44229	89687	45
16	645962	42.6	952669	10.4	693293	53.0	306707	44255	89674	44
17	646218	42.6	952606	10.4	693612	53.0	306388	44281	89662	43
18	646474	42.6	952544	10.4	693930	53.0	306070	44307	89649	42
19	646729	42.5	952481	10.4	694248	53.0	305752	44333	89636	41
20	646984	42.5	952419	10.4	694566	52.9	305434	44359	89623	40
21	9.647240	42.5	9.952356	10.4	9.694883	52.9	10.305117	44385	89610	39
22	647494	42.4	952294	10.4	695201	52.9	304799	44411	89597	38
23	647749	42.4	952231	10.4	695518	52.9	304482	44437	89584	37
24	648004	42.4	952168	10.5	695836	52.9	304164	44464	89571	36
25	648258	42.4	952106	10.5	696153	52.8	303847	44490	89558	35
26	648512	42.3	952043	10.5	696470	52.8	303530	44516	89545	34
27	648766	42.3	951980	10.5	696787	52.8	303213	44542	89532	33
28	649020	42.3	951917	10.5	697103	52.8	302897	44568	89519	32
29	649274	42.2	951854	10.5	697420	52.7	302580	44594	89506	31
30	649527	42.2	951791	10.5	697736	52.7	302264	44620	89493	30
31	9.649781	42.2	9.951728	10.5	9.698053	52.7	10.301947	44646	89480	29
32	650034	42.2	951665	10.5	698369	52.7	301631	44672	89467	28
33	650287	42.1	951602	10.5	698685	52.6	301315	44698	89454	27
34	650539	42.1	951539	10.5	699001	52.6	300999	44724	89441	26
35	650792	42.1	951476	10.5	699316	52.6	300684	44750	89428	25
36	651044	42.0	951412	10.5	699632	52.6	300368	44776	89415	24
37	651297	42.0	951349	10.6	699947	52.6	300053	44802	89402	23
38	651549	42.0	951286	10.6	700263	52.5	299737	44828	89389	22
39	651800	41.9	951222	10.6	700578	52.5	299422	44854	89376	21
40	652052	41.9	951159	10.6	700893	52.5	299107	44880	89363	20
41	9.652304	41.9	9.951096	10.6	9.701208	52.4	10.298792	44906	89350	19
42	652555	41.8	951032	10.6	701523	52.4	298477	44932	89337	18
43	652806	41.8	950968	10.6	701837	52.4	298163	44958	89324	17
44	653057	41.8	950905	10.6	702152	52.4	297848	44984	89311	16
45	653308	41.8	950841	10.6	702466	52.4	297534	45010	89298	15
46	653558	41.7	950778	10.6	702780	52.3	297220	45036	89285	14
47	653808	41.7	950714	10.6	703095	52.3	296905	45062	89272	13
48	654059	41.7	950650	10.6	703409	52.3	296591	45088	89259	12
49	654309	41.6	950586	10.6	703723	52.3	296277	45114	89245	11
50	654558	41.6	950522	10.7	704036	52.2	295964	45140	89232	10
51	9.654808	41.6	9.950458	10.7	9.704350	52.2	10.295650	45166	89219	9
52	655058	41.6	950394	10.7	704663	52.2	295337	45192	89206	8
53	655307	41.5	950330	10.7	704977	52.2	295023	45218	89193	7
54	655556	41.5	950366	10.7	705290	52.2	294710	45243	89180	6
55	655805	41.5	950202	10.7	705603	52.1	294397	45269	89167	5
56	656054	41.4	950138	10.7	705916	52.1	294084	45295	89153	4
57	656302	41.4	950074	10.7	706228	52.1	293772	45321	89140	3
58	656551	41.4	950010	10.7	706541	52.1	293459	45347	89127	2
59	656799	41.3	949945	10.7	706854	52.1	293146	45373	89114	1
60	657047		949881		707166		292834	45399	89101	0
	Cosine.		Sine.		Cotang.		Tang.	N. cos.	N. sine.	′

′	Sine.	D. 10″	Cosine.	D. 10″	Tang.	D. 10″	Cotang.	N. sine.	N. cos.	
0	9.657047	41.3	9.949881	10.7	9.707166	52.0	10.292834	45399	89101	60
1	657295	41.3	949816	10.7	707478	52.0	292522	45425	89087	59
2	657542	41.2	949752	10.7	707790	52.0	292210	45451	89074	58
3	657790	41.2	949688	10.8	708102	52.0	291898	45477	89061	57
4	658037	41.2	949623	10.8	708414	51.9	291586	45503	89048	56
5	658284	41.2	949558	10.8	708726	51.9	291274	45529	89035	55
6	658531	41.1	949494	10.8	709037	51.9	290963	45554	89021	54
7	658778	41.1	949429	10.8	709349	51.9	290651	45580	89008	53
8	659025	41.1	949364	10.8	709660	51.9	290340	45606	88995	52
9	659271	41.0	949300	10.8	709971	51.8	290029	45632	88981	51
10	659517	41.0	949235	10.8	710282	51.8	289718	45658	88968	50
11	9.659763	41.0	9.949170	10.8	9.710593	51.8	10.289407	45684	88955	49
12	660009	40.9	949105	10.8	710904	51.8	289096	45710	88942	48
13	660255	40.9	949040	10.8	711215	51.8	288785	45736	88928	47
14	660501	40.9	948975	10.8	711525	51.7	288475	45762	88915	46
15	660746	40.9	948910	10.8	711836	51.7	288164	45787	88902	45
16	660991	40.8	948845	10.8	712146	51.7	287854	45813	88888	44
17	661236	40.8	948780	10.9	712456	51.7	287544	45839	88875	43
18	661481	40.8	948715	10.9	712766	51.6	287234	45865	88862	42
19	661726	40.7	948650	10.9	713076	51.6	286924	45891	88848	41
20	661970	40.7	948584	10.9	713386	51.6	286614	45917	88835	40
21	9.662214	40.7	9.948519	10.9	9.713696	51.6	10.286304	45942	88822	39
22	662459	40.7	948454	10.9	714005	51.6	285995	45968	88808	38
23	662703	40.6	948388	10.9	714314	51.5	285686	45994	88795	37
24	662946	40.6	948323	10.9	714624	51.5	285376	46020	88782	36
25	663190	40.6	948257	10.9	714933	51.5	285067	46046	88768	35
26	663433	40.5	948192	10.9	715242	51.5	284758	46072	88755	34
27	663677	40.5	948126	10.9	715551	51.4	284449	46097	88741	33
28	663920	40.5	948060	10.9	715860	51.4	284140	46123	88728	32
29	664163	40.5	947995	11.0	716168	51.4	283832	46149	88715	31
30	664406	40.4	947929	11.0	716477	51.4	283523	46175	88701	30
31	9.664648	40.4	9.947863	11.0	9.716785	51.4	10.283215	46201	88688	29
32	664891	40.4	947797	11.0	717093	51.3	282907	46226	88674	28
33	665133	40.3	947731	11.0	717401	51.3	282599	46252	88661	27
34	665375	40.3	947665	11.0	717709	51.3	282291	46278	88647	26
35	665617	40.3	947600	11.0	718017	51.3	281983	46304	88634	25
36	665859	40.2	947533	11.0	718325	51.3	281675	46330	88620	24
37	666100	40.2	947467	11.0	718633	51.2	281367	46355	88607	23
38	666342	40.2	947401	11.0	718940	51.2	281060	46381	88593	22
39	666583	40.2	947335	11.0	719248	51.2	280752	46407	88580	21
40	666824	40.1	947269	11.0	719555	51.2	280445	46433	88566	20
41	9.667065	40.1	9.947203	11.0	9.719862	51.2	10.280138	46458	88553	19
42	667305	40.1	947136	11.1	720169	51.1	279831	46484	88539	18
43	667546	40.1	947070	11.1	720476	51.1	279524	46510	88526	17
44	667786	40.0	947004	11.1	720783	51.1	279217	46536	88512	16
45	668027	40.0	946937	11.1	721089	51.1	278911	46561	88499	15
46	668267	40.0	946871	11.1	721396	51.1	278604	46587	88485	14
47	668506	39.9	946804	11.1	721702	51.0	278298	46613	88472	13
48	668746	39.9	946738	11.1	722009	51.0	277991	46639	88458	12
49	668986	39.9	946671	11.1	722315	51.0	277685	46664	88445	11
50	669225	39.9	946604	11.1	722621	51.0	277379	46690	88431	10
51	9.669464	39.8	9.946538	11.1	9.722927	51.0	10.277073	46716	88417	9
52	669703	39.8	946471	11.1	723232	50.9	276768	46742	88404	8
53	669942	39.8	946404	11.1	723538	50.9	276462	46767	88390	7
54	670181	39.7	946337	11.1	723844	50.9	276156	46793	88377	6
55	670419	39.7	946270	11.2	724149	50.9	275851	46819	88363	5
56	670658	39.7	946203	11.2	724454	50.9	275546	46844	88349	4
57	670896	39.7	946136	11.2	724759	50.8	275241	46870	88336	3
58	671134	39.6	946069	11.2	725065	50.8	274935	46896	88322	2
59	671372	39.6	946002	11.2	725369	50.8	274631	46921	88308	1
60	671609		945935		725674		274326	46947	88295	0
	Cosine.		Sine.		Cotang.		Tang.	N. cos.	N. sine.	′

62 Degrees.

′	Sine.	D. 10″	Cosine.	D. 10″	Tang.	D. 10″	Cotang.	N. sine.	N. cos.	
0	9.671609	39.6	9.945935	11.2	9.725674	50.8	10.274326	46947	88295	60
1	671847	39.5	945868	11.2	725979	50.8	274021	46973	88281	59
2	672084	39.5	945800	11.2	726284	50.7	273716	46999	88267	58
3	672321	39.5	945733	11.2	726588	50.7	273412	47024	88254	57
4	672558	39.5	945666	11.2	726892	50.7	273108	47050	88240	56
5	672795	39.4	945598	11.2	727197	50.7	272803	47076	88226	55
6	673032	39.4	945531	11.2	727501	50.7	272499	47101	88213	54
7	673268	39.4	945464	11.3	727805	50.6	272195	47127	88199	53
8	673505	39.4	945396	11.3	728109	50.6	271891	47153	88185	52
9	673741	39.3	945328	11.3	728412	50.6	271588	47178	88172	51
10	673977	39.3	945261	11.3	728716	50.6	271284	47204	88158	50
11	9.674213	39.3	9.945193	11.3	9.729020	50.6	10.270980	47229	88144	49
12	674448	39.2	945125	11.3	729323	50.5	270677	47255	88130	48
13	674684	39.2	945058	11.3	729626	50.5	270374	47281	88117	47
14	674919	39.2	944990	11.3	729929	50.5	270071	47306	88103	46
15	675155	39.2	944922	11.3	730233	50.5	269767	47332	88089	45
16	675390	39.1	944854	11.3	730535	50.5	269465	47358	88075	44
17	675624	39.1	944786	11.3	730838	50.4	269162	47383	88062	43
18	675859	39.1	944718	11.3	731141	50.4	268859	47409	88048	42
19	676094	39.1	944650	11.3	731444	50.4	268556	47434	88034	41
20	676328	39.0	944582	11.4	731746	50.4	268254	47460	88020	40
21	9.676562	39.0	9.944514	11.4	9.732048	50.4	10.267952	47486	88006	39
22	676796	39.0	944446	11.4	732351	50.3	267649	47511	87993	38
23	677030	39.0	944377	11.4	732653	50.3	267347	47537	87979	37
24	677264	38.9	944309	11.4	732955	50.3	267045	47562	87965	36
25	677498	38.9	944241	11.4	733257	50.3	266743	47588	87951	35
26	677731	38.9	944172	11.4	733558	50.3	266442	47614	87937	34
27	677964	38.8	944104	11.4	733860	50.2	266140	47639	87923	33
28	678197	38.8	944036	11.4	734162	50.2	265838	47665	87909	32
29	678430	38.8	943967	11.4	734463	50.2	265537	47690	87896	31
30	678663	38.8	943899	11.4	734764	50.2	265236	47716	87882	30
31	9.678895	38.7	9.943830	11.4	9.735066	50.2	10.264934	47741	87868	29
32	679128	38.7	943761	11.4	735367	50.2	264633	47767	87854	28
33	679360	38.7	943693	11.5	735668	50.1	264332	47793	87840	27
34	679592	38.7	943624	11.5	735969	50.1	264031	47818	87826	26
35	679824	38.6	943555	11.5	736269	50.1	263731	47844	87812	25
36	680056	38.6	943486	11.5	736570	50.1	263430	47869	87798	24
37	680288	38.6	943417	11.5	736871	50.1	263129	47895	87784	23
38	680519	38.5	943348	11.5	737171	50.0	262829	47920	87770	22
39	680750	38.5	943279	11.5	737471	50.0	262529	47946	87756	21
40	680982	38.5	943210	11.5	737771	50.0	262229	47971	87743	20
41	9.681213	38.5	9.943141	11.5	9.738071	50.0	10.261929	47997	87729	19
42	681443	38.4	943072	11.5	738371	50.0	261629	48022	87715	18
43	681674	38.4	943003	11.5	738671	49.9	261329	48048	87701	17
44	681905	38.4	942934	11.5	738971	49.9	261029	48073	87687	16
45	682135	38.4	942864	11.5	739271	49.9	260729	48099	87673	15
46	682365	38.3	942795	11.6	739570	49.9	260430	48124	87659	14
47	682595	38.3	942726	11.6	739870	49.9	260130	48150	87645	13
48	682825	38.3	942656	11.6	740169	49.9	259831	48175	87631	12
49	683055	38.3	942587	11.6	740468	49.8	259532	48201	87617	11
50	683284	38.2	942517	11.6	740767	49.8	259233	48226	87603	10
51	9.683514	38.2	9.942448	11.6	9.741066	49.8	10.258934	48252	87589	9
52	683743	38.2	942378	11.6	741365	49.8	258635	48277	87575	8
53	683972	38.2	942308	11.6	741664	49.8	258336	48303	87561	7
54	684201	38.1	942239	11.6	741962	49.7	258038	48328	87546	6
55	684430	38.1	942169	11.6	742261	49.7	257739	48354	87532	5
56	684658	38.1	942099	11.6	742559	49.7	257441	48379	87518	4
57	684887	38.0	942029	11.6	742858	49.7	257142	48405	87504	3
58	685115	38.0	941959	11.6	743156	49.7	256844	48430	87490	2
59	685343	38.0	941889	11.7	743454	49.7	256546	48456	87476	1
60	685571		941819		743752		256248	48481	87462	0
	Cosine.		Sine.		Cotang.		Tang.	N. cos.	N. sine.	′

61 Degrees.

′	Sine.	D. 10″	Cosine.	D. 10″	Tang.	D. 10″	Cotang.	N. sine.	N. cos.	
0	9.685571	38.0	9.941819	11.7	9.743752	49.6	10.256248	48481	87462	60
1	685799	37.9	941749	11.7	744050	49.6	255950	48506	87448	59
2	686027	37.9	941679	11.7	744348	49.6	255652	48532	87434	58
3	686254	37.9	941609	11.7	744645	49.6	255355	48557	87420	57
4	686482	37.9	941539	11.7	744943	49.6	255057	48583	87406	56
5	686709	37.8	941469	11.7	745240	49.6	254760	48608	87391	55
6	686936	37.8	941398	11.7	745538	49.5	254462	48634	87377	54
7	687163	37.8	941328	11.7	745835	49.5	254165	48659	87363	53
8	687389	37.8	941258	11.7	746132	49.5	253868	48684	87349	52
9	687616	37.7	941187	11.7	746429	49.5	253571	48710	87335	51
10	687843	37.7	941117	11.7	746726	49.5	253274	48735	87321	50
11	9.688069	37.7	9.941046	11.8	9.747023	49.4	10.252977	48761	87306	49
12	688295	37.7	940975	11.8	747319	49.4	252681	48786	87292	48
13	688521	37.6	940905	11.8	747616	49.4	252384	48811	87278	47
14	688747	37.6	940834	11.8	747913	49.4	252087	48837	87264	46
15	688972	37.6	940763	11.8	748209	49.4	251791	48862	87250	45
16	689198	37.6	940693	11.8	748505	49.3	251495	48888	87235	44
17	689423	37.5	940622	11.8	748801	49.3	251199	48913	87221	43
18	689648	37.5	940551	11.8	749097	49.3	250903	48938	87207	42
19	689873	37.5	940480	11.8	749393	49.3	250607	48964	87193	41
20	690098	37.5	940409	11.8	749689	49.3	250311	48989	87178	40
21	9.690323	37.4	9.940338	11.8	9.749985	49.3	10.250015	49014	87164	39
22	690548	37.4	940267	11.8	750281	49.2	249719	49040	87150	38
23	690772	37.4	940196	11.8	750576	49.2	249424	49065	87136	37
24	690996	37.4	940125	11.9	750872	49.2	249128	49090	87121	36
25	691220	37.3	940054	11.9	751167	49.2	248833	49116	87107	35
26	691444	37.3	939982	11.9	751462	49.2	248538	49141	87093	34
27	691668	37.3	939911	11.9	751757	49.2	248243	49166	87079	33
28	691892	37.3	939840	11.9	752052	49.1	247948	49192	87064	32
29	692115	37.2	939768	11.9	752347	49.1	247653	49217	87050	31
30	692339	37.2	939697	11.9	752642	49.1	247358	49242	87036	30
31	9.692562	37.2	9.939625	11.9	9.752937	49.1	10.247063	49268	87021	29
32	692785	37.1	939554	11.9	753231	49.1	246769	49293	87007	28
33	693008	37.1	939482	11.9	753526	49.1	246474	49318	86993	27
34	693231	37.1	939410	11.9	753820	49.0	246180	49344	86978	26
35	693453	37.1	939339	11.9	754115	49.0	245885	49369	86964	25
36	693676	37.0	939267	12.0	754409	49.0	245591	49394	86949	24
37	693898	37.0	939195	12.0	754703	49.0	245297	49419	86935	23
38	694120	37.0	939123	12.0	754997	49.0	245003	49445	86921	22
39	694342	37.0	939052	12.0	755291	49.0	244709	49470	86906	21
40	694564	36.9	938980	12.0	755585	48.9	244415	49495	86892	20
41	9.694786	36.9	9.938908	12.0	9.755878	48.9	10.244122	49521	86878	19
42	695007	36.9	938836	12.0	756172	48.9	243828	49546	86863	18
43	695229	36.9	938763	12.0	756465	48.9	243535	49571	86849	17
44	695450	36.8	938691	12.0	756759	48.9	243241	49596	86834	16
45	695671	36.8	938619	12.0	757052	48.9	242948	49622	86820	15
46	695892	36.8	938547	12.0	757345	48.8	242655	49647	86805	14
47	696113	36.8	938475	12.0	757638	48.8	242362	49672	86791	13
48	696334	36.7	938402	12.1	757931	48.8	242069	49697	86777	12
49	696554	36.7	938330	12.1	758224	48.8	241776	49723	86762	11
50	696775	36.7	938258	12.1	758517	48.8	241483	49748	86748	10
51	9.696995	36.7	9.938185	12.1	9.758810	48.8	10.241190	49773	86733	9
52	697215	36.6	938113	12.1	759102	48.7	240898	49798	86719	8
53	697435	36.6	938040	12.1	759395	48.7	240605	49824	86704	7
54	697654	36.6	937967	12.1	759687	48.7	240313	49849	86690	6
55	697874	36.6	937895	12.1	759979	48.7	240021	49874	86675	5
56	698094	36.5	937822	12.1	760272	48.7	239728	49899	86661	4
57	698313	36.5	937749	12.1	760564	48.7	239436	49924	86646	3
58	698532	36.5	937676	12.1	760856	48.6	239144	49950	86632	2
59	698751	36.5	937604	12.1	761148	48.6	238852	49975	86617	1
60	698970		937531		761439		238561	50000	86603	0
	Cosine.		Sine.		Cotang.		Tang.	N. cos.	N. sine.	′

60 Degrees.

′	Sine.	D. 10″	Cosine.	D. 10″	Tang.	D. 10″	Cotang.	N. sine.	N. cos.	
0	9.698970	36.4	9.937531	12.1	9.761439	48.6	10.238561	50000	86603	60
1	699189	36.4	937458	12.2	761731	48.6	238269	50025	86588	59
2	699407	36.4	937385	12.2	762023	48.6	237977	50050	86573	58
3	699626	36.4	937312	12.2	762314	48.6	237686	50076	86559	57
4	699844	36.3	937238	12.2	762606	48.5	237394	50101	86544	56
5	700062	36.3	937165	12.2	762897	48.5	237103	50126	86530	55
6	700280	36.3	937092	12.2	763188	48.5	236812	50151	86515	54
7	700498	36.3	937019	12.2	763479	48.5	236521	50176	86501	53
8	700716	36.3	936946	12.2	763770	48.5	236230	50201	86486	52
9	700933	36.2	936872	12.2	764061	48.5	235939	50227	86471	51
10	701151	36.2	936799	12.2	764352	48.4	235648	50252	86457	50
11	9.701368	36.2	9.936725	12.2	9.764643	48.4	10.235357	50277	86442	49
12	701585	36.2	936652	12.3	764933	48.4	235067	50302	86427	48
13	701802	36.1	936578	12.3	765224	48.4	234776	50327	86413	47
14	702019	36.1	936505	12.3	765514	48.4	234486	50352	86398	46
15	702236	36.1	936431	12.3	765805	48.4	234195	50377	86384	45
16	702452	36.1	936357	12.3	766095	48.4	233905	50403	86369	44
17	702669	36.0	936284	12.3	766385	48.3	233615	50428	86354	43
18	702885	36.0	936210	12.3	766675	48.3	233325	50453	86340	42
19	703101	36.0	936136	12.3	766965	48.3	233035	50478	86325	41
20	703317	36.0	936062	12.3	767255	48.3	232745	50503	86310	40
21	9.703533	35.9	9.935988	12.3	9.767545	48.3	10.232455	50528	86295	39
22	703749	35.9	935914	12.3	767834	48.3	232166	50553	86281	38
23	703964	35.9	935840	12.3	768124	48.2	231876	50578	86266	37
24	704179	35.9	935766	12.4	768413	48.2	231587	50603	86251	36
25	704395	35.9	935692	12.4	768703	48.2	231297	50628	86237	35
26	704610	35.8	935618	12.4	768992	48.2	231008	50654	86222	34
27	704825	35.8	935543	12.4	769281	48.2	230719	50679	86207	33
28	705040	35.8	935469	12.4	769570	48.2	230430	50704	86192	32
29	705254	35.8	935395	12.4	769860	48.1	230140	50729	86178	31
30	705469	35.7	935320	12.4	770148	48.1	229852	50754	86163	30
31	9.705683	35.7	9.935246	12.4	9.770437	48.1	10.229563	50779	86148	29
32	705898	35.7	935171	12.4	770726	48.1	229274	50804	86133	28
33	706112	35.7	935097	12.4	771015	48.1	228985	50829	86119	27
34	706326	35.6	935022	12.4	771303	48.1	228697	50854	86104	26
35	706539	35.6	934948	12.4	771592	48.1	228408	50879	86089	25
36	706753	35.6	934873	12.4	771880	48.0	228120	50904	86074	24
37	706967	35.6	934798	12.5	772168	48.0	227832	50929	86059	23
38	707180	35.5	934723	12.5	772457	48.0	227543	50954	86045	22
39	707393	35.5	934649	12.5	772745	48.0	227255	50979	86030	21
40	707606	35.5	934574	12.5	773033	48.0	226967	51004	86015	20
41	9.707819	35.5	9.934499	12.5	9.773321	48.0	10.226679	51029	86000	19
42	708032	35.4	934424	12.5	773608	47.9	226392	51054	85985	18
43	708245	35.4	934349	12.5	773896	47.9	226104	51079	85970	17
44	708458	35.4	934274	12.5	774184	47.9	225816	51104	85956	16
45	708670	35.4	934199	12.5	774471	47.9	225529	51129	85941	15
46	708882	35.3	934123	12.5	774759	47.9	225241	51154	85926	14
47	709094	35.3	934048	12.5	775046	47.9	224954	51179	85911	13
48	709306	35.3	933973	12.5	775333	47.9	224667	51204	85896	12
49	709518	35.3	933898	12.6	775621	47.8	224379	51229	85881	11
50	709730	35.3	933822	12.6	775908	47.8	224092	51254	85866	10
51	9.709941	35.2	9.933747	12.6	9.776195	47.8	10.223805	51279	85851	9
52	710153	35.2	933671	12.6	776482	47.8	223518	51304	85836	8
53	710364	35.2	933596	12.6	776769	47.8	223231	51329	85821	7
54	710575	35.2	933520	12.6	777055	47.8	222945	51354	85806	6
55	710786	35.1	933445	12.6	777342	47.8	222658	51379	85792	5
56	710967	35.1	933369	12.6	777628	47.7	222372	51404	85777	4
57	711208	35.1	933293	12.6	777915	47.7	222085	51429	85762	3
58	711419	35.1	933217	12.6	778201	47.7	221799	51454	85747	2
59	711629	35.0	933141	12.6	778487	47.7	221512	51479	85732	1
60	711839		933066		778774		221226	51504	85717	0
	Cosine.		Sine		Cotang.		Tang.	N. cos.	N.sine.	′

59 Degrees.

′	Sine.	D. 10″	Cosine.	D. 10″	Tang.	D. 10″	Cotang.	N. sine.	N. cos.	
0	9.711839	35.0	9.933066	12.6	9.778774	47.7	10.221226	51504	85717	60
1	712050	35.0	932990	12.7	779060	47.7	220940	51529	85702	59
2	712260	35.0	932914	12.7	779346	47.6	220654	51554	85687	58
3	712469	34.9	932838	12.7	779632	47.6	220368	51579	85672	57
4	712679	34.9	932762	12.7	779918	47.6	220082	51604	85657	56
5	712889	34.9	932685	12.7	780203	47.6	219797	51628	85642	55
6	713098	34.9	932609	12.7	780489	47.6	219511	51653	85627	54
7	713308	34.9	932533	12.7	780775	47.6	219225	51678	85612	53
8	713517	34.8	932457	12.7	781060	47.6	218940	51703	85597	52
9	713726	34.8	932380	12.7	781346	47.5	218654	51728	85582	51
10	713935	34.8	932304	12.7	781631	47.5	218369	51753	85567	50
11	9.714144	34.8	9.932228	12.7	9.781916	47.5	10.218084	51778	85551	49
12	714352	34.7	932151	12.7	782201	47.5	217799	51803	85536	48
13	714561	34.7	932075	12.8	782486	47.5	217514	51828	85521	47
14	714769	34.7	931998	12.8	782771	47.5	217229	51852	85506	46
15	714978	34.7	931921	12.8	783056	47.5	216944	51877	85491	45
16	715186	34.7	931845	12.8	783341	47.5	216659	51902	85476	44
17	715394	34.6	931768	12.8	783626	47.4	216374	51927	85461	43
18	715602	34.6	931691	12.8	783910	47.4	216090	51952	85446	42
19	715809	34.6	931614	12.8	784195	47.4	215805	51977	85431	41
20	716017	34.6	931537	12.8	784479	47.4	215521	52002	85416	40
21	9.716224	34.5	9.931460	12.8	9.784764	47.4	10.215236	52026	85401	39
22	716432	34.5	931383	12.8	785048	47.4	214952	52051	85385	38
23	716639	34.5	931306	12.8	785332	47.3	214668	52076	85370	37
24	716846	34.5	931229	12.9	785616	47.3	214384	52101	85355	36
25	717053	34.5	931152	12.9	785900	47.3	214100	52126	85340	35
26	717259	34.4	931075	12.9	786184	47.3	213816	52151	85325	34
27	717466	34.4	930998	12.9	786468	47.3	213532	52175	85310	33
28	717673	34.4	930921	12.9	786752	47.3	213248	52200	85294	32
29	717879	34.4	930843	12.9	787036	47.3	212964	52225	85279	31
30	718085	34.3	930766	12.9	787319	47.2	212681	52250	85264	30
31	9.718291	34.3	9.930688	12.9	9.787603	47.2	10.212397	52275	85249	29
32	718497	34.3	930611	12.9	787886	47.2	212114	52299	85234	28
33	718703	34.3	930533	12.9	788170	47.2	211830	52324	85218	27
34	718909	34.3	930456	12.9	788453	47.2	211547	52349	85203	26
35	719114	34.2	930378	12.9	788736	47.2	211264	52374	85188	25
36	719320	34.2	930300	13.0	789019	47.2	210981	52399	85173	24
37	719525	34.2	930223	13.0	789302	47.1	210698	52423	85157	23
38	719730	34.2	930145	13.0	789585	47.1	210415	52448	85142	22
39	719935	34.1	930067	13.0	789868	47.1	210132	52473	85127	21
40	720140	34.1	929989	13.0	790151	47.1	209849	52498	85112	20
41	9.720345	34.1	9.929911	13.0	9.790433	47.1	10.209567	52522	85096	19
42	720549	34.1	929833	13.0	790716	47.1	209284	52547	85081	18
43	720754	34.0	929755	13.0	790999	47.1	209001	52572	85066	17
44	720958	34.0	929677	13.0	791281	47.1	208719	52597	85051	16
45	721162	34.0	929599	13.0	791563	47.0	208437	52621	85035	15
46	721366	34.0	929521	13.0	791846	47.0	208154	52646	85020	14
47	721570	34.0	929442	13.0	792128	47.0	207872	52671	85005	13
48	721774	33.9	929364	13.1	792410	47.0	207590	52696	84989	12
49	721978	33.9	929286	13.1	792692	47.0	207308	52720	84974	11
50	722181	33.9	929207	13.1	792974	47.0	207026	52745	84959	10
51	9.722385	33.9	9.929129	13.1	9.793256	47.0	10.206744	52770	84943	9
52	722588	33.9	929050	13.1	793538	46.9	206462	52794	84928	8
53	722791	33.8	928972	13.1	793819	46.9	206181	52819	84913	7
54	722994	33.8	928893	13.1	794101	46.9	205899	52844	84897	6
55	723197	33.8	928815	13.1	794383	46.9	205617	52869	84882	5
56	723400	33.8	928736	13.1	794664	46.9	205336	52893	84866	4
57	723603	33.7	928657	13.1	794945	46.9	205055	52918	84851	3
58	723805	33.7	928578	13.1	795227	46.9	204773	52943	84836	2
59	724007	33.7	928499	13.1	795508	46.8	204492	52967	84820	1
60	724210		928420		795789		204211	52992	84805	0
	Cosine.		Sine.		Cotang.		Tang.	N. cos.	N. sine.	′

′	Sine.	D. 10″	Cosine.	D. 10″	Tang.	D. 10″	Cotang.	N. sine.	N. cos.	
0	9.724210	33.7	9.928420	13.2	9.795789	46.8	10.204211	52992	84805	60
1	724412	33.7	928342	13.2	796070	46.8	203930	53017	84789	59
2	724614	33.6	928263	13.2	796351	46.8	203649	53041	84774	58
3	724816	33.6	928183	13.2	796632	46.8	203368	53066	84759	57
4	725017	33.6	928104	13.2	796913	46.8	203087	53091	84743	56
5	725219	33.6	928025	13.2	797194	46.8	202806	53115	84728	55
6	725420	33.5	927946	13.2	797475	46.8	202525	53140	84712	54
7	725622	33.5	927867	13.2	797755	46.8	202245	53164	84697	53
8	725823	33.5	927787	13.2	798036	46.7	201964	53189	84681	52
9	726024	33.5	927708	13.2	798316	46.7	201684	53214	84666	51
10	726225	33.5	927629	13.2	798596	46.7	201404	53238	84650	50
11	9.726426	33.4	9.927549	13.2	9.798877	46.7	10.201123	53263	84635	49
12	726626	33.4	927470	13.3	799157	46.7	200843	53288	84619	48
13	726827	33.4	927390	13.3	799437	46.7	200563	53312	84604	47
14	727027	33.4	927310	13.3	799717	46.7	200283	53337	84588	46
15	727228	33.4	927231	13.3	799997	46.6	200003	53361	84573	45
16	727428	33.3	927151	13.3	800277	46.6	199723	53386	84557	44
17	727628	33.3	927071	13.3	800557	46.6	199443	53411	84542	43
18	727828	33.3	926991	13.3	800836	46.6	199164	53435	84526	42
19	728027	33.3	926911	13.3	801116	46.6	198884	53460	84511	41
20	728227	33.3	926831	13.3	801396	46.6	198604	53484	84495	40
21	9.728427	33.2	9.926751	13.3	9.801675	46.6	10.198325	53509	84480	39
22	728626	33.2	926671	13.3	801955	46.6	198045	53534	84464	38
23	728825	33.2	926591	13.3	802234	46.5	197766	53558	84448	37
24	729024	33.2	926511	13.4	802513	46.5	197487	53583	84433	36
25	729223	33.1	926431	13.4	802792	46.5	197208	53607	84417	35
26	729422	33.1	926351	13.4	803072	46.5	196928	53632	84402	34
27	729621	33.1	926270	13.4	803351	46.5	196649	53656	84386	33
28	729820	33.1	926190	13.4	803630	46.5	196370	53681	84370	32
29	730018	33.0	926110	13.4	803908	46.5	196092	53705	84355	31
30	730216	33.0	926029	13.4	804187	46.5	195813	53730	84339	30
31	9.730415	33.0	9.925949	13.4	9.804466	46.4	10.195534	53754	84324	29
32	730613	33.0	925868	13.4	804745	46.4	195255	53779	84308	28
33	730811	33.0	925788	13.4	805023	46.4	194977	53804	84292	27
34	731009	32.9	925707	13.4	805302	46.4	194698	53828	84277	26
35	731206	32.9	925626	13.4	805580	46.4	194420	53853	84261	25
36	731404	32.9	925545	13.5	805859	46.4	194141	53877	84245	24
37	731602	32.9	925465	13.5	806137	46.4	193863	53902	84230	23
38	731799	32.9	925384	13.5	806415	46.3	193585	53926	84214	22
39	731996	32.8	925303	13.5	806693	46.3	193307	53951	84198	21
40	732193	32.8	925222	13.5	806971	46.3	193029	53975	84182	20
41	9.732390	32.8	9.925141	13.5	9.807249	46.3	10.192751	54000	84167	19
42	732587	32.8	925060	13.5	807527	46.3	192473	54024	84151	18
43	732784	32.8	924979	13.5	807805	46.3	192195	54049	84135	17
44	732980	32.7	924897	13.5	808083	46.3	191917	54073	84120	16
45	733177	32.7	924816	13.5	808361	46.3	191639	54097	84104	15
46	733373	32.7	924735	13.6	808638	46.2	191362	54122	84088	14
47	733569	32.7	924654	13.6	808916	46.2	191084	54146	84072	13
48	733765	32.7	924572	13.6	809193	46.2	190807	54171	84057	12
49	733961	32.6	924491	13.6	809471	46.2	190529	54195	84041	11
50	734157	32.6	924409	13.6	809748	46.2	190252	54220	84025	10
51	9.734353	32.6	9.924328	13.6	9.810025	46.2	10.189975	54244	84009	9
52	734549	32.6	924246	13.6	810302	46.2	189698	54269	83994	8
53	734744	32.5	924164	13.6	810580	46.2	189420	54293	83978	7
54	734939	32.5	924083	13.6	810857	46.2	189143	54317	83962	6
55	735135	32.5	924001	13.6	811134	46.1	188866	54342	83946	5
56	735330	32.5	923919	13.6	811410	46.1	188590	54366	83930	4
57	735525	32.5	923837	13.6	811687	46.1	188313	54391	83915	3
58	735719	32.4	923755	13.7	811964	46.1	188036	54415	83899	2
59	735914	32.4	923673	13.7	812241	46.1	187759	54440	83883	1
60	736109		923591		812517		187483	54464	83867	0
	Cosine.		Sine.		Cotang.		Tang.	N. cos.	N.sine.	′

′	Sine.	D. 10″	Cosine.	D. 10″	Tang.	D. 10″	Cotang.	N. sine.	N. cos.	
0	9.736109	32.4	9.923591	13.7	9.812517	46.1	10.187482	54464	83867	60
1	736303	32.4	923509	13.7	812794	46.1	187206	54488	83851	59
2	736498	32.4	923427	13.7	813070	46.1	186930	54513	83835	58
3	736692	32.3	923345	13.7	813347	46.0	186653	54537	83819	57
4	736886	32.3	923263	13.7	813623	46.0	186377	54561	83804	56
5	737080	32.3	923181	13.7	813899	46.0	186101	54586	83788	55
6	737274	32.3	923098	13.7	814175	46.0	185825	54610	83772	54
7	737467	32.3	923016	13.7	814452	46.0	185548	54635	83756	53
8	737661	32.2	922933	13.7	814728	46.0	185272	54659	83740	52
9	737855	32.2	922851	13.7	815004	46.0	184996	54683	83724	51
10	738048	32.2	922768	13.8	815279	46.0	184721	54708	83708	50
11	9.738241	32.2	9.922686	13.8	9.815555	45.9	10.184445	54732	83692	49
12	738434	32.2	922603	13.8	815831	45.9	184169	54756	83676	48
13	738627	32.1	922520	13.8	816107	45.9	183893	54781	83660	47
14	738820	32.1	922438	13.8	816382	45.9	183618	54805	83645	46
15	739013	32.1	922355	13.8	816658	45.9	183342	54829	83629	45
16	739206	32.1	922272	13.8	816933	45.9	183067	54854	83613	44
17	739398	32.1	922189	13.8	817209	45.9	182791	54878	83597	43
18	739590	32.0	922106	13.8	817484	45.9	182516	54902	83581	42
19	739783	32.0	922023	13.8	817759	45.9	182241	54927	83565	41
20	739975	32.0	921940	13.8	818035	45.8	181965	54951	83549	40
21	9.740167	32.0	9.921857	13.9	9.818310	45.8	10.181690	54975	83533	39
22	740359	32.0	921774	13.9	818585	45.8	181415	54999	83517	38
23	740550	31.9	921691	13.9	818860	45.8	181140	55024	83501	37
24	740742	31.9	921607	13.9	819135	45.8	180865	55048	83485	36
25	740934	31.9	921524	13.9	819410	45.8	180590	55072	83469	35
26	741125	31.9	921441	13.9	819684	45.8	180316	55097	83453	34
27	741316	31.9	921357	13.9	819959	45.8	180041	55121	83437	33
28	741508	31.8	921274	13.9	820234	45.8	179766	55145	83421	32
29	741699	31.8	921190	13.9	820508	45.7	179492	55169	83405	31
30	741889	31.8	921107	13.9	820783	45.7	179217	55194	83389	30
31	9.742080	31.8	9.921023	13.9	9.821057	45.7	10.178943	55218	83373	29
32	742271	31.8	920939	14.0	821332	45.7	178668	55242	83356	28
33	742462	31.7	920856	14.0	821606	45.7	178394	55266	83340	27
34	742652	31.7	920772	14.0	821880	45.7	178120	55291	83324	26
35	742842	31.7	920688	14.0	822154	45.7	177846	55315	83308	25
36	743033	31.7	920604	14.0	822429	45.7	177571	55339	83292	24
37	743223	31.7	920520	14.0	822703	45.7	177297	55363	83276	23
38	743413	31.6	920436	14.0	822977	45.6	177023	55388	83260	22
39	743602	31.6	920352	14.0	823250	45.6	176750	55412	83244	21
40	743792	31.6	920268	14.0	823524	45.6	176476	55436	83228	20
41	9.743982	31.6	9.920184	14.0	9.823798	45.6	10.176202	55460	83212	19
42	744171	31.6	920099	14.0	824072	45.6	175928	55484	83195	18
43	744361	31.5	920015	14.0	824345	45.6	175655	55509	83179	17
44	744550	31.5	919931	14.1	824619	45.6	175381	55533	83163	16
45	744739	31.5	919846	14.1	824893	45.6	175107	55557	83147	15
46	744928	31.5	919762	14.1	825166	45.6	174834	55581	83131	14
47	745117	31.5	919677	14.1	825439	45.5	174561	55605	83115	13
48	745306	31.4	919593	14.1	825713	45.5	174287	55630	83098	12
49	745494	31.4	919508	14.1	825986	45.5	174014	55654	83082	11
50	745683	31.4	919424	14.1	826259	45.5	173741	55678	83066	10
51	9.745871	31.4	9.919339	14.1	9.826532	45.5	10.173468	55702	83050	9
52	746059	31.4	919254	14.1	826805	45.5	173195	55726	83034	8
53	746248	31.3	919169	14.1	827078	45.5	172922	55750	83017	7
54	746436	31.3	919085	14.1	827351	45.5	172649	55775	83001	6
55	746624	31.3	919000	14.1	827624	45.5	172376	55799	82985	5
56	746812	31.3	918915	14.2	827897	45.4	172103	55823	82969	4
57	746999	31.3	918830	14.2	828170	45.4	171830	55847	82953	3
58	747187	31.2	918745	14.2	828442	45.4	171558	55871	82936	2
59	747374	31.2	918659	14.2	828715	45.4	171285	55895	82920	1
60	747562		918574		828987		171013	55919	82904	0
	Cosine.		Sine.		Cotang.		Tang.	N. cos.	N.sine.	′

56 Degrees.

′	Sine.	D. 10″	Cosine.	D. 10″	Tang.	D. 10″	Cotang.	N. sine	N. cos.	
0	9.747562	31.2	9.918574	14.2	9.828987	45.4	10.171013	55919	82904	60
1	747749	31.2	918489	14.2	829260	45.4	170740	55943	82887	59
2	747936	31.2	918404	14.2	829532	45.4	170468	55968	82871	58
3	748123	31.1	918318	14.2	829805	45.4	170195	55992	82855	57
4	748310	31.1	918233	14.2	830077	45.4	169923	56016	82839	56
5	748497	31.1	918147	14.2	830349	45.3	169651	56040	82822	55
6	748683	31.1	918062	14.2	830621	45.3	169379	56064	82806	54
7	748870	31.1	917976	14.3	830893	45.3	169107	56088	82790	53
8	749056	31.0	917891	14.3	831165	45.3	168835	56112	82773	52
9	749243	31.0	917805	14.3	831437	45.3	168563	56136	82757	51
10	749426	31.0	917719	14.3	831709	45.3	168291	56160	82741	50
11	9.749615	31.0	9.917634	14.3	9.831981	45.3	10.168019	56184	82724	49
12	749801	31.0	917548	14.3	832253	45.3	167747	56208	82708	48
13	749987	30.9	917462	14.3	832525	45.3	167475	56232	82692	47
14	750172	30.9	917376	14.3	832796	45.3	167204	56256	82675	46
15	750358	30.9	917290	14.3	833068	45.2	166932	56280	82659	45
16	750543	30.9	917204	14.3	833339	45.2	166661	56305	82643	44
17	750729	30.9	917118	14.4	833611	45.2	166389	56329	82626	43
18	750914	30.8	917032	14.4	833882	45.2	166118	56353	82610	42
19	751099	30.8	916946	14.4	834154	45.2	165846	56377	82593	41
20	751284	30.8	916859	14.4	834425	45.2	165575	56401	82577	40
21	9.751469	30.8	9.916773	14.4	9.834696	45.2	10.165304	56425	82561	39
22	751654	30.8	916687	14.4	834967	45.2	165033	56449	82544	38
23	751839	30.8	916600	14.4	835238	45.2	164762	56473	82528	37
24	752023	30.7	916514	14.4	835509	45.2	164491	56497	82511	36
25	752208	30.7	916427	14.4	835780	45.1	164220	56521	82495	35
26	752392	30.7	916341	14.4	836051	45.1	163949	56545	82478	34
27	752576	30.7	916254	14.4	836322	45.1	163678	56569	82462	33
28	752760	30.7	916167	14.5	836593	45.1	163407	56593	82446	32
29	752944	30.6	916081	14.5	836864	45.1	163136	56617	82429	31
30	753128	30.6	915994	14.5	837134	45.1	162866	56641	82413	30
31	9.753312	30.6	9.915907	14.5	9.837405	45.1	10.162595	56665	82396	29
32	753495	30.6	915820	14.5	837675	45.1	162325	56689	82380	28
33	753679	30.6	915733	14.5	837946	45.1	162054	56713	82363	27
34	753862	30.5	915646	14.5	838216	45.1	161784	56736	82347	26
35	754046	30.5	915559	14.5	838487	45.0	161513	56760	82330	25
36	754229	30.5	915472	14.5	838757	45.0	161243	56784	82314	24
37	754412	30.5	915385	14.5	839027	45.0	160973	56808	82297	23
38	754595	30.5	915297	14.5	839297	45.0	160703	56832	82281	22
39	754778	30.4	915210	14.5	839568	45.0	160432	56856	82264	21
40	754960	30.4	915123	14.6	839838	45.0	160162	56880	82248	20
41	9.755143	30.4	9.915035	14.6	9.840108	45.0	10.159892	56904	82231	19
42	755326	30.4	914948	14.6	840378	45.0	159622	56928	82214	18
43	755508	30.4	914860	14.6	840647	45.0	159353	56952	82198	17
44	755690	30.4	914773	14.6	840917	44.9	159083	56976	82181	16
45	755872	30.3	914685	14.6	841187	44.9	158813	57000	82165	15
46	756054	30.3	914598	14.6	841457	44.9	158543	57024	82148	14
47	756236	30.3	914510	14.6	841726	44.9	158274	57047	82132	13
48	756418	30.3	914422	14.6	841996	44.9	158004	57071	82115	12
49	756600	30.3	914334	14.6	842266	44.9	157734	57095	82098	11
50	756782	30.2	914246	14.7	842535	44.9	157465	57119	82082	10
51	9.756963	30.2	9.914158	14.7	9.842805	44.9	10.157195	57143	82065	9
52	757144	30.2	914070	14.7	843074	44.9	156926	57167	82048	8
53	757326	30.2	913982	14.7	843343	44.9	156657	57191	82032	7
54	757507	30.2	913894	14.7	843612	44.9	156388	57215	82015	6
55	757688	30.1	913806	14.7	843882	44.8	156118	57238	81999	5
56	757869	30.1	913718	14.7	844151	44.8	155849	57262	81982	4
57	758050	30.1	913630	14.7	844420	44.8	155580	57286	81965	3
58	758230	30.1	913541	14.7	844689	44.8	155311	57310	81949	2
59	758411	30.1	913453	14.7	844958	44.8	155042	57334	81932	1
60	758591		913365		845227		154773	57358	81915	0
	Cosine.		Sine.		Cotang.		Tang.	N. cos.	N. sine.	′

55 Degrees.

′	Sine.	D. 10″	Cosine.	D. 10″	Tang.	D. 10″	Cotang.	N. sine.	N. cos.	
0	9.758591	30.1	9.913365	14.7	9.845227	44.8	10.154773	57358	81915	60
1	758772	30.0	913276	14.7	845496	44.8	154504	57381	81899	59
2	758952	30.0	913187	14.8	845764	44.8	154236	57405	81882	58
3	759132	30.0	913099	14.8	846033	44.8	153967	57429	81865	57
4	759312	30.0	913010	14.8	846302	44.8	153698	57453	81848	56
5	759492	30.0	912922	14.8	846570	44.7	153430	57477	81832	55
6	759672	29.9	912833	14.8	846839	44.7	153161	57501	81815	54
7	759852	29.9	912744	14.8	847107	44.7	152893	57524	81798	53
8	760031	29.9	912655	14.8	847376	44.7	152624	57548	81782	52
9	760211	29.9	912566	14.8	847644	44.7	152356	57572	81765	51
10	760390	29.9	912477	14.8	847913	44.7	152087	57596	81748	50
11	9.760569	29.8	9.912388	14.8	9.848181	44.7	10.151819	57619	81731	49
12	760748	29.8	912299	14.9	848449	44.7	151551	57643	81714	48
13	760927	29.8	912210	14.9	848717	44.7	151283	57667	81698	47
14	761106	29.8	912121	14.9	848986	44.7	151014	57691	81681	46
15	761285	29.8	912031	14.9	849254	44.7	150746	57715	81664	45
16	761464	29.8	911942	14.9	849522	44.7	150478	57738	81647	44
17	761642	29.7	911853	14.9	849790	44.6	150210	57762	81631	43
18	761821	29.7	911763	14.9	850058	44.6	149942	57786	81614	42
19	761999	29.7	911674	14.9	850325	44.6	149675	57810	81597	41
20	762177	29.7	911584	14.9	850593	44.6	149407	57833	81580	40
21	9.762356	29.7	9.911495	14.9	9.850861	44.6	10.149139	57857	81563	39
22	762534	29.6	911405	14.9	851129	44.6	148871	57881	81546	38
23	762712	29.6	911315	15.0	851396	44.6	148604	57904	81530	37
24	762889	29.6	911226	15.0	851664	44.6	148336	57928	81513	36
25	763067	29.6	911136	15.0	851931	44.6	148069	57952	81496	35
26	763245	29.6	911046	15.0	852199	44.6	147801	57976	81479	34
27	763422	29.6	910956	15.0	852466	44.6	147534	57999	81462	33
28	763600	29.5	910866	15.0	852733	44.5	147267	58023	81445	32
29	763777	29.5	910776	15.0	853001	44.5	146999	58047	81428	31
30	763954	29.5	910686	15.0	853268	44.5	146732	58070	81412	30
31	9.764131	29.5	9.910596	15.0	9.853535	44.5	10.146465	58094	81395	29
32	764308	29.5	910506	15.0	853802	44.5	146198	58118	81378	28
33	764485	29.4	910415	15.0	854069	44.5	145931	58141	81361	27
34	764662	29.4	910325	15.1	854336	44.5	145664	58165	81344	26
35	764838	29.4	910235	15.1	854603	44.5	145397	58189	81327	25
36	765015	29.4	910144	15.1	854870	44.5	145130	58212	81310	24
37	765191	29.4	910054	15.1	855137	44.5	144863	58236	81293	23
38	765367	29.4	909963	15.1	855404	44.5	144596	58260	81276	22
39	765544	29.3	909873	15.1	855671	44.4	144329	58283	81259	21
40	765720	29.3	909782	15.1	855938	44.4	144062	58307	81242	20
41	9.765896	29.3	9.909691	15.1	9.856204	44.4	10.143796	58330	81225	19
42	766072	29.3	909601	15.1	856471	44.4	143529	58354	81208	18
43	766247	29.3	909510	15.1	856737	44.4	143263	58378	81191	17
44	766423	29.3	909419	15.1	857004	44.4	142996	58401	81174	16
45	766598	29.2	909328	15.2	857270	44.4	142730	58425	81157	15
46	766774	29.2	909237	15.2	857537	44.4	142463	58449	81140	14
47	766949	29.2	909146	15.2	857803	44.4	142197	58472	81123	13
48	767124	29.2	909055	15.2	858069	44.4	141931	58496	81106	12
49	767300	29.2	908964	15.2	858336	44.4	141664	58519	81089	11
50	767475	29.1	908873	15.2	858602	44.3	141398	58543	81072	10
51	9.767649	29.1	9.908781	15.2	9.858868	44.3	10.141132	58567	81055	9
52	767824	29.1	908690	15.2	859134	44.3	140866	58590	81038	8
53	767999	29.1	908599	15.2	859400	44.3	140600	58614	81021	7
54	768173	29.1	908507	15.2	859666	44.3	140334	58637	81004	6
55	768348	29.0	908416	15.3	859932	44.3	140068	58661	80987	5
56	768522	29.0	908324	15.3	850198	44.3	139802	58684	80970	4
57	768697	29.0	908233	15.3	850464	44.3	139536	58708	80953	3
58	768871	29.0	908141	15.3	850730	44.3	139270	58731	80936	2
59	769045	29.0	908049	15.3	850995	44.3	139005	58755	80919	1
60	769219		907958		851261		138739	58779	80902	0
	Cosine.		Sine.		Cotang.		Tang.	N. cos.	N.sine.	′

54 Degrees.

′	Sine.	D. 10″	Cosine.	D. 10″	Tang.	D. 10″	Cotang.	N. sine.	N. cos.	
0	9.769219	29.0	9.907958	15.3	9.861261	44.3	10.138739	58779	80902	60
1	769393	28.9	907866	15.3	861527	44.3	138473	58802	80885	59
2	769566	28.9	907774	15.3	861792	44.2	138208	58826	80867	58
3	769740	28.9	907682	15.3	862058	44.2	137942	58849	80850	57
4	769913	28.9	907590	15.3	862323	44.2	137677	58873	80833	56
5	770087	28.9	907498	15.3	862589	44.2	137411	58896	80816	55
6	770260	28.8	907406	15.3	862854	44.2	137146	58920	80799	54
7	770433	28.8	907314	15.4	863119	44.2	136881	58943	80782	53
8	770606	28.8	907222	15.4	863385	44.2	136615	58967	80765	52
9	770779	28.8	907129	15.4	863650	44.2	136350	58990	80748	51
10	770952	28.8	907037	15.4	863915	44.2	136085	59014	80730	50
11	9.771125	28.8	9.906945	15.4	9.864180	44.2	10.135820	59037	80713	49
12	771298	28.7	906852	15.4	864445	44.2	135555	59061	80696	48
13	771470	28.7	906760	15.4	864710	44.2	135290	59084	80679	47
14	771643	28.7	906667	15.4	864975	44.2	135025	59108	80662	46
15	771815	28.7	906575	15.4	865240	44.1	134760	59131	80644	45
16	771987	28.7	906482	15.4	865505	44.1	134495	59154	80627	44
17	772159	23.7	906389	15.5	865770	44.1	134230	59178	80610	43
18	772331	28.6	906296	15.5	866035	44.1	133965	59201	80593	42
19	772503	28.6	906204	15.5	866300	44.1	133700	59225	80576	41
20	772675	28.6	906111	15.5	866564	44.1	133436	59248	80558	40
21	9.772847	28.6	9.906018	15.5	9.866829	44.1	10.133171	59272	80541	39
22	773018	28.6	905925	15.5	867094	44.1	132906	59295	80524	38
23	773190	28.6	905832	15.5	867358	44.1	132642	59318	80507	37
24	773361	28.5	905739	15.5	867623	44.1	132377	59342	80489	36
25	773533	28.5	905645	15.5	867887	44.1	132113	59365	80472	35
26	773704	28.5	905552	15.5	868152	44.1	131848	59389	80455	34
27	773875	28.5	905459	15.5	868416	44.0	131584	59412	80438	33
28	774046	28.5	905366	15.6	868680	44.0	131320	59436	80422	32
29	774217	28.5	905272	15.6	868945	44.0	131055	59459	80403	31
30	774388	28.4	905179	15.6	869209	44.0	130791	59482	80386	30
31	9.774558	28.4	9.905085	15.6	9.869473	44.0	10.130527	59506	80368	29
32	774729	28.4	904992	15.6	869737	44.0	130263	59529	80351	28
33	774899	28.4	904898	15.6	870001	44.0	129999	59552	80334	27
34	775070	28.4	904804	15.6	870265	44.0	129735	59576	80316	26
35	775240	28.4	904711	15.6	870529	44.0	129471	59599	80299	25
36	775410	28.3	904617	15.6	870793	44.0	129207	59622	80282	24
37	775580	28.3	904523	15.6	871057	44.0	128943	59646	80264	23
38	775750	28.3	904429	15.7	871321	44.0	128679	59669	80247	22
39	775920	28.3	904335	15.7	871585	44.0	128415	59693	80230	21
40	776090	28.3	904241	15.7	871849	44.0	128151	59716	80212	20
41	9.776259	28.3	9.904147	15.7	9.872112	43.9	10.127888	59739	80195	19
42	776429	28.2	904053	15.7	872376	43.9	127624	59763	80178	18
43	776598	28.2	903959	15.7	872640	43.9	127360	59786	80160	17
44	776768	28.2	903864	15.7	872903	43.9	127097	59809	80143	16
45	776937	28.2	903770	15.7	873167	43.9	126833	59832	80125	15
46	777106	28.2	903676	15.7	873430	43.9	126570	59856	80108	14
47	777275	28.1	903581	15.7	873694	43.9	126306	59879	80091	13
48	777444	28.1	903487	15.7	873957	43.9	126043	59902	80073	12
49	777613	28.1	903392	15.8	874220	43.9	125780	59926	80056	11
50	777781	28.1	903298	15.8	874484	43.9	125516	59949	80038	10
51	9.777950	28.1	9.903202	15.8	9.874747	43.9	10.125253	59972	80021	9
52	778119	28.1	903108	15.8	875010	43.9	124990	59995	80003	8
53	778287	28.0	903014	15.8	875273	43.8	124727	60019	79986	7
54	778455	28.0	902919	15.8	875536	43.8	124464	60042	79968	6
55	778624	28.0	902824	15.8	875800	43.8	124200	60065	79951	5
56	778792	28.0	902729	15.8	876063	43.8	123937	60089	79934	4
57	778960	28.0	902634	15.8	876326	43.8	123674	60112	79916	3
58	779128	28.0	902539	15.9	876589	43.8	123411	60135	79899	2
59	779295	27.9	902444	15.9	876851	43.8	123149	60158	79881	1
60	779463		902349		877114		122886	60182	79864	0
.	Cosine.		Sine.		Cotang.		Tang.	N. cos.	N.sine.	′

53 Degrees.

′	Sine.	D. 10″	Cosine.	D. 10″	Tang.	D. 10″	Cotang.	N.sine.	N. cos.	
0	9.779463	27.9	9.902349	15.9	9.877114	43.8	10.122886	60182	79864	60
1	779631	27.9	902253	15.9	877377	43.8	122623	60205	79846	59
2	779798	27.9	902158	15.9	877640	43.8	122360	60228	79829	58
3	779966	27.9	902063	15.9	877903	43.8	122097	60251	79811	57
4	780133	27.9	901967	15.9	878165	43.8	121835	60274	79793	56
5	780300	27.8	901872	15.9	878428	43.8	121572	60298	79776	55
6	780467	27.8	901776	15.9	878691	43.8	121309	60321	79758	54
7	780634	27.8	901681	15.9	878953	43.7	121047	60344	79741	53
8	780801	27.8	901585	15.9	879216	43.7	120784	60367	79723	52
9	780968	27.8	901490	15.9	879478	43.7	120522	60390	79706	51
10	781134	27.8	901394	15.9	879741	43.7	120259	60414	79688	50
11	9.781301	27.7	9.901298	16.0	9.880003	43.7	10.119997	60437	79671	49
12	781468	27.7	901202	16.0	880265	43.7	119735	60460	79658	48
13	781634	27.7	901106	16.0	880528	43.7	119472	60483	79635	47
14	781800	27.7	901010	16.0	880790	43.7	119210	60506	79618	46
15	781966	27.7	900914	16.0	881052	43.7	118948	60529	79600	45
16	782132	27.7	900818	16.0	881314	43.7	118686	60553	79583	44
17	782298	27.6	900722	16.0	881576	43.7	118424	60576	79565	43
18	782464	27.6	900626	16.0	881839	43.7	118161	60599	79547	42
19	782630	27.6	900529	16.0	882101	43.7	117899	60622	79530	41
20	782796	27.6	900433	16.0	882363	43.7	117637	60645	79512	40
21	9.782961	27.6	9.900337	16.1	9.882625	43.6	10.117375	60668	79494	39
22	783127	27.6	900242	16.1	882887	43.6	117113	60691	79477	38
23	783282	27.5	900144	16.1	883148	43.6	116852	60714	79459	37
24	783458	27.5	900047	16.1	883410	43.6	116590	60738	79441	36
25	783623	27.5	899951	16.1	883672	43.6	116328	60761	79424	35
26	783788	27.5	899854	16.1	883934	43.6	116066	60784	79406	34
27	783953	27.5	899757	16.1	884196	43.6	115804	60807	79388	33
28	784118	27.5	899660	16.1	884457	43.6	115543	60830	79371	32
20	784282	27.4	899564	16.1	884719	43.6	115281	60853	79353	31
39	784447	27.4	899467	16.1	884980	43.6	115020	60876	79335	30
31	9.784612	27.4	9.899370	16.2	9.885242	43.6	10.114758	60899	79318	29
32	784776	27.4	899273	16.2	885503	43.6	114497	60922	79300	28
33	784941	27.4	899176	16.2	885765	43.6	114235	60945	79282	27
34	785105	27.4	899078	16.2	886026	43.6	113974	60968	79264	26
35	785269	27.3	898981	16.2	886288	43.6	113712	60991	79247	25
36	785433	27.3	898884	16.2	886549	43.5	113451	61015	79229	24
37	785597	27.3	898787	16.2	886810	43.5	113190	61038	79211	23
38	785761	27.3	898689	16.2	887072	43.5	112928	61061	79193	22
39	785925	27.3	898592	16.2	887333	43.5	112667	61084	79176	21
40	786089	27.3	898494	16.2	887594	43.5	112406	61107	79158	20
41	9.786252	27.2	9.898397	16.3	9.887855	43.5	10.112145	61130	79140	19
42	786416	27.2	898299	16.3	888116	43.5	111884	61153	79122	18
43	786579	27.2	898202	16.3	888377	43.5	111623	61176	79105	17
44	786742	27.2	898104	16.3	888639	43.5	111361	61199	79087	16
45	786906	27.2	898006	16.3	888900	43.5	111100	61222	79069	15
46	787069	27.2	897908	16.3	889160	43.5	110840	61245	79051	14
47	787232	27.1	897810	16.3	889421	43.5	110579	61268	79033	13
48	787395	27.1	897712	16.3	889682	43.5	110318	61291	79016	12
49	787557	27.1	897614	16.3	889943	43.5	110057	61314	78998	11
50	787720	27.1	897516	16.3	890204	43.4	109796	61337	78980	10
51	9.787883	27.1	9.897418	16.3	9.890465	43.4	10.109535	61360	78962	9
52	788045	27.1	897320	16.4	890725	43.4	109275	61383	78944	8
53	788208	27.1	897222	16.4	890986	43.4	109014	61406	78926	7
54	788370	27.0	897123	16.4	891247	43.4	108753	61429	78908	6
55	788532	27.0	897025	16.4	891507	43.4	108493	61451	78891	5
56	788694	27.0	896926	16.4	891768	43.4	108232	61474	78873	4
57	788856	27.0	896828	16.4	892028	43.4	107972	61497	78855	3
58	789018	27.0	896729	16.4	892289	43.4	107711	61520	78837	2
59	789180	27.0	896631	16.4	892549	43.4	107451	61543	78819	1
60	789342		896532		892810		107190	61566	78801	0
	Cosine.		Sine.		Cotang.		Tang.	N. cos.	N.sine.	′

′	Sine.	D. 10″	Cosine.	D. 10″	Tang.	D. 10″	Cotang.	N. sine.	N. cos.	
0	9.789342	26.9	9.896532	16.4	9.892810	43.4	10.107190	61566	78801	60
1	789504	26.9	896433	16.5	893070	43.4	106930	61589	78783	59
2	789665	26.9	896335	16.5	893331	43.4	106669	61612	78765	58
3	789827	26.9	896236	16.5	893591	43.4	106409	61635	78747	57
4	789988	26.9	896137	16.5	893851	43.4	106149	61658	78729	56
5	790149	26.9	896038	16.5	894111	43.4	105889	61681	78711	55
6	790310	26.8	895939	16.5	894371	43.4	105629	61704	78694	54
7	790471	26.8	895840	16.5	894632	43.3	105368	61726	78676	53
8	790632	26.8	895741	16.5	894892	43.3	105108	61749	78658	52
9	790793	26.8	895641	16.5	895152	43.3	104848	61772	78640	51
10	790954	26.8	895542	16.5	895412	43.3	104588	61795	78622	50
11	9.791115	26.8	9.895443	16.6	9.895672	43.3	10.104328	61818	78604	49
12	791275	26.7	895343	16.6	895932	43.3	104068	61841	78586	48
13	791436	26.7	895244	16.6	896192	43.3	103808	61864	78568	47
14	791596	26.7	895145	16.6	896452	43.3	103548	61887	78550	46
15	791757	26.7	895045	16.6	896712	43.3	103288	61909	78532	45
16	791917	26.7	894945	16.6	896971	43.3	103029	61932	78514	44
17	792077	26.7	894846	16.6	897231	43.3	102769	61955	78496	43
18	792237	26.6	894746	16.6	897491	43.3	102509	61978	78478	42
19	792397	26.6	894646	16.6	897751	43.3	102249	62001	78460	41
20	792557	26.6	894546	16.6	898010	43.3	101990	62024	78442	40
21	9.792716	26.6	9.894446	16.7	9.898270	43.3	10.101730	62046	78424	39
22	792876	26.6	894346	16.7	898530	43.3	101470	62069	78405	38
23	793035	26.6	894246	16.7	898789	43.3	101211	62092	78387	37
24	793195	26.5	894146	16.7	899049	43.2	100951	62115	78369	36
25	793354	26.5	894046	16.7	899308	43.2	100692	62138	78351	35
26	793514	26.5	893946	16.7	899568	43.2	100432	62160	78333	34
27	793673	26.5	893846	16.7	899827	43.2	100173	62183	78315	33
28	793832	26.5	893745	16.7	900086	43.2	099914	62206	78297	32
29	793991	26.5	893645	16.7	900346	43.2	099654	62229	78279	31
30	794150	26.4	893544	16.7	900605	43.2	099395	62251	78261	30
31	9.794308	26.4	9.893444	16.8	9.900864	43.2	10.099136	62274	78243	29
32	794467	26.4	893343	16.8	901124	43.2	098876	62297	78225	28
33	794626	26.4	893243	16.8	901383	43.2	098617	62320	78206	27
34	794784	26.4	893142	16.8	901642	43.2	098358	62342	78188	26
35	794942	26.4	893041	16.8	901901	43.2	098099	62365	78170	25
36	795101	26.4	892940	16.8	902160	43.2	097840	62388	78152	24
37	795259	26.3	892839	16.8	902419	43.2	097581	62411	78134	23
38	795417	26.3	892739	16.8	902679	43.2	097321	62433	78116	22
39	795575	26.3	892638	16.8	902938	43.2	097062	62456	78098	21
40	795733	26.3	892536	16.8	903197	43.1	096803	62479	78079	20
41	9.795891	26.3	9.892435	16.9	9.903455	43.1	10.096545	62502	78061	19
42	796049	26.3	892334	16.9	903714	43.1	096286	62524	78043	18
43	796206	26.3	892233	16.9	903973	43.1	096027	62547	78025	17
44	796364	26.2	892132	16.9	904232	43.1	095768	62570	78007	16
45	796521	26.2	892030	16.9	904491	43.1	095509	62592	77988	15
46	796679	26.2	891929	16.9	904750	43.1	095250	62615	77970	14
47	796836	26.2	891827	16.9	905008	43.1	094992	62638	77952	13
48	796993	26.2	891726	16.9	905267	43.1	094733	62660	77934	12
49	797150	26.1	891624	16.9	905526	43.1	094474	62683	77916	11
50	797307	26.1	891523	17.0	905784	43.1	094216	62706	77897	10
51	9.797464	26.1	9.891421	17.0	9.906043	43.1	10.093957	62728	77879	9
52	797621	26.1	891319	17.0	906302	43.1	093698	62751	77861	8
53	797777	26.1	891217	17.0	906560	43.1	093440	62774	77843	7
54	797934	26.1	891115	17.0	906819	43.1	093181	62796	77824	6
55	798091	26.1	891013	17.0	907077	43.1	092923	62819	77806	5
56	798247	26.1	890911	17.0	907336	43.1	092664	62842	77788	4
57	798403	26.0	890809	17.0	907594	43.1	092406	62864	77769	3
58	798560	26.0	890707	17.0	907852	43.1	092148	62887	77751	2
59	798716	26.0	890605	17.0	908111	43.0	091889	62909	77733	1
60	798872		890503		908369		091631	62932	77715	0
	Cosine.		Sine.		Cotang.		Tang.	N. cos.	N.sine.	′

51 Degrees.

′	Sine.	D. 10″	Cosine.	D. 10″	Tang.	D. 10″	Cotang.	N. sine.	N. cos.	
0	9.798772	26.0	9.890503	17.0	9.908369	43.0	10.091631	62932	77715	60
1	799028	26.0	890400	17.1	908628	43.0	091372	62955	77696	59
2	799184	26.0	890298	17.1	908886	43.0	091114	62977	77678	58
3	799339	25.9	890195	17.1	909144	43.0	090856	63000	77660	57
4	799495	25.9	890093	17.1	909402	43.0	090598	63022	77641	56
5	799651	25.9	889990	17.1	909660	43.0	090340	63045	77623	55
6	799806	25.9	889888	17.1	909918	43.0	090082	63068	77605	54
7	799962	25.9	889785	17.1	910177	43.0	089823	63090	77586	53
8	800117	25.9	889682	17.1	910435	43.0	089565	63113	77568	52
9	800272	25.8	889579	17.1	910693	43.0	089307	63135	77550	51
10	800427	25.8	889477	17.1	910951	43.0	089049	63158	77531	50
11	9.800582	25.8	9.889374	17.2	9.911209	43.0	10.088791	93180	77513	49
12	800737	25.8	889271	17.2	911467	43.0	088533	63203	77494	48
13	800892	25.8	889168	17.2	911724	43.0	088276	63225	77476	47
14	801047	25.8	889064	17.2	911982	43.0	088018	63248	77458	46
15	801201	25.8	888961	17.2	912240	43.0	087760	63271	77439	45
16	801356	25.7	888858	17.2	912498	43.0	087502	63293	77421	44
17	801511	25.7	888755	17.2	912756	43.0	087244	63316	77402	43
18	801665	25.7	888651	17.2	913014	42.9	086986	63338	77384	42
19	801819	25.7	888548	17.2	913271	42.9	086729	63361	77366	41
20	801973	25.7	888444	17.3	913529	42.9	086471	63383	77347	40
21	9.802128	25.7	9.888341	17.3	9.913787	42.9	10.086213	63406	77329	39
22	802282	25.6	888237	17.3	914044	42.9	085956	63428	77310	38
23	802436	25.6	888134	17.3	914302	42.9	085698	63451	77292	37
24	802589	25.6	888030	17.3	914560	42.9	085440	63473	77273	36
25	802743	25.6	887926	17.3	914817	42.9	085183	63496	77255	35
26	802897	25.6	887822	17.3	915075	42.9	084925	63518	77236	34
27	803050	25.6	887718	17.3	915332	42.9	084668	63540	77218	33
28	803204	25.6	887614	17.3	915590	42.9	084410	63563	77199	32
29	803357	25.5	887510	17.3	915847	42.9	084153	63585	77181	31
30	803511	25.5	887406	17.4	916104	42.9	083896	63608	77162	30
31	9.803664	25.5	9.887302	17.4	9.916362	42.9	10.083638	63630	77144	29
32	803817	25.5	887198	17.4	916619	42.9	083381	63653	77125	28
33	803970	25.5	887093	17.4	916877	42.9	083123	63675	77107	27
34	804123	25.5	886989	17.4	917134	42.9	082866	63698	77088	26
35	804276	25.4	886885	17.4	917391	42.9	082609	63720	77070	25
36	804428	25.4	886780	17.4	917648	42.9	082352	63742	77051	24
37	804581	25.4	886676	17.4	917905	42.9	082095	63765	77033	23
38	804734	25.4	886571	17.4	918163	42.8	081837	63787	77014	22
39	804886	25.4	886466	17.4	918420	42.8	081580	63810	76996	21
40	805039	25.4	886362	17.5	918677	42.8	081323	63832	76977	20
41	9.805191	25.4	9.886257	17.5	9.918934	42.8	10.081066	63854	76959	19
42	805343	25.3	886152	17.5	919191	42.8	080809	63877	76940	18
43	805495	25.3	886047	17.5	919448	42.8	080552	63899	76921	17
44	805647	25.3	885942	17.5	919705	42.8	080295	63922	76903	16
45	805799	25.3	885837	17.5	919962	42.8	080038	63944	76884	15
46	805951	25.3	885732	17.5	920219	42.8	079781	63966	76866	14
47	806103	25.3	885627	17.5	920476	42.8	079524	63989	76847	13
48	806254	25.3	885522	17.5	920733	42.8	079267	64011	76828	12
49	806406	25.2	885416	17.5	920990	42.8	079010	64083	76810	11
50	806557	25.2	885311	17.6	921247	42.8	078753	64056	76791	10
51	9.806709	25.2	9.885205	17.6	9.921503	42.8	10.078497	64078	76772	9
52	806860	25.2	885100	17.6	921760	42.8	078240	64100	76754	8
53	807011	25.2	884994	17.6	922017	42.8	077983	64123	76735	7
54	807163	25.2	884889	17.6	922274	42.8	077726	64145	76717	6
55	807314	25.2	884783	17.6	922530	42.8	077470	64167	76698	5
56	807465	25.1	884677	17.6	922787	42.8	077213	64190	76679	4
57	807615	25.1	884572	17.6	923044	42.8	076956	64212	76661	3
58	807766	25.1	884466	17.6	923300	42.8	076700	64234	76642	2
59	807917	25.1	884360	17.6	923557	42.7	076443	64256	76623	1
60	808067		884254		923813		076187	64279	76604	0
	Cosine.		Sine.		Cotang.		Tang.	N. cos.	N.sine.	′

′	Sine.	D. 10″	Cosine.	D. 10″	Tang.	D. 10″	Cotang.	N. sine.	N. cos.	
0	9.808067	25.1	9.884254	17.7	9.923813	42.7	10.076187	64279	76604	60
1	808218	25.1	884148	17.7	924070	42.7	075930	64301	76586	59
2	808368	25.1	884042	17.7	924327	42.7	075673	64323	76567	58
3	808519	25.0	883936	17.7	924583	42.7	075417	64346	76548	57
4	808669	25.0	883829	17.7	924840	42.7	075160	64368	76530	56
5	808819	25.0	883723	17.7	925096	42.7	074904	64390	76511	55
6	808969	25.0	883617	17.7	925352	42.7	074648	64412	76492	54
7	809119	25.0	883510	17.7	925609	42.7	074391	64435	76473	53
8	809269	25.0	883404	17.7	925865	42.7	074135	64457	76455	52
9	809419	24.9	883297	17.8	926122	42.7	073878	64479	76436	51
10	809569	24.9	883191	17.8	926378	42.7	073622	64501	76417	50
11	9.809718	24.9	9.883084	17.8	9.926634	42.7	10.073366	64524	76398	49
12	809868	24.9	882977	17.8	926890	42.7	073110	64546	76380	48
13	810017	24.9	882871	17.8	927147	42.7	072853	64568	76361	47
14	810167	24.9	882764	17.8	927403	42.7	072597	64590	76342	46
15	810316	24.8	882657	17.8	927659	42.7	072341	64612	76323	45
16	810465	24.8	882550	17.8	927915	42.7	072085	64635	76304	44
17	810614	24.8	882443	17.8	928171	42.7	071829	64657	76286	43
18	810763	24.8	882336	17.9	928427	42.7	071573	64679	76267	42
19	810912	24.8	882229	17.9	928683	42.7	071317	64701	76248	41
20	811061	24.8	882121	17.9	928940	42.7	071060	64723	76229	40
21	9.811210	24.8	9.882014	17.9	9.929196	42.7	10.070804	64746	76210	39
22	811358	24.7	881907	17.9	929452	42.7	070548	64768	76192	38
23	811507	24.7	881799	17.9	929708	42.7	070292	64790	76173	37
24	811655	24.7	881692	17.9	929964	42.6	070036	64812	76154	36
25	811804	24.7	881584	17.9	930220	42.6	069780	64834	76135	35
26	811952	24.7	881477	17.9	930475	42.6	069525	64856	76116	34
27	812100	24.7	881369	17.9	930731	42.6	069269	64878	76097	33
28	812248	24.7	881261	18.0	930987	42.6	069013	64901	76078	32
29	812396	24.6	881153	18.0	931243	42.6	068757	64923	76059	31
30	812544	24.6	881046	18.0	931499	42.6	068501	64945	76041	30
31	9.812692	24.6	9.880938	18.0	9.931755	42.6	10.068245	64967	76022	29
32	812840	24.6	880830	18.0	932010	42.6	067990	64989	76003	28
33	812988	24.6	880722	18.0	932266	42.6	067734	65011	75984	27
34	813135	24.6	880613	18.0	932522	42.6	067478	65033	75965	26
35	813283	24.6	880505	18.0	932778	42.6	067222	65055	75946	25
36	813430	24.5	880397	18.0	933033	42.6	066967	65077	75927	24
37	813578	24.5	880289	18.1	933289	42.6	066711	65100	75908	23
38	813725	24.5	880180	18.1	933545	42.6	066455	65122	75889	22
39	813872	24.5	880072	18.1	933800	42.6	066200	65144	75870	21
40	814019	24.5	879963	18.1	934056	42.6	065944	65166	75851	20
41	9.814166	24.5	9.879855	18.1	9.934311	42.6	10.065689	65188	75832	19
42	814313	24.5	879746	18.1	934567	42.6	065433	65210	75813	18
43	814460	24.4	879637	18.1	934823	42.6	065177	65232	75794	17
44	814607	24.4	879529	18.1	935078	42.6	064922	65254	75775	16
45	814753	24.4	879420	18.1	935333	42.6	064667	65276	75756	15
46	814900	24.4	879311	18.1	935589	42.6	064411	65298	75738	14
47	815046	24.4	879202	18.2	935844	42.6	064156	65320	75719	13
48	815193	24.4	879093	18.2	936100	42.6	063900	65342	75700	12
49	815339	24.4	878984	18.2	936355	42.6	063645	65364	75680	11
50	815485	24.3	878875	18.2	936610	42.6	063390	65386	75661	10
51	9.815631	24.3	9.878766	18.2	9.936866	42.5	10.063134	65408	75642	9
52	815778	24.3	878656	18.2	937121	42.5	062879	65430	75623	8
53	815924	24.3	878547	18.2	937376	42.5	062624	65452	75604	7
54	816069	24.3	878438	18.2	937632	42.5	062368	65474	75585	6
55	816215	24.3	878328	18.2	937887	42.5	062113	65496	75566	5
56	816361	24.3	878219	18.3	938142	42.5	061858	65518	75547	4
57	816507	24.2	878109	18.3	938398	42.5	061602	65540	75528	3
58	816652	24.2	877999	18.3	938653	42.5	061347	65562	75509	2
59	816798	24.2	877890	18.3	938908	42.5	061092	65584	75490	1
60	816943		877780		939163		060837	65606	75471	0
	Cosine.		Sine.		Cotang.		Tang.	N. cos.	N. sine.	′

′	Sine.	D. 10″	Cosine.	D. 10″	Tang.	D. 10″	Cotang.	N. sine.	N. cos.	
0	9.816943	24.2	9.877780	18.3	9.939163	42.5	10.060837	65606	75471	60
1	817088	24.2	877670	18.3	939418	42.5	060582	65628	75452	59
2	817233	24.2	877560	18.3	939673	42.5	060327	65650	75433	58
3	817379	24.2	877450	18.3	939928	42.5	060072	65672	75414	57
4	817524	24.1	877340	18.3	940183	42.5	059817	65694	75395	56
5	817668	24.1	877230	18.4	940438	42.5	059562	65716	75375	55
6	817813	24.1	877120	18.4	940694	42.5	059306	65738	75356	54
7	817958	24.1	877010	18.4	940949	42.5	059051	65759	75337	53
8	818103	24.1	876899	18.4	941204	42.5	058796	65781	75318	52
9	818247	24.1	876789	18.4	941458	42.5	058542	65803	75299	51
10	818392	24.1	876678	18.4	941714	42.5	058286	65825	75280	50
11	9.818536	24.0	9.876568	18.4	9.941968	42.5	10.058032	65847	75261	49
12	818681	24.0	876457	18.4	942223	42.5	057777	65869	75241	48
13	818825	24.0	876347	18.4	942478	42.5	057522	65891	75222	47
14	818969	24.0	876236	18.5	942733	42.5	057267	65913	75203	46
15	819113	24.0	876125	18.5	942988	42.5	057012	65935	75184	45
16	819257	24.0	876014	18.5	943243	42.5	056757	65956	75165	44
17	819401	24.0	875904	18.5	943498	42.5	056502	65978	75146	43
18	819545	23.9	875793	18.5	943752	42.5	056248	66000	75126	42
19	819689	23.9	875682	18.5	944007	42.5	055993	66022	75107	41
20	819832	23.9	875571	18.5	944262	42.5	055738	66044	75088	40
21	9.819976	23.9	9.875459	18.5	9.944517	42.5	10.055483	66066	75069	39
22	820120	23.9	875348	18.5	944771	42.4	055229	66088	75050	38
23	820263	23.9	875237	18.5	945026	42.4	054974	66109	75030	37
24	820406	23.9	875126	18.6	945281	42.4	054719	66131	75011	36
25	820550	23.8	875014	18.6	945535	42.4	054465	66153	74992	35
26	820693	23.8	874903	18.6	945790	42.4	054210	66175	74973	34
27	820836	23.8	874791	18.6	946045	42.4	053955	66197	74953	33
28	820979	23.8	874680	18.6	946299	42.4	053701	66218	74934	32
29	821122	23.8	874568	18.6	946554	42.4	053446	66240	74915	31
30	821265	23.8	874456	18.6	946808	42.4	053192	66262	74896	30
31	9.821407	23.8	9.874344	18.6	9.947063	42.4	10.052937	66284	74876	29
32	821550	23.8	874232	18.7	947318	42.4	052682	66306	74857	28
33	821693	23.7	874121	18.7	947572	42.4	052428	66327	74838	27
34	821835	23.7	874009	18.7	947826	42.4	052174	66349	74818	26
35	821977	23.7	873896	18.7	948081	42.4	051919	66371	74799	25
36	822120	23.7	873784	18.7	948336	42.4	051664	66393	74780	24
37	822262	23.7	873672	18.7	948590	42.4	051410	66414	74760	23
38	822404	23.7	873560	18.7	948844	42.4	051156	66436	74741	22
39	822546	23.7	873448	18.7	949099	42.4	050901	66458	74722	21
40	822688	23.6	873335	18.7	949353	42.4	050647	66480	74703	20
41	9.822830	23.6	9.873223	18.7	9.949607	42.4	10.050393	66501	74683	19
42	822972	23.6	873110	18.8	949862	42.4	050138	66523	74663	18
43	823114	23.6	872998	18.8	950116	42.4	049884	66545	74644	17
44	823255	23.6	872885	18.8	950370	42.4	049630	66566	74625	16
45	823397	23.6	872772	18.8	950625	42.4	049375	66588	74606	15
46	823539	23.6	872659	18.8	950879	42.4	049121	66610	74586	14
47	823680	23.5	872547	18.8	951133	42.4	048867	66632	74567	13
48	823821	23.5	872434	18.8	951388	42.4	048612	66653	74548	12
49	823963	23.5	872321	18.8	951642	42.4	048358	66675	74522	11
50	824104	23.5	872208	18.8	951896	42.4	048104	66697	74509	10
51	9.824245	23.5	9.872095	18.9	9.952150	42.4	10.047850	66718	74489	9
52	824386	23.5	871981	18.9	952405	42.4	047595	66740	74470	8
53	824527	23.5	871868	18.9	952659	42.4	047341	66762	74451	7
54	824668	23.4	871755	18.9	952913	42.4	047087	66783	74431	6
55	824808	23.4	871641	18.9	953167	42.3	046833	66805	74412	5
56	824949	23.4	871528	18.9	953421	42.3	046579	66827	74392	4
57	825090	23.4	871414	18.9	953675	42.3	046325	66848	74373	3
58	825230	23.4	871301	18.9	953929	42.3	046071	66870	74353	2
59	825371	23.4	871187	18.9	954183	42.3	045817	66891	74334	1
60	825511		871073		954437		045563	66913	74314	0
	Cosine.		Sine.		Cotang.		Tang.	N. cos.	N.sine.	′

48 Degrees.

′	Sine.	D. 10″	Cosine.	D. 10″	Tang.	D. 10″	Cotang.	N. sine.	N. cos.	
0	9.825511	23.4	9.871073	19.0	9.954437	42.3	10.045563	66913	74314	60
1	825651	23.3	870960	19.0	954691	42.3	045309	66935	74295	59
2	825791	23.3	870846	19.0	954945	42.3	045055	66956	74276	58
3	825931	23.3	870732	19.0	955200	42.3	044800	66978	74256	57
4	826071	23.3	870618	19.0	955454	42.3	044546	66999	74237	56
5	826211	23.3	870504	19.0	955707	42.3	044293	67021	74217	55
6	826351	23.3	870390	19.0	955961	42.3	044039	67043	74198	54
7	826491	23.3	870276	19.0	956215	42.3	043785	67064	74178	53
8	826631	23.3	870161	19.0	956469	42.3	043531	67086	74159	52
9	826770	23.2	870047	19.1	956723	42.3	043277	67107	74139	51
10	826910	23.2	869933	19.1	956977	42.3	043023	67129	74120	50
11	9.827049	23.2	9.869818	19.1	9.957231	42.3	10.042769	67151	74100	49
12	827189	23.2	869704	19.1	957485	42.3	042515	67172	74080	48
13	827328	23.2	869589	19.1	957739	42.3	042261	67194	74061	47
14	827467	23.2	869474	19.1	957993	42.3	042007	67215	74041	46
15	827606	23.2	869360	19.1	958246	42.3	041754	67237	74022	45
16	827745	23.2	869245	19.1	958500	42.3	041500	67258	74002	44
17	827884	23.1	869130	19.1	958754	42.3	041246	67280	73983	43
18	828023	23.1	869015	19.2	959008	42.3	040992	67301	73963	42
19	828162	23.1	868900	19.2	959262	42.3	040738	67323	73944	41
20	828301	23.1	868785	19.2	959516	42.3	040484	67344	73924	40
21	9.828439	23.1	9.868670	19.2	9.959769	42.3	10.040231	67366	73904	39
22	828578	23.1	868555	19.2	960023	42.3	039977	67387	73885	38
23	828716	23.1	868440	19.2	960277	42.3	039723	67409	73865	37
24	828855	23.0	868324	19.2	960531	42.3	039469	67430	73846	36
25	828993	23.0	868209	19.2	960784	42.3	039216	67452	73826	35
26	829131	23.0	868093	19.2	961038	42.3	038962	67473	73806	34
27	829269	23.0	867978	19.3	961291	42.3	038709	67495	73787	33
28	829407	23.0	867862	19.3	961545	42.3	038455	67516	73767	32
29	829545	23.0	867747	19.3	961799	42.3	038201	67538	73747	31
30	829683	23.0	867631	19.3	962052	42.3	037948	67559	73728	30
31	9.829821	22.9	9.867515	19.3	9.962306	42.3	10.037694	67580	73708	29
32	829959	22.9	867399	19.3	962560	42.3	037440	67602	73688	28
33	830097	22.9	867283	19.3	962813	42.3	037187	67623	73669	27
34	830234	22.9	867167	19.3	963067	42.3	036933	67645	73649	26
35	830372	22.9	867051	19.3	963320	42.3	036680	67666	73629	25
36	830509	22.9	866935	19.4	963574	42.3	036426	67688	73610	24
37	830646	22.9	866819	19.4	963827	42.3	036173	67709	73590	23
38	830784	22.9	866703	19.4	964081	42.3	035919	67730	73570	22
39	830921	22.8	866586	19.4	964335	42.3	035665	67752	73551	21
40	831058	22.8	866470	19.4	964588	42.2	035412	67773	73531	20
41	9.831195	22.8	9.866353	19.4	9.964842	42.2	10.035158	67795	73511	19
42	831332	22.8	866237	19.4	965095	42.2	034905	67816	73491	18
43	831469	22.8	866120	19.4	965349	42.2	034651	67837	73472	17
44	831606	22.8	866004	19.5	965602	42.2	034398	67859	73452	16
45	831742	22.8	865887	19.5	965855	42.2	034145	67880	73432	15
46	831879	22.8	865770	19.5	966109	42.2	033891	67901	73413	14
47	832015	22.7	865653	19.5	966362	42.2	033638	67923	73393	13
48	832152	22.7	865536	19.5	966616	42.2	033384	67944	73373	12
49	832288	22.7	865419	19.5	966869	42.2	033131	67965	73353	11
50	832425	22.7	865302	19.5	967123	42.2	032877	67987	73333	10
51	9.832561	22.7	9.865185	19.5	9.967376	42.2	10.032624	68008	73314	9
52	832697	22.7	865068	19.5	967629	42.2	032371	68029	73294	8
53	832833	22.7	864950	19.5	967883	42.2	032117	68051	73274	7
54	832969	22.6	864833	19.6	968136	42.2	031864	68072	73254	6
55	833105	22.6	864716	19.6	968389	42.2	031611	68093	73234	5
56	833241	22.6	864598	19.6	968643	42.2	031357	68115	73215	4
57	833377	22.6	864481	19.6	968896	42.2	031104	68136	73195	3
58	833512	22.6	864363	19.6	969149	42.2	030851	68157	73175	2
59	833648	22.6	864245	19.6	969403	42.2	030597	68179	73155	1
60	833783		864127		969656		030344	68200	73135	0
	Cosine.		Sine.		Cotang.		Tang.	N. cos.	N. sine.	′

47 Degrees.

′	Sine.	D. 10″	Cosine.	D. 10″	Tang.	D. 10″	Cotang.	N. sine.	N. cos.	
0	9.833783	22.6	9.864127	19.6	9.969656	42.2	10.030344	68200	73135	60
1	833919	22.5	864010	19.6	969909	42.2	030091	68221	73116	59
2	834054	22.5	863892	19.7	970162	42.2	029838	68242	73096	58
3	834189	22.5	863774	19.7	970416	42.2	029584	68264	73076	57
4	834325	22.5	863656	19.7	970669	42.2	029331	68285	73056	56
5	834460	22.5	863538	19.7	970922	42.2	029078	68306	73036	55
6	834595	22.5	863419	19.7	971175	42.2	028825	68327	73016	54
7	834730	22.5	863301	19.7	971429	42.2	028571	68349	72996	53
8	834865	22.5	863183	19.7	971682	42.2	028318	68370	72976	52
9	834999	22.4	863064	19.7	971935	42.2	028065	68391	72957	51
10	835134	22.4	862946	19.8	972188	42.2	027812	68412	72937	50
11	9.835269	22.4	9.862827	19.8	9.972441	42.2	10.027559	68434	72917	49
12	835403	22.4	862709	19.8	972694	42.2	027306	68455	72897	48
13	835538	22.4	862590	19.8	972948	42.2	027052	68476	72877	47
14	835672	22.4	862471	19.8	973201	42.2	026799	68497	72857	46
15	835807	22.4	862353	19.8	973454	42.2	026546	68518	72837	45
16	835941	22.4	862234	19.8	973707	42.2	026293	68539	72817	44
17	836075	22.3	862115	19.8	973960	42.2	026040	68561	72797	43
18	836209	22.3	861996	19.8	974213	42.2	025787	68582	72777	42
19	836343	22.3	861877	19.8	974466	42.2	025534	68603	72757	41
20	836477	22.3	861758	19.9	974719	42.2	025281	68624	72737	40
21	9.836611	22.3	9.861638	19.9	9.974973	42.2	10.025027	68645	72717	39
22	836745	22.3	861519	19.9	975226	42.2	024774	68666	72697	38
23	836878	22.3	861400	19.9	975479	42.2	024521	68688	72677	37
24	837012	22.2	861280	19.9	975732	42.2	024268	68709	72657	36
25	837146	22.2	861161	19.9	975985	42.2	024015	68730	72637	35
26	837279	22.2	861041	19.9	976238	42.2	023762	68751	72617	34
27	837412	22.2	860922	19.9	976491	42.2	023509	68772	72597	33
28	837546	22.2	860802	19.9	976744	42.2	023256	68793	72577	32
29	837679	22.2	860682	20.0	976997	42.2	023003	68814	72557	31
30	837812	22.2	860562	20.0	977250	42.2	022750	68835	72537	30
31	9.837945	22.2	9.860442	20.0	9.977503	42.2	10.022497	68857	72517	29
32	838078	22.1	860322	20.0	977756	42.2	022244	68878	72497	28
33	838211	22.1	860202	20.0	978009	42.2	021991	68899	72477	27
34	838344	22.1	860082	20.0	978262	42.2	021738	68920	72457	26
35	838477	22.1	859962	20.0	978515	42.2	021485	68941	72437	25
36	838610	22.1	859842	20.0	978768	42.2	021232	68962	72417	24
37	838742	22.1	859721	20.1	979021	42.2	020979	68983	72397	23
38	838875	22.1	859601	20.1	979274	42.2	020726	69004	72377	22
39	839007	22.1	859480	20.1	979527	42.2	020473	69025	72357	21
40	839140	22.0	859360	20.1	979780	42.2	020220	69046	72337	20
41	9.839272	22.0	9.859239	20.1	9.980033	42.2	10.019967	69067	72317	19
42	839404	22.0	859119	20.1	980286	42.2	019714	69088	72297	18
43	839536	22.0	858998	20.1	980538	42.2	019462	69109	72277	17
44	839668	22.0	858877	20.1	980791	42.1	019209	69130	72257	16
45	839800	22.0	858756	20.2	981044	42.1	018956	69151	72236	15
46	839932	22.0	858635	20.2	981297	42.1	018703	69172	72216	14
47	840064	21.9	858514	20.2	981550	42.1	018450	69193	72196	13
48	840196	21.9	858393	20.2	981803	42.1	018197	69214	72176	12
49	840328	21.9	858272	20.2	982056	42.1	017944	69235	72156	11
50	840459	21.9	858151	20.2	982309	42.1	017691	69256	72136	10
51	9.840591	21.9	9.858029	20.2	9.982562	42.1	10.017438	69277	72116	9
52	840722	21.9	857908	20.2	982814	42.1	017186	69298	72095	8
53	840854	21.9	857786	20.2	983067	42.1	016933	69319	72075	7
54	840985	21.9	857665	20.3	983320	42.1	016680	69340	72055	6
55	841116	21.8	857543	20.3	983573	42.1	016427	69361	72035	5
56	841247	21.8	857422	20.3	983826	42.1	016174	69382	72015	4
57	841378	21.8	857300	20.3	984079	42.1	015921	69403	71995	3
58	841509	21.8	857178	20.3	984331	42.1	015669	69424	71974	2
59	841640	21.8	857056	20.3	984584	42.1	015416	69445	71954	1
60	841771		856934		984837		015163	69466	71934	0
	Cosine.		Sine.		Cotang.		Tang.	N. cos.	N. sine.	′

′	Sine.	D. 10″	Cosine.	D. 10″	Tang.	D. 10″	Cotang.	N. sine.	N. cos.	
0	9.841771	21.8	9.856934	20.3	9.984837	42.1	10.015163	69466	71934	60
1	841902	21.8	856812	20.3	985090	42.1	014910	69487	71914	59
2	842033	21.8	856690	20.4	985343	42.1	014657	69508	71894	58
3	842163	21.7	856568	20.4	985596	42.1	014404	69529	71873	57
4	842294	21.7	856446	20.4	985848	42.1	014152	69549	71853	56
5	842424	21.7	856323	20.4	986101	42.1	013899	69570	71833	55
6	842555	21.7	856201	20.4	986354	42.1	013646	69591	71813	54
7	842685	21.7	856078	20.4	986607	42.1	013393	69612	71792	53
8	842815	21.7	855956	20.4	986860	42.1	013140	69633	71772	52
9	842946	21.7	855833	20.4	987112	42.1	012888	69654	71752	51
10	843076	21.7	855711	20.5	987365	42.1	012635	69675	71732	50
11	9.843206	21.6	9.855588	20.5	9.987618	42.1	10.012382	69696	71711	49
12	843336	21.6	855465	20.5	987871	42.1	012129	69717	71691	48
13	843466	21.6	855342	20.5	988123	42.1	011877	69737	71671	47
14	843595	21.6	855219	20.5	988376	42.1	011624	69758	71650	46
15	843725	21.6	855096	20.5	988629	42.1	011371	69779	71630	45
16	843855	21.6	854973	20.5	988882	42.1	011118	69800	71610	44
17	843984	21.6	854850	20.5	989134	42.1	010866	69821	71590	43
18	844114	21.5	854727	20.6	989387	42.1	010613	69842	71569	42
19	844243	21.5	854603	20.6	989640	42.1	010360	69862	71549	41
20	844372	21.5	854480	20.6	989893	42.1	010107	69883	71529	40
21	9.844502	21.5	9.854356	20.6	9.990145	42.1	10.009855	69904	71508	39
22	844631	21.5	854233	20.6	990398	42.1	009602	69925	71488	38
23	844760	21.5	854109	20.6	990651	42.1	009349	69946	71468	37
24	844889	21.5	853986	20.6	990903	42.1	009097	69966	71447	36
25	845018	21.5	853862	20.6	991156	42.1	008844	69987	71427	35
26	845147	21.5	853738	20.6	991409	42.1	008591	70008	71407	34
27	845276	21.4	853614	20.7	991662	42.1	008338	70029	71386	33
28	845405	21.4	853490	20.7	991914	42.1	008086	70049	71366	32
29	845533	21.4	853366	20.7	992167	42.1	007833	70070	71345	31
30	845662	21.4	853242	20.7	992420	42.1	007580	70091	71325	30
31	9.845790	21.4	9.853118	20.7	9.992672	42.1	10.007328	70112	71305	29
32	845919	21.4	852994	20.7	992925	42.1	007075	70132	71284	28
33	846047	21.4	852869	20.7	993178	42.1	006822	70153	71264	27
34	846175	21.4	852745	20.7	993430	42.1	006570	70174	71243	26
35	846304	21.4	852620	20.7	993683	42.1	006317	70195	71223	25
36	846432	21.3	852496	20.8	993936	42.1	006064	70215	71203	24
37	846560	21.3	852371	20.8	994189	42.1	005811	70236	71182	23
38	846688	21.3	852247	20.8	994441	42.1	005559	70257	71162	22
39	846816	21.3	852122	20.8	994694	42.1	005306	70277	71141	21
40	846944	21.3	851997	20.8	994947	42.1	005053	70298	71121	20
41	9.847071	21.3	9.851872	20.8	9.995199	42.1	10.004801	70319	71100	19
42	847199	21.3	851747	20.8	995452	42.1	004548	70339	71080	18
43	847327	21.3	851622	20.8	995705	42.1	004295	70360	71059	17
44	847454	21.2	851497	20.9	995957	42.1	004043	70381	71039	16
45	847582	21.2	851372	20.9	996210	42.1	003790	70401	71019	15
46	847709	21.2	851246	20.9	996463	42.1	003537	70422	70998	14
47	847836	21.2	851121	20.9	996715	42.1	003285	70443	70978	13
48	847964	21.2	850996	20.9	996968	42.1	003032	70463	70957	12
49	848091	21.2	850870	20.9	997221	42.1	002779	70484	70937	11
50	848218	21.2	850745	20.9	997473	42.1	002527	70505	70916	10
51	9.848345	21.2	9.850619	20.9	9.997726	42.1	10.002274	70525	70896	9
52	848472	21.1	850493	21.0	997979	42.1	002021	70546	70875	8
53	848599	21.1	850368	21.0	998231	42.1	001769	70567	70855	7
54	848726	21.1	850242	21.0	998484	42.1	001516	70587	70834	6
55	848852	21.1	850116	21.0	998737	42.1	001263	70608	70813	5
56	848979	21.1	849990	21.0	998989	42.1	001011	70628	70793	4
57	849106	21.1	849864	21.0	999242	42.1	000758	70649	70772	3
58	849232	21.1	849738	21.0	999495	42.1	000505	70670	70752	2
59	849359	21.1	849611	21.0	999748	42.1	000253	70690	70731	1
60	849485		849485		10.000000		000000	70711	70711	0
	Cosine.		Sine.		Cotang.		Tang.	N. cos.	N. sine.	′

45 Degrees.

TABLE III.

LOGARITHMS OF NUMBERS,

FROM 1 TO 110,

INCLUDING TWELVE DECIMAL PLACES.

N.	Log.	N.	Log.
1	0. 000 000 000 000	36	1. 556 302 500 767
2	0. 301 029 995 644	37	1. 568 201 724 067
3	0. 477 121 254 720	38	1. 579 783 596 617
4	0. 602 059 991 328	39	1. 591 064 607 264
5	0. 698 970 004 336	40	1. 602 059 991 328
6	0. 778 151 250 384	41	1. 612 783 846 720
7	0. 845 098 040 014	42	1. 623 249 290 398
8	0. 903 089 986 992	43	1. 633 468 455 579
9	0. 954 242 509 440	44	1. 643 452 676 486
10	1. 000 000 000 000	45	1. 653 212 513 775
11	1. 041 392 685 158	46	1. 662 757 831 682
12	1. 079 181 246 048	47	1. 672 097 857 936
13	1. 113 943 352 309	48	1. 681 241 237 376
14	1. 146 128 035 678	49	1. 690 196 080 028
15	1. 176 091 259 059	50	1. 698 970 004 336
16	1. 204 119 982 656	51	1. 707 570 176 098
17	1. 230 448 921 378	52	1. 716 003 243 635
18	1. 255 272 505 103	53	1. 724 275 869 601
19	1. 278 753 600 953	54	1. 732 393 759 823
20	1. 301 029 995 664	55	1. 740 362 689 494
21	1. 322 219 294 734	56	1 748 188 027 006
22	1. 342 422 680 822	57	1. 755 874 855 672
23	1. 361 727 836 076	58	1. 763 427 993 563
24	1. 380 211 241 712	59	1. 770 852 011 642
25	1. 397 940 008 672	60	1. 778 151 250 384
26	1. 414 973 347 971	61	1. 785 329 835 011
27	1. 431 363 764 159	62	1. 792 391 689 492
28	1. 447 158 031 342	63	1. 799 340 549 454
29	1. 462 397 997 899	64	1. 806 179 973 984
30	1. 477 121 254 720	65	1. 812 913 356 643
31	1. 491 361 693 834	66	1. 819 543 935 542
32	1. 505 149 978 320	67	1. 826 074 302 701
33	1. 518 513 939 878	68	1. 832 508 912 706
34	1. 531 478 917 042	69	1. 838 849 090 737
35	1. 544 068 044 350	70	1. 845 098 040 014

N.	Log.					N.	Log.				
71	1.	851	258	348	719	91	1.	959	041	392	321
72	1.	857	332	496	431	92	1.	963	787	827	346
73	1.	863	322	860	120	93	1.	968	482	948	554
74	1.	869	231	719	731	94	1.	973	127	853	600
75	1.	875	061	263	392	95	1.	977	723	605	289
76	1.	880	813	592	281	96	1.	982	271	233	040
77	1.	886	490	725	172	97	1.	986	771	734	266
78	1.	892	094	602	690	98	1.	991	226	075	692
79	1.	897	627	091	290	99	1.	995	635	194	598
80	1.	903	089	986	992	100	2.	000	000	000	000
81	1.	908	485	018	879	101	2.	004	321	373	783
82	1.	913	813	852	384	102	2.	008	600	171	762
83	1.	919	078	092	376	103	2.	012	837	224	705
84	1.	924	279	286	062	104	2.	017	033	339	299
85	1.	929	418	925	714	105	2.	021	189	299	070
86	1.	934	498	451	244	106	2.	025	305	865	265
87	1.	939	519	252	619	107	2.	029	383	777	685
88	1.	944	482	672	150	108	2.	033	423	755	487
89	1.	949	390	006	645	109	2.	037	426	497	941
90	1.	954	242	509	439	110	2.	041	392	685	158

LOGARITHMS OF THE PRIME NUMBERS

FROM 110 TO 1129.

INCLUDING TWELVE DECIMAL PLACES.

N.	Log.					N.	Log.				
113	2.	053	078	443	483	197	2.	294	466	266	162
127	2.	103	803	720	956	199	2.	298	853	076	410
131	2.	117	271	295	656	211	2.	324	282	455	298
137	2.	136	720	567	156	223	2.	348	304	863	222
139	2.	143	014	800	254	227	2.	356	025	857	189
149	3.	173	186	268	412	229	2.	359	835	482	343
151	2.	178	976	947	293	233	2.	367	355	922	471
157	2.	195	899	653	409	239	2.	378	397	902	352
163	2.	212	187	604	404	241	2.	382	017	042	576
167	2.	222	716	471	148	251	2.	399	673	721	509
173	2.	238	046	103	129	257	2.	409	933	123	332
179	2.	252	853	030	980	263	2.	419	955	748	490
181	2.	257	678	574	869	269	2.	429	752	261	993
191	2.	281	033	367	248	271	2.	432	969	290	877
193	2.	285	557	309	008	277	2.	442	479	768	999

N.	Log.	N.	Log.
281	2. 448 706 319 906	601	2. 778 874 471 998
283	2. 451 786 435 523	607	2. 783 188 691 074
293	2. 466 867 523 562	613	2. 787 460 556 130
307	2. 487 138 375 477	617	2. 790 285 164 033
311	2. 492 760 389 026	619	2. 791 690 648 987
313	2. 495 544 337 550	631	2. 800 029 359 232
317	2. 501 059 267 324	641	2. 806 858 879 634
331	2. 519 827 993 783	643	2. 808 210 973 921
337	2. 527 629 883 034	647	2. 810 904 280 666
347	2. 540 329 475 079	653	2. 814 912 981 274
349	2. 542 826 426 673	659	2. 818 885 490 409
353	2. 547 774 138 015	661	2. 820 201 459 485
359	2. 555 094 447 578	673	2. 828 015 064 225
367	2. 564 666 064 254	677	2. 830 588 667 946
373	2. 571 708 831 809	683	2. 834 420 703 630
379	2. 578 639 209 957	691	2. 839 477 902 551
383	2. 583 198 773 980	701	2. 845 718 017 237
389	2. 589 949 601 323	709	2. 850 646 235 112
397	2. 598 790 506 763	719	2. 856 728 890 383
401	2. 603 144 372 687	727	2. 861 534 410 855
409	2. 611 723 298 019	733	2. 865 103 970 639
419	2. 622 214 022 971	739	2. 868 643 643 162
421	2. 624 282 095 835	743	2. 870 988 813 759
431	2. 634 477 268 999	751	2. 875 639 937 004
433	2. 636 488 015 871	757	2. 879 095 879 497
439	2. 642 464 520 242	761	2. 881 384 656 769
443	2. 646 403 726 235	769	2. 885 926 339 800
449	2. 652 246 388 777	773	2. 888 179 493 917
457	2. 659 916 200 054	787	2. 895 974 732 358
461	2. 663 700 925 389	797	2. 901 458 321 400
463	2. 665 580 991 012	809	2. 977 948 459 773
467	2. 669 317 881 008	811	2. 909 020 854 210
479	2. 680 335 513 415	821	2. 914 343 157 120
487	2. 687 528 961 120	823	2. 915 399 835 203
491	2. 691 081 487 026	827	2. 917 505 509 487
499	2. 698 100 545 623	829	2. 918 554 530 558
503	2. 701 567 985 083	839	2. 923 761 960 830
509	2. 706 717 782 345	853	2. 930 949 031 163
521	2. 716 837 623 304	857	2. 932 980 821 917
523	2. 718 502 688 873	859	2. 933 993 163 838
541	2. 733 197 265 134	863	2. 936 010 794 546
547	2. 737 987 326 358	877	2. 942 999 593 360
557	2. 745 855 195 192	881	2. 944 975 908 412
563	2. 750 508 395 940	883	2. 945 960 703 512
569	2. 755 112 178 598	887	2. 947 923 619 839
571	2. 756 636 108 333	907	2. 957 607 287 059
577	2. 761 175 813 171	911	2. 959 518 376 972
587	2. 768 638 004 455	919	2. 963 315 513 609
593	2. 773 054 693 364	929	2. 968 015 713 997
599	2. 777 427 303 257	937	2. 971 739 590 780

N.	Log.					N.	Log.				
941	2.	973	589	620	234	1039	3.	016	615	547	558
947	2.	976	349	979	055	1049	3.	020	775	488	195
953	2.	979	092	900	639	1051	3.	021	602	716	026
967	2.	985	426	474	084	1061	3.	025	715	383	898
971	2.	987	219	229	907	1063	3.	026	533	264	523
977	2.	989	894	559	717	1069	3.	028	977	705	205
983	2.	992	553	512	733	1087	3.	036	229	513	712
991	2.	996	073	604	003	1091	3.	037	824	749	671
997	2.	998	695	158	313	1093	3.	038	620	157	372
1009	3.	003	891	170	203	1097	3.	040	206	627	571
1013	3.	005	609	445	427	1103	3.	042	575	512	437
1019	3.	008	174	244	007	1109	3.	044	931	546	149
1021	3.	009	025	742	086	1117	3.	048	053	173	103
1031	3.	013	258	660	430	1123	3.	050	379	756	239
1033	3.	014	100	321	518	1129	3.	052	693	942	370

It is not necessary to extend this table, as the logarithm of any one of the higher numbers can be readily computed by the following formula, which may be found in any of the standard works on algebra, namely:

$$\text{Log. }(z+1)=\text{log. }z+0.8685889638\left(\frac{1}{2z+1}\right)$$

The result will be true to ten decimal places for all numbers over 1000, and true to twelve decimals for all numbers over 2000.

The logarithms of composite numbers can be determined by the combination of logarithms already in the table, and the prime numbers from the formula.

Thus, the number 3083 is a prime number, find its logarithm, true to ten places of decimals.

We first find the logarithm of 3082. By factoring this number, we find that it may be composed by the multiplication of 46 into 67.

Log. 46	1.	662 757 8316
Log. 67	1.	826 074 3027
Log. 3082	3.	488 832 1343

Now, $\text{Log. } 3083 = 3.4888321343 + \frac{0.8685889638}{6165}$

We give a few additional prime numbers:

1151	1223	1291	1373	1451	1511
1153	1229	1297	1381	1453	1523
1163	1231	1301	1399	1459	1531
1171	1237	1303	1409	1471	1543
1181	1249	1307	1423	1481	1549
1187	1259	1319	1427	1483	1553
1193	1277	1321	1429	1487	1559
1201	1279	1327	1433	1489	1567
1213	1283	1361	1439	1493	1571
1217	1289	1367	1447	1499	1579

AUXILIARY LOGARITHMS.

N.	Log.		N.	Log.	
1. 009	0. 003 891 170 203	A	1. 0009	0. 000 390 576 304	B
1. 008	0. 003 461 527 188		1. 0008	0. 000 347 233 698	
1. 007	0. 003 030 465 635		1. 0007	0. 000 303 836 798	
1. 006	0. 002 597 985 739		1. 0006	0. 000 260 435 561	
1. 005	0. 002 166 071 750		1. 0005	0. 000 217 099 966	
1. 004	0. 001 733 722 804		1. 0004	0. 000 173 690 053	
1. 003	0. 001 300 943 017		1. 0003	0. 000 130 268 803	
1. 002	0. 000 867 721 529		1. 0002	0. 000 086 850 213	
1. 001	0. 000 434 077 479		1. 0001	0. 000 043 427 277	

N.	Log.	
1. 00009	0. 000 039 084 741	C
1. 00008	0. 000 034 742 166	
1. 00007	0. 000 030 399 546	
1. 00006	0. 000 026 056 884	
1. 00005	0. 000 021 714 178	
1. 00004	0. 000 017 371 430	
1. 00003	0. 000 013 028 638	
1. 00002	0. 000 008 685 802	
1. 00001	0. 000 004 342 923 (a)	
1. 000001	0. 000 000 434 294 (b)	
1. 0000001	0. 000 000 043 429 (c)	
1. 00000001	0. 000 000 004 343 (d)	
1. 000000001	0. 000 000 000 434 (e)	
1. 0000000001	0. 000 000 000 043 (f)	

Number.	Log.
0. 4342944819	—1. 637 784 298

This decimal number is the modulus of our system of logarithms. Its logarithm is very useful in correcting other logarithms, as may be seen in the Chapter on Logarithms.

Distance.	½ Deg.		1 Deg.		1½ Deg.		2 Deg.	
	Lat.	Dep.	Lat.	Dep.	Lat.	Dep.	Lat.	Dep.
1	1. 00	0. 01	1. 00	0. 02	1. 00	0. 03	1. 00	0. 03
2	2. 00	0. 02	2. 00	0. 03	2. 00	0. 05	2. 00	0. 07
3	3. 00	0. 03	3. 00	0. 05	3. 00	0. 08	3. 00	0. 10
4	4. 00	0. 03	4. 00	0. 07	4. 00	0. 10	4. 00	0. 14
5	5. 00	0. 04	5. 00	0. 09	5. 00	0. 13	5. 00	0. 17
6	6. 00	0. 05	6. 90	0. 10	6. 00	0. 16	6. 00	0. 21
7	7. 00	0. 06	7. 00	0. 12	7. 00	0. 18	7. 00	0. 24
8	8. 00	0. 07	8. 00	0. 14	8. 00	0. 21	7. 99	0. 28
9	9. 00	0. 08	9. 00	0. 16	9. 00	0. 24	8. 99	0. 31
10	10. 00	0. 09	10. 00	0. 17	10. 00	0. 26	9. 99	0. 35
11	11. 00	0. 10	11. 00	0. 19	11. 00	0. 28	10. 99	0. 38
12	12. 00	0. 10	12. 00	0. 21	12. 00	0. 31	11. 99	0. 42
13	13. 00	0. 11	13. 00	0. 23	13. 00	0. 34	12. 99	0. 45
14	14. 00	0. 12	14. 00	0. 24	14. 00	0. 37	13. 99	0. 49
15	15. 00	0. 13	15. 00	0. 26	14. 99	0. 39	14. 99	0. 52
16	16. 00	0. 14	16. 00	0. 28	15. 99	0. 42	15. 99	0. 56
17	17. 00	0. 15	17. 00	0. 30	16. 99	0. 45	16. 99	0. 59
18	18. 00	0. 16	18. 00	0. 31	17. 99	0. 47	17. 99	0. 63
19	19. 00	0. 17	19. 00	0. 33	18. 99	0. 50	18. 99	0. 66
20	20. 00	0. 17	20. 00	0. 35	19. 99	0. 52	19. 99	0. 70
21	21. 00	0. 18	21. 00	0. 37	20. 99	0. 55	20. 99	0. 73
22	22. 00	0. 19	22. 00	0. 38	21. 99	0. 58	21. 99	0. 77
23	23. 00	0. 20	23. 00	0. 40	22. 99	0. 60	22. 99	0. 80
24	24. 00	0. 21	24. 00	0. 42	23. 99	0. 63	23. 99	0. 84
25	25. 00	0. 22	25. 00	0. 44	24. 99	0. 65	24. 98	0. 87
26	26. 00	0. 23	26. 00	0. 45	25. 99	0. 68	25. 98	0. 91
27	27. 00	0. 24	27. 00	0. 47	26. 99	0. 71	26. 98	0. 94
28	28. 00	0. 24	28. 00	0. 49	27. 99	0. 73	27. 98	0. 98
29	29. 00	0. 25	29. 00	0. 51	28. 99	0. 76	28. 98	1. 01
30	30. 00	0. 26	30. 00	0. 52	29. 99	0. 79	29. 98	1. 05
35	35. 00	0. 31	34. 99	0. 61	34. 99	0. 92	34. 98	1. 22
40	40. 00	0. 35	39. 99	0. 70	39. 99	1. 05	39. 98	1. 40
45	45. 00	0. 39	44. 90	0. 79	44. 99	1. 18	44. 97	1. 57
50	50. 00	0. 44	49. 99	0. 87	49. 98	1. 31	49. 97	1. 74
55	55. 00	0. 48	54. 99	0. 96	54. 98	1. 44	54. 97	1. 92
60	60. 00	0. 52	59. 90	0. 05	59. 98	1. 57	59. 96	2. 09
65	65. 00	0. 57	64. 99	1. 13	64. 98	1. 70	64. 96	2. 27
70	70. 00	0. 61	69. 99	1. 22	69. 98	1. 83	69. 96	2. 44
75	75. 00	0. 65	74. 99	1. 31	74. 97	1. 96	74. 95	2. 62
80	80. 00	0. 70	79. 99	1. 40	79. 97	2. 09	79. 95	2. 79
85	85. 00	0. 74	84. 99	1. 48	84. 97	2. 23	84. 95	2. 97
90	90. 00	0. 79	89. 99	1. 57	89. 97	2. 36	89. 95	3. 14
95	90. 00	0. 83	94. 99	1. 66	94. 97	2. 49	94. 94	3. 32
100	100. 00	0. 87	99. 98	1. 75	99. 97	2. 62	99. 94	3. 49
	Dep.	Lat.	Dep.	Lat.	Dep.	Lat.	Dep.	Lat.
	89½ Deg.		89 Deg.		88½ Deg.		88 Deg.	

Distance.	2½ Deg.		3 Deg.		3½ Deg.		4 Deg.	
	Lat.	Dep.	Lat.	Dep.	Lat.	Dep.	Lat.	Dep.
1	1.00	0.04	1.00	0.05	1.00	0.06	1.00	0.07
2	2.00	0.09	2.00	0.10	2.00	0.12	2.00	0.14
3	3.00	0.13	3.00	0.16	2.99	0.18	2.99	0.21
4	4.00	0.17	3.99	0.21	3.99	0.24	3.99	0.28
5	5.00	0.22	4.99	0.26	4.99	0.31	4.99	0.35
6	5.99	0.26	5.99	0.31	5.99	0.37	5.99	0.42
7	6.99	0.31	6.99	0.37	6.99	0.43	6.98	0.49
8	7.99	0.35	7.99	0.42	7.99	0.49	7.98	0.56
9	8.99	0.39	8.99	0.47	8.98	0.55	8.98	0.63
10	9.99	0.44	9.99	0.52	9.98	0.61	9.98	0.70
11	10.99	0.48	10.98	0.58	10.98	0.67	10.97	0.77
12	11.99	0.52	11.98	0.63	11.98	0.73	11.97	0.84
13	12.99	0.57	12.98	0.68	12.99	0.79	12.97	0.91
14	13.99	0.61	13.98	0.73	13.97	0.85	13.97	0.98
15	14.99	0.65	14.98	0.79	14.97	0.92	14.96	1.05
16	15.99	0.70	15.98	0.84	15.97	0.98	15.96	1.12
17	16.98	0.74	16.98	0.89	16.97	1.04	16.96	1.19
18	17.98	0.79	17.98	0.94	17.97	1.10	17.96	1.26
19	18.98	0.83	18.98	0.99	18.96	1.16	18.95	1.33
20	19.98	0.87	19.97	1.05	19.96	1.22	19.95	1.40
21	20.98	0.92	20.97	1.10	20.96	1.28	20.95	1.46
22	21.98	0.96	21.97	1.15	21.96	1.34	21.95	1.53
23	22.98	1.00	22.97	1.20	22.96	1.40	22.94	1.60
24	23.98	1.05	23.97	1.26	23.96	1.47	23.94	1.67
25	24.98	1.09	24.97	1.31	24.95	1.53	24.94	1.74
26	25.98	1.13	25.96	1.36	25.95	1.59	25.94	1.81
27	26.97	1.18	26.96	1.41	26.95	1.65	26.93	1.88
28	27.97	1.22	27.96	1.47	27.95	1.71	27.93	1.95
29	28.97	1.26	28.96	1.52	28.95	1.77	28.93	2.02
30	29.97	1.31	29.96	1.57	29.94	1.83	29.93	2.09
35	34.97	1.53	34.95	1.83	34.93	2.14	34.91	2.44
40	39.96	1.75	39.95	2.09	39.93	2.44	39.90	2.79
45	44.96	1.96	44.94	2.36	44.92	2.75	44.89	3.14
50	49.95	2.18	49.93	2.62	49.91	3.05	49.88	3.49
55	54.95	2.40	54.92	2.88	54.90	3.36	54.87	3.84
60	59.94	2.62	59.92	3.14	59.89	3.66	59.83	4.19
65	64.94	2.84	64.91	3.40	64.88	3.97	64.84	4.53
70	69.93	3.05	69.90	3.66	69.87	4.27	69.83	4.88
75	74.93	3.27	74.90	3.93	74.86	4.58	74.82	5.23
80	79.92	3.49	79.89	4.19	79.85	4.88	79.81	5.58
85	84.92	3.71	84.88	4.45	84.84	5.19	84.79	5.93
90	89.91	3.93	89.98	4.71	89.83	5.49	89.78	6.28
95	94.91	4.14	94.87	4.97	94.82	5.80	94.77	6.63
100	99.91	4.36	99.86	5.23	99.81	6.10	99.76	6.98
	Dep.	Lat.	Dep.	Lat.	Dep.	Lat.	Dep.	Lat.
	87½ Deg.		87 Deg.		86½ Deg.		86 Deg.	

Distance.	4½ Deg.		5 Deg.		5½ Deg.		6 Deg.	
	Lat.	Dep.	Lat.	Dep.	Lat.	Dep.	Lat.	Dep.
1	1. 00	0. 08	1. 00	0. 09	1. 00	0. 10	0. 99	0. 10
2	1. 99	0. 16	1. 99	0. 17	1. 99	0. 19	1. 99	0. 21
3	2. 99	0. 24	2. 99	0. 26	2. 99	0. 29	2. 98	0. 31
4	3. 99	0. 31	3. 98	0. 35	3. 98	0. 38	3. 98	0. 41
5	4. 98	0. 39	4. 98	0. 44	4. 98	0. 48	4. 97	0. 52
6	5. 98	0. 47	5. 98	0. 52	5. 97	0. 58	5. 97	0. 63
7	6. 98	0. 55	6. 97	0. 61	6. 97	0. 67	6. 96	0. 73
8	7. 98	0. 63	7. 97	0. 70	7. 96	0. 76	7. 96	0. 84
9	8. 97	0. 71	8. 97	0. 78	8. 96	0. 86	8. 95	0. 94
10	9. 97	0. 78	9. 96	0. 87	9. 95	0. 96	9. 95	1. 05
11	10. 97	0. 86	10. 96	0. 96	10. 95	1. 05	10. 94	1. 15
12	11. 96	0. 94	11. 95	1. 05	11. 94	1. 15	11. 93	1. 25
13	12. 96	1. 02	12. 95	1. 13	12. 94	1. 25	12. 93	1. 36
14	13. 96	1. 10	13. 95	1. 22	13. 94	1. 34	13. 92	1. 46
15	14. 95	1. 18	14. 94	1. 31	14. 93	1. 44	14. 92	1. 57
16	15. 95	1. 26	15. 94	1. 39	15. 93	1. 53	15. 91	1. 67
17	16. 95	1. 33	16. 94	1. 48	16. 92	1. 63	16. 91	1. 78
18	17. 94	1. 41	17. 93	1. 57	17. 92	1. 73	17. 90	1. 88
19	18. 94	1. 49	18. 93	1. 66	18. 91	1. 82	18. 90	1. 99
20	19. 94	1. 57	19. 92	1. 74	19. 91	1. 92	19. 89	2. 09
21	20. 94	1. 65	20. 92	1. 83	20. 90	2. 01	20. 88	2. 20
22	21. 93	1. 73	21. 92	1. 92	21. 90	2. 11	21. 88	2. 30
23	22. 93	1. 80	22. 91	2. 00	22. 89	2. 20	22. 87	2. 40
24	23. 93	1. 88	23. 91	2. 09	23. 89	2. 30	23. 87	2. 51
25	24. 92	1. 96	24. 90	2. 18	24. 88	2. 40	24. 86	2. 61
26	25. 92	2. 04	25. 90	2. 27	25. 88	2. 49	25. 86	2. 72
27	26. 92	2. 12	26. 90	2. 35	26. 88	2. 59	26. 85	2. 82
28	27. 91	2. 20	27. 89	2. 44	27. 87	2. 68	27. 85	2. 93
29	28. 91	2. 28	28. 89	2. 53	28. 87	2. 78	28. 84	3. 03
30	29. 21	2. 35	29. 89	2. 61	29. 86	2. 88	29. 84	3. 14
35	34. 89	2. 75	34. 87	3. 05	34. 84	3. 35	34. 81	3. 66
40	39. 88	3. 14	39. 85	3. 49	39. 82	3. 83	39. 78	4. 18
45	44. 86	3. 53	44. 83	3. 92	44. 79	4. 31	44. 75	4. 70
50	49. 85	3. 92	49. 81	4. 36	49. 77	4. 79	49. 73	5. 23
55	54. 85	4. 08	54. 79	4. 79	54. 75	5. 27	54. 70	5. 75
60	59. 84	4. 45	59. 77	5. 23	59. 72	5. 75	59. 67	6. 27
65	64. 82	4. 82	64. 75	5. 67	64. 70	6. 23	64. 64	6. 79
70	69. 81	5. 19	69. 73	6. 10	69. 68	6. 71	69. 62	7. 32
75	74. 79	5. 65	74. 71	6. 54	74. 65	7. 19	74. 59	7. 84
80	79. 78	5. 93	79. 70	6. 97	79. 63	7. 67	76. 56	8. 36
85	84. 77	6. 30	84. 68	7. 41	84. 61	8. 15	84. 53	8. 88
90	89. 75	6. 67	89. 66	7. 84	89. 59	8. 63	89. 51	9. 41
95	94. 74	7. 04	94. 64	8. 28	94. 56	9. 11	94. 48	9. 93
100	99. 73	7. 41	99. 62	8. 72	99. 54	9. 58	99. 45	10. 43
	Dep.	Lat.	Dep.	Lat.	Dep.	Lat.	Dep.	Lat.
	85½ Deg.		85 Deg.		84½ Deg.		84 Deg.	

Distance.	6½ Deg.		7 Deg.		7½ Deg.		8 Deg.	
	Lat.	Dep.	Lat.	Dep.	Lat.	Dep.	Lat.	Dep.
1	0. 99	0. 11	0. 99	0. 12	0. 99	0. 13	0. 99	0. 14
2	1. 99	0. 23	1. 99	0. 24	1. 98	0. 26	1. 98	0. 28
3	2. 98	0. 34	2. 98	0. 37	2 97	0. 39	2. 97	0. 42
4	3. 97	0. 45	3. 97	0. 49	3. 97	0. 52	3. 96	0. 56
5	4. 97	0. 57	4. 96	0. 61	4. 96	0. 65	4. 95	0. 70
6	5. 96	0. 68	5. 96	0. 73	5. 95	0. 78	5. 94	0. 84
7	6. 96	0. 79	6. 95	0. 85	6. 94	0. 91	6. 93	0. 97
8	7. 95	0. 91	7. 94	0. 97	7. 93	1. 04	7. 92	1. 11
9	8. 94	1. 02	8. 93	1. 10	8. 92	1. 17	8. 91	1. 25
10	9. 94	1. 13	9. 93	1. 22	9. 91	1. 31	9. 90	1. 39
11	10. 93	1. 25	10. 92	1. 34	10. 91	1. 44	10. 89	1. 53
12	11. 92	1. 36	11. 91	1. 46	11. 90	1. 57	11. 88	1. 67
13	12. 92	1. 47	12. 90	1. 58	12. 89	1. 70	12. 87	1. 81
14	13. 91	1. 59	13. 90	1. 71	13. 88	1. 83	13. 86	1. 95
15	14. 90	1. 70	14. 89	1. 83	14. 87	1. 96	14. 85	2. 09
16	15. 90	1. 81	15. 88	1. 95	15. 86	2. 09	15. 84	2. 23
17	16. 89	1. 92	16. 87	2. 07	16. 85	2. 22	16. 83	2. 37
18	17. 88	2. 04	17. 87	2. 19	17. 85	2. 35	17. 82	2. 51
19	18. 88	2. 15	18. 86	2. 32	18. 84	2. 48	18. 82	2. 64
20	19. 87	2. 26	19. 85	2. 44	19. 83	2. 61	19. 81	2. 78
21	20. 87	2. 38	20. 84	2. 56	20. 82	2. 74	20. 80	2. 92
22	21. 86	2. 49	21. 84	2. 68	21. 81	2. 87	21. 79	3. 06
23	22. 85	2. 60	22. 83	2. 80	22. 80	3. 00	22. 78	3. 20
24	23. 85	2. 72	23. 82	2. 92	23. 79	3. 13	23. 77	3. 34
25	24. 84	2. 83	24. 81	3. 05	24. 79	3. 26	24. 76	3. 48
26	25. 83	2. 94	25. 81	3. 17	25. 78	3. 39	25. 75	3. 62
27	26. 83	3. 06	26. 80	3. 29	26. 77	3. 52	26. 74	3. 76
28	27. 82	3. 17	27. 79	3. 41	27. 76	3. 65	27. 73	3. 90
29	28. 81	3. 28	28. 78	3. 53	28. 75	3. 79	28. 72	4. 04
30	29. 81	3. 40	29. 78	3. 66	29. 74	3. 92	29. 71	4. 18
35	34. 78	3. 96	34. 74	4. 27	34. 70	4. 57	34. 66	4. 87
40	39. 74	4. 53	39. 70	4. 87	39. 66	5. 22	39. 61	5. 57
45	44. 71	5. 09	44. 67	5. 48	44. 62	5. 87	44. 56	6. 26
50	49. 68	5. 66	49. 63	6. 09	49. 57	6. 53	49. 51	6. 96
55	54. 65	6. 23	54. 59	6. 70	55. 58	6. 70	54. 46	7. 65
60	59. 61	6. 79	59. 55	7. 31	59. 55	7. 31	59. 42	8. 35
65	64. 58	7. 36	64. 52	7. 92	64. 52	7. 92	64. 37	9 05
70	69. 55	7. 92	69. 48	8. 53	69. 48	8. 53	69. 32	9. 74
75	74. 52	8. 49	74. 44	9. 14	74. 44	9. 14	74. 27	10. 44
80	79. 49	9. 06	79. 40	9. 75	79. 40	9. 75	79. 22	11. 13
85	84. 45	9. 62	84. 37	10. 36	84. 37	10. 36	84. 17	11. 83
90	89. 42	10. 19	89. 33	10. 97	89. 33	10. 97	89. 12	12. 53
95	94. 39	10. 75	94. 29	11. 58	94. 29	11. 58	94. 08	13. 22
100	99. 36	11. 32	99. 25	12. 19	99. 25	12. 19	99. 03	13. 92
	Dep.	Lat.	Dep.	Lat.	Dep.	Lat.	Dep.	Lat.
	83½ Deg.		83 Deg.		82½ Deg.		82 Deg.	

Distance.	8½ Deg.		9 Deg.		9 ½ Deg.		10 Deg.	
	Lat.	Dep.	Lat.	Dep.	Lat.	Dep.	Lat.	Dep.
1	0. 99	0. 15	0. 99	0. 16	0. 99	0. 17	0. 98	0. 17
2	1. 98	0. 30	1. 98	0. 31	1. 97	0. 33	1. 97	0. 35
3	2. 97	0. 44	2. 96	0. 47	2. 96	0. 50	2. 95	0. 52
4	3. 96	0. 59	3. 95	0. 63	3. 95	0. 66	3. 94	0. 69
5	4. 95	0. 74	4. 94	0. 78	4. 93	0. 83	4. 92	0. 87
6	5. 93	0. 89	5. 93	0. 94	5. 92	0. 99	5. 91	1. 04
7	6. 92	1. 03	6. 91	1. 10	6. 90	1. 16	6. 89	1. 22
8	7. 91	1. 18	7. 90	1. 25	7. 89	1. 32	7. 88	1. 39
9	8. 90	1. 33	8. 89	1. 41	8. 88	1. 49	8. 86	1. 56
10	9. 89	1. 48	9. 88	1. 56	9. 86	1. 65	9. 85	1. 74
11	10. 88	1. 63	10. 86	1. 72	10. 85	1. 82	10. 83	1. 91
12	11. 87	1. 77	11. 85	1. 88	11. 84	1. 98	11. 82	2. 08
13	12. 86	1. 92	12. 84	2. 03	12. 82	2. 15	12. 80	2. 26
14	13. 85	2. 07	13. 83	2. 19	13. 81	2. 31	13. 79	2. 43
15	14. 84	2. 22	14. 82	2. 35	14. 79	2. 48	14. 77	2. 60
16	15. 82	2. 36	15. 80	2. 50	15. 78	2. 64	15. 76	2. 78
17	16. 81	2. 51	16. 79	2. 66	16. 77	2. 81	16. 74	2. 95
18	17. 80	2. 66	17. 78	2. 82	17. 75	2. 97	17. 73	3. 13
19	18. 79	2. 81	18. 77	2. 97	18. 74	3. 14	18. 71	3. 30
20	19. 78	2. 96	19. 75	3. 13	19. 73	3. 30	19. 70	3. 47
21	20. 77	3. 10	20. 74	3. 29	20. 71	3. 47	20. 68	3. 65
22	21. 76	3. 25	21. 73	3. 44	21. 70	3. 63	21. 67	3. 82
23	22. 75	3. 40	22. 72	3. 60	22. 68	3. 80	22. 65	3. 99
24	23. 74	3. 55	23. 70	3. 75	23. 67	3. 96	23. 64	4. 17
25	24. 73	3. 70	24. 69	3. 91	24. 66	4. 13	24. 62	4. 34
26	25. 71	3. 84	25. 68	4. 07	25. 64	4. 29	25. 61	4. 51
27	26. 70	3. 99	26. 67	4. 22	26. 63	4. 46	26. 59	4. 69
28	27. 69	4. 14	27. 66	4. 38	27. 62	4. 62	27. 57	4. 86
29	28. 68	4. 29	28. 64	4. 54	28. 60	4. 79	28. 56	5. 04
30	29. 67	4. 43	29. 63	4. 69	29. 59	4. 95	29. 54	5. 21
35	34. 62	5. 17	34. 57	5. 48	34. 52	5. 78	34. 47	6. 08
40	39. 56	5. 91	39. 51	6. 26	39. 45	6. 60	39. 39	6. 95
45	44. 51	6. 65	44. 45	7. 04	44. 38	7. 43	44. 32	7. 81
50	49. 45	7. 39	49. 38	7. 82	49. 32	8. 25	49. 24	8. 68
55	54. 40	8. 13	54. 32	8. 60	54. 25	9. 08	54. 16	9. 95
60	59. 34	8. 87	59. 26	9. 39	59. 18	9. 90	59. 09	10. 42
65	64. 29	9. 61	64. 20	10. 17	64. 11	10. 73	64. 01	11. 29
70	69. 23	10. 35	69. 14	10. 95	69. 04	11. 55	68. 94	12. 16
75	74. 18	11. 09	74. 08	11. 73	73. 97	12. 38	73. 86	13. 02
80	79. 12	11. 82	79. 02	12. 51	78. 90	13. 20	78. 78	13. 89
85	84. 07	12. 56	83. 95	13. 30	83. 83	14. 03	83. 71	14. 76
90	89. 01	13. 30	88. 89	14. 08	88. 77	14. 85	88. 63	15. 63
95	93. 96	14. 04	93. 83	14. 86	93. 70	15. 68	93. 56	16. 50
100	98. 90	14. 78	98. 77	15. 64	98. 63	16. 50	98. 48	17. 36
	Dep.	Lat.	Dep.	Lat.	Dep.	Lat.	Dep.	Lat.
	81½ Deg.		81 Deg.		80½ Deg.		80 Deg.	

Distance.	10½ Deg.		11 Deg.		11½ Deg.		12 Deg.	
	Lat.	Dep.	Lat.	Dep.	Lat.	Dep.	Lat.	Dep.
1	0.98	0.18	0.98	0.19	0.98	0.20	0.98	0.21
2	1.97	0.36	1.96	0.38	1.96	0.40	1.96	0.42
3	2.95	0.55	2.94	0.57	2.94	0.60	2.93	0.62
4	3.93	0.73	3.93	0.76	3.92	0.80	3.91	0.83
5	4.92	0.91	4.91	0.95	4.90	1.00	4.89	1.04
6	5.90	1.09	5.89	1.14	5.88	1.20	5.87	1.25
7	6.88	1.28	6.87	1.34	6.86	1.40	6.85	1.46
8	7.87	1.46	7.85	1.53	7.84	1.59	7.83	1.66
9	8.85	1.64	8.83	1.72	8.82	1.79	8.80	1.87
10	9.83	1.82	9.82	1.91	9.80	1.99	9.78	2.08
11	10.82	2.00	10.80	2.10	10.78	2.19	10.76	2.29
12	11.80	2.19	11.78	2.29	11.76	2.39	11.74	2.49
13	12.78	2.37	12.76	2.48	12.74	2.59	12.72	2.70
14	13.77	2.55	13.74	2.67	13.72	2.79	13.69	2.91
15	14.75	2.73	14.72	2.87	14.70	2.99	14.67	3.12
16	15.73	2.92	15.71	2.86	15.68	3.19	15.65	3.33
17	16.72	3.10	16.69	3.05	16.66	3.39	16.63	3.53
18	17.70	3.28	17.67	3.24	17.64	3.59	17.61	3.74
19	18.68	3.46	18.65	3.43	18.62	3.79	18.58	3.95
20	19.67	3.64	19.63	3.63	19.60	3.99	19.56	4.16
21	20.65	3.83	20.61	3.82	20.58	4.13	20.54	4.37
22	21.63	4.01	21.60	4.01	21.56	4.39	21.52	4.57
23	22.61	4.19	22.58	4.20	22.54	4.59	22.50	4.78
24	23.60	4.37	23.56	4.39	23.52	4.78	23.48	4.99
25	24.58	4.56	24.54	4.58	24.50	4.98	24.45	5.20
26	25.56	4.74	25.52	4.77	25.48	5.18	25.43	5.41
27	26.55	4.92	26.50	4.96	26.46	5.38	26.41	5.61
28	27.53	5.10	27.49	5.15	27.44	5.58	27.39	5.82
29	28.51	5.28	28.47	5.34	28.42	5.78	28.37	6.03
30	29.50	5.47	29.45	5.72	29.40	5.98	29.34	6.24
35	34.41	6.38	34.36	6.68	34.30	6.98	34.24	7.28
40	39.33	7.29	39.27	7.63	39.20	7.97	39.13	8.32
45	44.25	8.20	44.17	8.59	44.10	8.97	44.02	9.36
50	49.16	9.11	49.08	9.54	49.00	9.97	48.91	10.40
55	54.08	10.02	53.99	10.49	53.90	10.97	53.80	11.44
60	59.00	10.93	58.90	11.45	58.80	11.96	58.69	12.47
65	63.91	11.85	63.81	12.40	63.70	12.96	63.58	13.51
70	68.83	12.76	68.71	13.36	68.59	13.96	68.47	14.55
75	73.74	13.67	73.62	14.31	73.49	14.95	73.36	15.59
80	78.66	14.58	78.53	15.26	78.39	15.95	78.25	16.63
85	83.58	15.49	83.44	16.22	83.29	16.95	83.14	17.67
90	88.49	16.40	88.35	17.17	88.19	17.94	88.03	18.71
95	93.41	17.31	93.25	18.13	93.09	18.94	92.92	19.75
100	98.33	18.22	98.16	19.08	97.99	19.94	97.81	20.79
	Dep.	Lat.	Dep.	Lat.	Dep.	Lat.	Dep.	Lat.
	79½ Deg.		79 Deg.		78½ Deg.		78 Deg.	

Distance.	12½ Deg.		13 Deg.		13½ Deg.		14 Deg.	
	Lat.	Dep.	Lat.	Dep.	Lat.	Dep.	Lat.	Dep.
1	0. 98	0. 22	0. 97	0. 23	0. 97	0. 23	0. 97	0. 24
2	1. 95	0. 43	1. 95	0. 46	1. 95	0. 47	1. 94	0. 48
3	2. 93	0. 65	2. 92	0. 67	2. 92	0. 70	2. 91	0. 73
4	3. 91	0. 87	3. 90	0. 90	3. 89	0. 93	3. 88	0. 97
5	4. 88	1. 08	4. 87	1. 12	4. 86	1. 17	4. 85	1. 21
6	5. 86	1. 30	5. 85	1. 35	5. 83	1. 40	5. 82	1. 45
7	6. 83	1. 52	6. 82	1. 57	6. 81	1. 63	6. 79	1. 69
8	7. 81	1. 73	7. 80	1. 80	7. 78	1. 87	7. 76	1. 94
9	8. 79	1. 95	8. 77	2. 02	8. 75	2. 10	8. 73	2. 18
10	9. 76	2. 16	9. 74	2. 25	9. 72	2. 33	9 70	2. 42
11	10. 74	2. 38	10. 72	2. 47	10. 70	2. 57	10. 67	2. 66
12	11. 72	2. 60	11. 69	2. 70	11. 67	2. 80	11. 64	2. 90
13	12. 69	2. 81	12. 67	2. 92	12. 64	3. 03	12. 61	3. 15
14	13. 67	3. 03	13. 64	3. 15	13. 61	3. 27	13. 58	3. 39
15	14. 64	3. 25	14. 62	3. 37	14. 59	3. 50	14. 55	3. 63
16	15. 62	3. 46	15. 59	3. 60	15. 56	3. 74	15. 52	3. 87
17	16. 60	3. 68	16. 57	3. 82	16. 53	3. 97	16. 50	4. 11
18	17. 57	3. 90	17. 54	4. 05	17. 50	4. 20	17. 47	4. 35
19	18. 55	4. 11	18. 51	4. 27	18. 48	4. 44	18. 44	4. 60
20	19. 53	4. 33	19. 49	4. 50	19. 45	4. 67	19. 41	4. 84
21	20. 50	4. 55	20. 46	4. 72	20. 42	4. 90	20. 38	5. 08
22	21. 48	4. 76	21. 44	4. 95	21. 39	5. 14	21. 35	5. 32
23	22. 45	4. 98	22. 41	5. 17	22. 36	5. 37	22. 32	5. 56
24	23. 43	5. 19	23. 38	5. 40	23. 34	5. 60	23. 29	5. 81
25	24. 41	5. 41	24. 36	5. 62	24. 31	5. 84	24. 26	6. 05
26	25. 38	5. 63	25. 33	5. 85	25. 28	6. 07	25. 23	6 29
27	26. 36	5. 84	26. 31	6. 07	26. 25	6. 30	26. 20	6. 53
28	27. 34	6. 06	27. 28	6. 30	27. 23	6. 54	27. 17	6. 77
29	28. 31	6. 28	28. 26	6. 52	28. 20	6. 77	28. 14	7. 02
30	29. 29	6. 49	29. 23	6. 75	29. 17	7. 00	29. 11	7. 26
35	34. 17	7. 58	34. 10	7. 87	34. 03	8. 17	33. 96	8. 47
40	39. 05	8. 66	38. 97	9. 00	38. 89	9. 34	38. 81	9. 68
45	43. 93	9. 74	43. 85	10. 12	43. 76	10. 51	43. 66	10. 89
50	48. 81	10. 82	48. 72	11. 25	48. 62	11. 67	48. 51	12. 10
55	53. 70	11. 90	53. 59	12. 37	53. 48	12. 84	53. 37	13. 31
60	58. 58	12. 99	58. 46	13. 50	58. 34	14. 01	58. 22	14 52
65	63. 46	14. 07	63. 33	14. 62	63. 20	15. 17	63. 07	15. 72
70	68. 34	15. 15	68. 21	15. 75	68. 07	16. 34	67. 92	16. 93
75	73. 22	16. 23	73. 08	16. 87	72. 93	17. 50	72. 77	18. 14
80	78. 10	17. 32	77. 95	18. 00	77. 79	18. 68	77. 62	19. 35
85	82. 99	18. 40	82. 82	19. 12	82. 65	19. 84	82. 48	20. 56
90	87. 87	19. 48	87. 69	20. 25	87. 51	21. 01	87. 33	21. 77
95	92. 75	20. 56	92. 57	21. 37	92. 38	22. 18	92. 18	22. 98
100	97. 63	21. 64	97. 44	22. 50	97. 24	23. 34	97. 03	24. 19
	Dep.	Lat.	Dep.	Lat.	Dep.	Lat.	Dep.	Lat.
	77½ Deg.		77 Deg.		76½ Deg.		76 Deg.	

Distance.	14½ Deg.		15 Deg.		15½ Deg.		16 Deg.	
	Lat.	Dep.	Lat.	Dep.	Lat.	Dep.	Lat.	Dep.
1	0. 97	0. 25	0. 97	0. 26	0. 96	0. 27	0. 96	0. 28
2	1. 94	0. 50	1. 93	0. 52	1. 93	0. 53	1. 92	0. 55
3	2. 90	0. 75	2. 90	0. 78	2. 89	0. 80	2. 88	0. 83
4	3. 87	1. 00	3. 86	1. 04	3. 85	1. 07	3. 85	1. 10
5	4. 84	1. 25	4. 83	1. 29	4. 82	1. 34	4. 81	1. 38
6	5. 81	1. 50	5. 80	1. 55	5. 78	1. 60	5. 77	1. 65
7	6. 78	1. 75	6. 76	1. 81	6. 75	1. 87	6. 73	1. 93
8	7. 75	2. 00	7. 73	2. 07	7. 71	2. 14	7. 69	2. 21
9	8. 71	2. 25	8. 69	2. 33	8. 67	2. 41	8. 65	2. 48
10	9. 68	2. 50	9. 66	2. 59	9. 64	2. 67	9. 61	2. 76
11	10. 65	2. 75	10. 63	2. 85	10. 60	2. 94	10. 57	3. 03
12	11. 62	3. 00	11. 59	3. 11	11. 56	3. 21	11. 54	3. 31
13	12. 59	3. 25	12. 56	3. 36	12. 53	3. 47	12. 50	3. 58
14	13. 55	3. 51	13. 52	3. 62	13. 49	3. 74	13. 46	3. 86
15	14. 52	3. 76	14. 49	3. 88	14. 45	4. 01	14. 42	4. 13
16	15. 49	4. 01	15. 45	4. 14	15. 42	4. 28	15. 38	4. 41
17	16. 46	4. 26	16. 42	4. 40	16. 38	4. 54	16. 34	4. 69
18	17. 43	4. 51	17. 39	4. 66	17. 35	4. 81	17. 30	4. 96
19	18. 39	4. 76	18. 35	4. 92	18. 31	5. 08	18. 26	5. 24
20	19. 36	5. 01	19. 32	5. 18	19. 27	5. 34	19. 23	5. 51
21	20. 33	5. 26	20. 28	5. 44	20. 24	5. 61	20. 19	5. 79
22	21. 30	5. 51	21. 25	5. 69	21. 20	5. 88	21. 15	6. 06
23	22. 27	5. 76	22. 22	5. 95	22. 16	6. 15	22. 11	6. 34
24	23. 24	6. 01	23. 18	6. 21	23. 13	6. 41	23. 07	6. 62
25	24. 20	6. 26	24. 15	6. 47	24. 09	6. 68	24. 03	6. 89
26	25. 17	6. 51	25. 11	6. 73	25. 05	6. 95	24. 99	7. 17
27	26. 14	6. 76	26. 08	6. 99	26. 02	7. 22	25. 95	7. 44
28	27. 11	7. 01	27. 05	7. 25	26. 98	7. 48	26. 92	7. 72
29	28. 08	7. 26	28. 01	7. 51	27. 95	7. 75	27 88	7. 99
30	29. 04	7. 51	28. 98	7. 76	28. 91	8. 02	28. 84	8. 27
35	33. 89	8. 76	33. 81	9. 06	33. 73	9. 35	33. 64	9. 65
40	38. 73	10. 02	38. 64	10. 35	38. 55	10. 69	38. 45	11. 03
45	43. 57	11. 27	43. 47	11. 65	43. 36	12. 03	43. 26	12. 40
50	48. 41	12. 52	48. 30	12. 94	48. 18	13. 36	48. 06	13. 78
55	53. 25	13. 77	53. 13	14. 24	53. 00	14. 70	52. 87	15. 16
60	58. 09	15. 02	57. 96	15. 53	57. 82	16. 03	57. 68	16. 54
65	62. 93	16. 27	62. 79	16. 82	62. 64	17. 37	62. 48	17. 92
70	67. 77	17. 53	67. 61	18. 12	67. 45	18. 71	67. 29	19. 29
75	72. 61	18. 78	72. 44	19. 41	72. 27	20. 04	72. 09	20. 67
80	77. 45	20. 03	77. 27	20. 71	77. 09	21. 38	76. 90	22. 05
85	82. 29	21. 28	82. 10	22. 00	81. 91	22. 72	81. 71	23. 43
90	87. 13	22. 53	86. 93	23. 29	86. 73	24. 05	86. 51	24. 81
95	91. 97	23. 79	91. 76	24. 59	91. 54	25. 39	91. 32	26. 19
100	96. 81	25. 04	96. 59	25. 88	96. 36	26. 72	96. 13	27. 56
	Dep.	Lat.	Dep.	Lat.	Dep.	Lat.	Dep.	Lat.
	75½ Deg.		75 Deg.		74½ Deg.		74 Deg.	

Distance.	16½ Deg.		17 Deg.		17½ Deg.		18 Deg.	
	Lat.	Dep.	Lat.	Dep.	Lat.	Dep.	Lat.	Dep.
1	0. 96	0. 28	0. 96	0. 29	0. 95	0. 30	0. 95	0. 31
2	1. 92	0. 57	1. 91	0. 58	1. 91	0. 60	1. 90	0. 62
3	2. 88	0. 85	2. 87	0. 88	2. 86	0. 90	2. 85	0. 93
4	3. 84	1. 14	3. 83	1. 17	3. 81	1. 20	3. 80	1. 24
5	4. 79	1. 42	4. 78	1. 46	4. 77	1. 50	4. 76	1. 55
6	5. 75	1. 70	5. 74	1. 75	5. 72	1. 80	5. 71	1. 85
7	6. 71	1. 99	6. 69	2. 05	6. 68	2. 10	6. 66	2. 16
8	7. 67	2. 27	7. 65	2. 34	7. 63	2. 41	7. 61	2. 47
9	8. 63	2. 56	8. 61	2. 63	8. 58	2. 71	8. 56	2. 78
10	9. 59	2. 84	9. 56	2. 92	9. 54	3. 01	9 51	3. 09
11	10. 55	3. 12	10. 52	3. 22	10. 49	3. 31	10. 46	3. 40
12	11. 51	3. 41	11. 48	3. 51	11. 44	3. 61	11. 41	3. 71
13	12. 46	3. 69	12. 43	3. 80	12. 40	3. 91	12. 36	4. 02
14	13. 42	3. 98	13. 39	4. 09	13. 35	4. 21	13. 31	4. 33
15	14. 38	4. 26	14. 34	4. 39	14. 31	4. 51	14. 27	4. 64
16	15. 34	4. 54	15. 30	4. 68	15. 26	4. 81	15. 22	4. 94
17	16. 30	4. 83	16. 26	4. 97	16. 21	5. 11	16. 17	5. 25
18	17. 26	5. 11	17. 21	5. 26	17. 17	5. 41	17. 12	5. 56
19	18. 22	5. 40	18. 17	5. 56	18. 12	5. 71	18. 07	5. 87
20	19. 18	5. 68	19. 13	5. 85	19. 07	6. 01	19. 02	6. 18
21	20. 14	5. 96	20. 08	6. 14	20. 03	6. 31	19. 97	6. 49
22	21. 09	6. 25	21. 04	6. 43	20. 98	6. 62	20. 92	6. 80
23	22. 05	6. 53	21. 99	6. 72	21. 94	6. 92	21. 87	7. 11
24	23. 01	6. 82	22. 95	7. 02	22. 89	7. 22	22. 83	7. 42
25	23. 97	7. 10	23. 91	7. 30	23. 84	7. 52	23. 78	7. 73
26	24. 93	7. 38	24. 86	7. 60	24. 80	7. 82	24. 73	8. 03
27	25. 89	7. 67	25. 82	7. 89	25. 75	8. 12	25. 68	8. 34
28	26. 85	7. 95	26. 78	8. 19	26. 70	8. 42	26. 63	8. 65
29	27. 81	8. 24	27. 73	8. 48	27. 66	8. 72	27. 58	8. 96
30	28. 76	8. 52	28. 69	8. 77	28. 61	9. 02	28. 53	9. 27
35	33. 56	9. 94	33. 47	10. 23	33. 38	10. 52	33. 29	10. 82
40	38. 35	11. 36	38. 25	11. 69	38. 15	12. 03	38. 04	12. 36
45	43. 15	12. 78	43. 03	13. 16	42. 92	13. 53	42. 80	13. 91
50	47. 94	14. 20	47. 82	14. 62	47. 69	15. 04	47. 55	15. 45
55	52. 74	15. 62	52. 60	16. 08	52. 45	16. 54	52. 31	17. 00
60	57. 53	17. 04	57. 38	17. 54	57. 22	18. 04	57. 06	18. 54
65	62. 32	18. 46	62. 16	19. 00	61. 99	19. 55	61. 82	20. 09
70	67. 12	19. 88	66. 94	20. 47	66. 76	21. 05	66. 57	21. 63
75	71. 91	21. 30	71. 72	21. 93	71. 53	22. 55	71. 33	23. 18
80	76. 71	22. 72	76. 50	23. 39	76. 30	24. 06	76. 08	24. 72
85	81. 50	24. 14	81. 29	24. 85	81. 07	25. 56	80. 84	26. 27
90	86. 29	25. 56	86. 07	26. 31	85. 83	27. 06	85. 60	27. 81
95	91. 09	26. 98	90. 85	27. 78	90. 60	28. 57	90. 35	29. 36
100	95. 88	28. 40	95. 63	29. 24	95. 37	30. 07	95. 11	30. 90
	Dep.	Lat.	Dep.	Lat.	Dep.	Lat.	Dep.	Lat.
	73½ Deg.		73 Deg.		72½ Deg.		72 Deg.	

Distance.	18½ Deg.		19 Deg.		19½ Deg.		20 Deg.	
	Lat.	Dep.	Lat.	Dep.	Lat.	Dep.	Lat.	Dep.
1	0.95	0.32	0.95	0.33	0.94	0.33	0.94	0.34
2	1.90	0.63	1.89	0.65	1.89	0.67	1.88	0.68
3	2.84	0.95	2.84	0.98	2.83	1.00	2.82	1.03
4	3.79	1.27	3.78	1.30	3.77	1.34	3.76	1.37
5	4.74	1.59	4.73	1.63	4.71	1.67	4.70	1.71
6	5.69	1.90	5.67	1.95	5.66	2.00	5.64	2.05
7	6.64	2.22	6.62	2.28	6.60	2.34	6.58	2.39
8	7.59	2.54	7.56	2.60	7.54	2.67	7.52	2.74
9	8.53	2.86	8.51	2.93	8.48	3.01	8.46	3.08
10	9.48	3.17	9.46	3.26	9.43	3.34	9.40	3.42
11	10.43	3.49	10.40	3.58	10.37	3.67	10.34	3.76
12	11.38	3.81	11.35	3.91	11.31	4.01	11.28	4.10
13	12.33	4.12	12.29	4.23	12.25	4.34	12.22	4.45
14	13.28	4.44	13.24	4.56	13.20	4.67	13.16	4.79
15	14.22	4.76	14.18	4.88	14.14	5.01	14.10	5.13
16	15.17	5.08	15.13	5.21	15.08	5.34	15.04	5.47
17	16.12	5.39	16.07	5.53	16.02	5.67	15.97	5.81
18	17.07	5.71	17.02	5.86	16.97	6.01	16.91	6.16
19	18.02	6.03	17.96	6.19	17.91	6.34	17.85	6.50
20	18.97	6.35	18.91	6.51	18.85	6.68	18.79	6.84
21	19.91	6.66	19.86	6.84	19.80	7.01	19.73	7.18
22	20.86	6.98	20.80	7.16	20.74	7.34	20.67	7.52
23	21.81	7.30	21.75	7.49	21.68	7.68	21.61	7.87
24	22.76	7.62	22.69	7.81	22.62	8.01	22.55	8.21
25	23.71	7.93	23.64	8.14	23.57	8.34	23.49	8.55
26	24.66	8.25	24.58	8.46	24.51	8.68	24.43	8.89
27	25.60	8.57	25.53	8.79	25.45	9.01	25.37	9.23
28	26.55	8.88	26.47	9.12	26.39	9.34	26.31	9.58
29	27.50	9.20	27.42	9.44	27.34	9.68	27.25	9.92
30	28.45	9.52	28.37	9.77	28.28	10.01	28.19	10.26
35	33.19	11.11	33.09	11.39	32.99	11.68	32.89	11.97
40	37.93	12.69	37.82	13.02	37.71	13.35	37.59	13.68
45	42.67	14.28	42.55	14.65	42.42	15.02	42.29	15.39
50	47.42	15.87	47.28	16.28	47.13	16.69	46.98	17.10
55	52.16	17.45	52.00	17.91	51.85	18.36	51.68	18.81
60	56.90	19.04	56.73	19.53	56.56	20.03	56.38	20.52
65	61.64	20.62	61.46	21.16	61.27	21.70	61.08	22.23
70	66.38	22.21	66.19	22.79	67.98	23.37	65.78	23.94
75	71.12	23.80	70.91	24.42	70.70	25.04	70.48	25.65
80	75.87	25.38	75.64	26.05	75.41	26.70	75.18	27.36
85	80.61	26.97	80.37	27.67	80.12	28.37	79.87	29.07
90	85.35	28.56	85.10	29.30	84.84	30.04	84.57	30.78
95	90.09	30.14	89.82	30.93	89.55	31.71	89.27	32.49
100	94.83	31.73	94.55	32.56	94.26	33.38	93.97	34.20
	Dep.	Lat.	Dep.	Lat.	Dep.	Lat.	Dep.	Lat.
	71½ Deg.		71 Deg.		70½ Deg.		70 Deg.	

Distance.	20½ Deg.		21 Deg.		21½ Deg.		22 Deg.	
	Lat.	Dep.	Lat.	Dep.	Lat.	Dep.	Lat.	Dep.
1	0.94	0.35	0.93	0.36	0.93	0.37	0.93	0.37
2	1.87	0.70	1.87	0.72	1.86	0.73	1.85	0.75
3	2.81	1.05	2.80	1.08	2.79	1.10	2.78	1.12
4	3.75	1.40	3.73	1.43	3.72	1.47	3.71	1.50
5	4.68	1.75	4.67	1.79	4.65	1.83	4.64	1.87
6	5.62	2.10	5.60	2.15	5.58	2.20	5.56	2.25
7	6.56	2.45	6.54	2.51	6.51	2.57	6.49	2.62
8	7.49	2.80	7.47	2.87	7.44	2.93	7.42	3.00
9	8.43	3.15	8.40	3.23	8.37	3.30	8.34	3.37
10	9.37	3.50	9.34	3.58	9.30	3.67	9 27	3.75
11	10.30	3.85	10.27	3.94	10.23	4.03	10.20	4.12
12	11.24	4.20	11.20	4.30	11.17	4.40	11.13	4.50
13	12.18	4.55	12.14	4.66	12.10	4.76	12.05	4.87
14	13.11	4.90	13.07	5.02	13.03	5.13	12.98	5.24
15	14.05	5.25	14.00	5.38	13.96	5.50	13.91	5.62
16	14.99	5.60	14.94	5.73	14.89	5.86	14.83	5.99
17	15.92	5.95	15.87	6.09	15.82	6.23	15.76	6.37
18	16.86	6.30	16.80	6.45	16.75	6.60	16.69	6.74
19	17.80	6.65	17.74	6.81	17.68	6.96	17.62	7.12
20	18.73	7.00	18.67	7.17	18.61	7.33	18.54	7.49
21	19.67	7.35	19.61	7.53	19.54	7.70	19.47	7.87
22	20.61	7.70	20.54	7.88	20.47	8.06	20.40	8.24
23	21.54	8.05	21.47	8.24	21.40	8.43	21.33	8.62
24	22.48	8.40	22.41	8.60	22.33	8.80	22.25	8.99
25	23.42	8.76	23.34	8.96	23.26	9.16	23.18	9.37
26	24.35	9.11	24.27	9.32	24.19	9.53	24.11	9.74
27	25.29	9.46	25.21	9.68	25.12	9.90	25.03	10.11
28	26.23	9.81	26.14	10.08	26.05	10.26	25.96	10.49
29	27.16	10.16	27.07	10.39	26.98	10.63	26.89	10.86
30	28.10	10.51	28.01	10.75	27.91	11.00	27.82	11.24
35	32.78	12.26	32.68	12.54	32.56	12.83	32.45	13.11
40	37.47	14.01	37.34	14.33	37.22	14.66	37.09	14.98
45	42.15	15.76	42.01	16.13	41.87	16.49	41.72	16.86
50	46.83	17.51	46.68	17.92	46.52	18.33	46.36	18.73
55	51.52	19.26	51.35	19.71	51.17	20.16	51.00	20.60
60	56.20	21.01	56.01	21.50	55.83	21.99	55.63	22.48
65	60.88	22.76	60.68	23.29	60.48	23.82	60.27	24.35
70	65.57	24.51	65.35	25.09	65.13	25.66	64.90	26.22
75	70.25	26.27	70.02	26.88	69.78	27.49	69.54	28.10
80	74.93	28.02	74.69	28.67	74.43	29.32	74.17	29.97
85	79.62	29.77	79.35	30.46	79.09	31.15	78.81	31.84
90	84.30	31.52	84.02	32.25	83.74	32.99	83.45	33.71
95	98.98	33.27	88.69	34.04	88.39	34.82	88.08	35.59
100	93.67	35.02	93.36	35.84	93.04	36.65	92.72	37.46
	Dep.	Lat.	Dep.	Lat.	Dep.	Lat.	Dep.	Lat.
	69½ Deg.		69 Deg.		68½ Deg.		68 Deg.	

Distance.	22½ Deg.		23 Deg.		23½ Deg.		24 Deg.	
	Lat.	Dep.	Lat.	Dep.	Lat.	Dep.	Lat.	Dep.
1	0. 90	0. 38	0. 92	0. 39	0. 92	0. 40	0. 91	0. 41
2	1. 85	0. 77	1. 84	0. 78	1. 83	0. 80	1. 83	0. 81
3	2. 77	1. 15	2. 76	1. 17	2. 75	1. 20	2. 74	1. 22
4	3. 70	1. 53	3. 68	1. 56	3. 67	1. 59	3. 65	1. 63
5	4. 62	1. 91	4. 60	1. 95	4. 59	1. 99	4. 57	2. 03
6	5. 54	2. 30	5. 52	2. 34	5. 50	2. 39	5. 48	2. 44
7	6. 47	2. 68	6. 44	2. 74	6. 42	2. 79	6. 39	2. 85
8	7. 39	3. 06	7. 36	3. 13	7. 34	3. 19	7. 31	3. 25
9	8. 31	3. 44	8. 28	3. 52	8. 25	3. 59	8. 22	3. 66
10	9. 24	3. 83	9. 20	3. 91	9. 17	3. 99	9. 14	4. 07
11	10. 16	4. 21	10. 13	4. 30	10. 09	4. 39	10. 05	4. 47
12	11. 09	4. 59	11. 05	4. 69	11. 00	4. 78	10. 96	4. 88
13	12. 01	4. 97	11. 97	5. 08	11. 92	5. 18	11. 88	5. 29
14	12. 93	5. 36	12. 89	5. 47	12. 84	5. 58	12. 79	5. 69
15	13. 86	5. 74	13. 81	5. 86	13. 76	5. 98	13. 70	6. 10
16	14. 78	6. 12	14. 73	6. 25	14. 67	6. 38	14. 62	6. 51
17	15. 71	6. 51	15. 65	6. 64	15. 59	6. 78	15. 53	6. 92
18	16. 63	6. 89	16. 57	7. 03	16. 51	7. 18	16. 44	7. 32
19	17. 55	7. 27	17. 49	7. 42	17. 42	7. 58	17. 36	7. 73
20	18. 48	7. 65	18. 41	7. 81	18. 34	7. 97	18. 27	8. 13
21	19. 40	8. 04	19. 33	8. 21	19. 26	8. 37	19. 18	8. 54
22	20. 33	8. 42	20. 25	8. 60	20. 18	8. 77	20. 10	8. 95
23	21. 25	8. 80	21. 17	8. 99	21. 09	9. 17	21. 01	9. 35
24	22. 17	9. 18	22. 09	9. 38	22. 01	9. 57	21. 93	9. 76
25	23. 10	9. 57	23. 01	9. 77	22. 93	9. 97	22. 84	10. 17
26	24. 02	9. 95	23. 93	10. 16	23. 84	10. 37	23. 75	10. 58
27	24. 94	10. 33	24. 85	10. 55	24. 76	10. 77	24. 67	10. 98
28	25. 87	10. 72	25. 77	10. 94	25. 68	11. 16	25. 58	11. 39
29	26. 79	11. 10	26. 69	11. 33	26. 59	11. 56	26. 49	11. 80
30	27. 72	11. 48	27. 62	11. 52	27. 51	11. 96	27. 41	12. 20
35	32. 34	13. 39	32. 22	13. 68	32. 10	13. 96	31. 97	14. 24
40	36. 96	15. 31	36. 82	15. 63	36. 68	15. 95	36. 54	16. 27
45	41. 57	17. 22	41. 42	17. 58	41. 27	17. 94	41. 11	18. 30
50	46. 19	19. 13	46. 03	19. 54	45. 85	19. 94	45. 68	20. 34
55	50. 81	21. 05	50. 63	21. 49	50. 44	21. 93	50. 24	22. 37
60	55. 43	22. 96	55. 23	23. 44	55. 02	23. 92	54. 81	24. 40
65	60. 05	24. 87	59. 83	25. 40	59. 61	25. 92	59. 38	26. 44
70	64. 67	26. 79	64. 44	27. 35	64. 19	27. 91	63. 95	28. 47
75	69. 29	28. 70	69. 04	29. 30	68. 78	29. 91	68. 52	30. 51
80	73. 91	30. 61	73. 64	31. 26	73. 36	31. 90	73. 08	32. 54
85	78. 53	32. 53	78. 24	33. 21	77. 95	33. 89	77. 65	34. 57
90	83. 15	34. 44	82. 85	35. 17	82. 54	35. 89	82. 22	36. 61
95	87. 77	36. 35	87. 45	37. 12	87. 12	37. 88	86. 79	38. 64
100	92. 39	38. 27	92. 05	39. 07	91. 71	39. 87	91. 35	40. 67
	Dep.	Lat.	Dep.	Lat.	Dep.	Lat.	Dep.	Lat.
	67½ Deg.		67 Deg.		66½ Deg.		66 Deg.	

Distance.	24½ Deg.		25 Deg.		25½ Deg.		26 Deg.	
	Lat.	Dep.	Lat.	Dep.	Lat.	Dep.	Lat.	Dep.
1	0. 91	0. 41	0. 91	0. 42	0. 90	0. 43	0. 90	0. 44
2	1. 82	0. 83	1. 81	0. 85	1. 81	0. 86	1. 80	0. 88
3	2. 73	1. 24	2. 72	1. 27	2. 71	1. 29	2. 70	1. 32
4	3. 64	1. 66	3. 63	1. 69	3. 61	1. 72	3. 60	1. 75
5	4. 55	2. 07	4. 53	2. 11	4. 51	2. 15	4. 49	2. 19
6	5. 46	2. 49	5. 44	2. 54	5. 42	2. 58	5. 39	2. 63
7	6. 37	2. 90	6. 34	2. 96	6. 32	3. 01	6. 29	3. 07
8	7. 28	3. 32	7. 25	3. 38	7. 22	3. 44	7. 19	3. 51
9	8. 19	3. 73	8. 16	3. 80	8. 12	3. 87	8. 09	3. 95
10	9. 10	4. 15	9. 06	4. 23	9. 03	4. 31	8. 99	4. 38
11	10. 01	4. 56	9. 97	4. 65	9. 93	4. 74	9. 89	4. 82
12	10. 92	4. 98	10. 88	5. 07	10. 83	5. 17	10. 79	5. 26
13	11. 83	5. 39	11. 78	5. 49	11. 73	5. 60	11. 68	5. 70
14	12. 74	5. 81	12. 69	5. 92	12. 64	6. 03	12. 58	6. 14
15	13. 65	6. 22	13. 59	6. 34	13. 54	6. 46	13. 48	6. 58
16	14. 56	6. 64	14. 50	6. 76	14. 44	6. 89	14. 38	7. 01
17	15. 47	7. 05	15. 41	7. 18	15. 34	7. 32	15. 28	7. 45
18	16. 38	7. 46	16. 31	7. 61	16. 25	7. 75	16. 18	7. 89
19	17. 19	7. 88	17. 22	8. 03	17. 15	8. 18	17. 08	8. 33
20	18. 20	8. 29	18. 13	8. 45	18. 05	8. 61	17. 98	8. 77
21	19. 11	8. 71	19. 03	8. 87	18. 95	9. 04	18. 87	9. 21
22	20. 02	9. 12	19. 94	9. 30	19. 86	9. 47	19. 77	9. 64
23	20. 93	9. 54	20. 85	9. 72	20. 76	9. 90	20. 67	10. 08
24	21. 84	9. 95	21. 75	10. 14	21. 66	10. 33	21. 57	10. 52
25	22. 75	10. 37	22. 66	10. 57	22. 56	10. 76	22. 47	10. 96
26	23. 66	10. 78	23. 56	10. 99	23. 47	11. 19	23. 37	11. 40
27	24. 57	11. 20	24. 47	11. 41	24. 37	11. 62	24. 27	11. 84
28	25. 48	11. 61	25. 38	11. 83	25. 27	12. 05	25. 17	12. 27
29	26. 39	12. 03	26. 28	12. 26	26. 17	12. 48	26. 06	12. 71
30	27. 30	12. 44	27. 19	12. 68	27. 08	12. 92	26. 96	13. 15
35	31. 85	14. 51	31. 72	14. 79	31. 59	15. 07	31. 46	15. 34
40	36. 40	16. 59	36. 25	16. 90	36. 10	17. 22	35. 95	17. 53
45	40. 95	18. 66	40. 78	19. 02	40. 62	19. 37	40. 45	19. 73
50	45. 50	20. 73	45. 32	21. 13	45. 13	21. 53	44. 94	21. 92
55	50. 05	22. 81	49. 85	23. 24	49. 64	23. 68	49. 43	24. 11
60	54. 60	24. 88	54. 38	25. 36	54. 16	25. 83	53. 93	26. 30
65	59. 15	26. 96	58. 91	27. 47	58. 67	27. 98	58. 42	28. 49
70	63. 70	29. 03	63. 44	29. 58	63. 18	30. 14	62. 92	30. 69
75	68. 25	31. 10	67. 97	31. 70	67. 69	32. 29	67. 41	32. 88
80	72. 80	33. 18	72. 50	33. 81	72. 21	34. 44	71. 90	35. 07
85	77. 35	35. 25	77. 04	35. 92	76. 72	36. 59	76. 40	37. 26
90	81. 90	37. 32	81. 57	38. 04	81. 23	38. 75	80. 89	39. 45
95	86. 45	39. 40	86. 10	40. 15	85. 75	40. 90	85. 39	41. 65
100	91. 00	41. 47	90. 63	42. 26	90. 26	43. 05	89. 88	43. 84
	Dep.	Lat.	Dep.	Lat.	Dep.	Lat.	Dep.	Lat.
	65½ Deg.		65 Deg.		64½ Deg.		64 Deg.	

Distance.	26½ Deg.		27 Deg.		27½ Deg.		28 Deg.	
	Lat.	Dep.	Lat.	Dep.	Lat.	Dep.	Lat.	Dep.
1	0.89	0.45	0.89	0.45	0.89	0.46	0.88	0.47
2	1.79	0.89	1.78	0.91	1.77	0.92	1.77	0.94
3	2.68	1.34	2.67	1.36	2.66	1.39	2.65	1.41
4	3.58	1.78	3.56	1.82	3.55	1.85	3.53	1.88
5	4.57	2.23	4.45	2.27	4.44	2.31	4.41	2.35
6	5.37	2.68	5.35	2.72	5.32	2.77	5.30	2.82
7	6.26	3.12	6.24	3.18	6.21	3.23	6.18	3.29
8	7.16	3.47	7.13	3.63	7.10	3.69	7.06	3.76
9	8.05	4.02	8.02	4.09	7.98	4.16	7.95	4.23
10	8.95	4.46	8.91	4.54	8.87	4.62	8.83	4.69
11	9.84	4.91	9.80	4.99	9.76	5.08	9.71	5.16
12	10.74	5.35	10.69	5.45	10.64	5.54	10.60	5.63
13	11.63	5.80	11.58	5.90	11.53	6.00	11.48	6.10
14	12.53	6.25	12.47	6.36	12.42	6.49	12.36	6.57
15	13.42	6.69	13.37	6.81	13.31	6.93	13.24	7.04
16	14.32	7.14	14.26	7.26	14.19	7.39	14.13	7.51
17	15.21	7.59	15.15	7.72	15.08	7.85	15.01	7.98
18	16.11	8.03	16.04	8.17	15.97	8.31	15.89	8.45
19	17.00	8.48	16.93	8.63	16.85	8.77	16.78	8.92
20	17.90	8.92	17.82	9.08	17.74	9.23	17.66	9.39
21	18.79	9.37	18.71	9.53	18.63	9.70	18.54	9.86
22	19.69	9.82	19.60	9.99	19.51	10.16	19.42	10.33
23	20.58	10.26	20.49	10.44	20.40	10.62	20.31	10.80
24	21.48	10.71	21.38	10.90	21.29	11.08	21.19	11.27
25	22.37	11.15	22.28	11.35	22.18	11.54	22.07	11.74
26	23.27	11.60	23.17	11.80	23.06	12.01	22.96	12.21
27	24.16	12.05	24.06	12.26	23.95	12.47	23.84	12.68
28	25.06	12.49	24.95	12.71	24.84	12.93	24.72	13.15
29	25.95	12.94	25.84	13.17	25.72	13.39	25.61	13.61
30	26.85	13.39	26.73	13.62	26.61	13.85	26.49	14.08
35	31.32	15.62	31.19	15.89	31.05	16.16	30.90	16.43
40	35.80	17.85	35.64	18.16	35.48	18.47	35.32	18.78
45	40.27	20.08	40.10	20.43	39.92	20.78	39.73	21.13
50	44.75	22.31	44.55	22.70	44.35	23.09	44.15	23.47
55	49.22	24.54	49.01	24.97	48.79	25.40	48.56	25.82
60	53.70	26.77	53.46	27.24	53.22	27.70	52.98	28.17
65	58.17	29.00	57.92	29.51	57.66	30.01	57.39	30.52
70	62.65	31.23	62.37	31.78	62.09	32.32	61.81	32.86
75	67.12	33.46	66.83	34.05	66.53	34.63	66.22	35.21
80	71.59	35.70	71.28	36.32	70.96	36.94	70.64	37.56
85	76.07	37.93	75.74	38.59	75.40	39.25	75.05	39.91
90	80.54	40.16	80.19	40.86	79.83	41.56	79.47	42.25
95	85.02	42.39	84.65	43.13	84.27	43.87	83.88	44.60
100	89.49	44.62	89.10	45.40	88.90	46.17	88.29	46.95
	Dep.	Lat.	Dep.	Lat.	Dep.	Lat.	Dep.	Lat.
	63½ Deg.		63 Deg.		62½ Deg.		62 Deg.	

Distance.	28½ Deg.		29 Deg.		29½ Deg.		30 Deg.	
	Lat.	Dep.	Lat.	Dep.	Lat.	Dep.	Lat.	Dep.
1	0. 88	0. 48	0. 87	0. 48	0. 87	0. 49	0. 87	0. 50
2	1. 76	0. 95	1. 75	0. 97	1. 74	0. 98	1. 73	1. 00
3	2. 64	1. 43	2. 62	1. 45	2. 61	1. 48	2. 60	1. 50
4	3. 52	1. 91	3. 50	1. 94	3. 48	1. 97	3. 46	2. 00
5	4. 39	2. 39	4. 37	2. 42	4. 35	2. 46	4. 33	2. 50
6	5. 27	2. 86	5. 25	2. 91	5. 22	2. 95	5. 20	3. 00
7	6. 15	3. 34	6. 12	3. 39	6. 09	3. 45	6. 06	3. 50
8	7. 03	3. 82	7. 00	3. 88	6. 96	3. 94	6. 93	4. 00
9	7. 91	4. 29	7. 87	4. 36	7. 83	4. 43	7. 79	4. 50
10	8. 79	4. 77	8. 75	4. 85	8. 70	4. 92	8. 66	5. 00
11	9. 67	5. 25	9. 62	5. 33	9. 57	5. 42	9. 53	5. 50
12	10. 55	5. 73	10. 50	5. 82	10. 44	5. 91	10. 39	6. 00
13	11. 42	6. 20	11. 37	6. 30	11. 31	6. 40	11. 26	6. 50
14	12. 30	6. 68	12. 24	6. 79	12. 18	6. 89	12. 12	7. 00
15	13. 18	7. 16	13. 12	7. 27	13. 06	7. 39	12. 99	7. 50
16	14. 06	7. 63	13. 99	7. 76	13. 93	7. 88	13. 86	8. 00
17	14. 94	8. 11	14. 87	8. 24	14. 80	8. 37	14. 72	8. 50
18	15. 82	8. 59	15. 74	8. 73	15. 67	8. 86	15. 59	9. 00
19	16. 70	9. 07	16. 62	9. 21	16. 54	9. 36	16. 45	9. 50
20	17. 58	9. 54	17. 49	9. 70	17. 41	9. 85	17. 32	10. 00
21	18. 46	10. 02	18. 37	10. 18	18. 28	10. 34	18. 19	10. 50
22	19. 33	10. 50	19. 24	10. 67	19. 15	10. 83	19. 05	11. 00
23	20. 21	10. 97	20. 12	11. 15	20. 02	11. 33	19. 92	11. 50
24	21. 09	11. 45	20. 99	11. 64	20. 89	11. 82	20. 78	12. 00
25	21. 97	11. 93	21. 87	12. 12	21. 76	12. 31	21. 65	12. 50
26	22. 85	12. 41	22. 74	12. 60	22. 63	12. 80	22. 52	13. 00
27	23. 73	12. 88	23. 61	13. 09	23. 50	13. 30	23. 38	13. 50
28	24. 61	13. 36	24. 49	13. 57	24. 37	13. 79	24. 25	14. 00
29	25. 49	13. 84	25. 36	14. 06	25. 24	14. 28	25. 11	14. 50
30	26. 36	14. 31	26. 24	14. 54	26. 11	14. 77	25. 98	15. 00
35	30. 76	16. 70	30. 61	16. 97	30. 46	17. 23	30. 31	17. 50
40	35. 15	19. 09	34. 98	19. 39	34. 81	19. 70	34. 64	20. 00
45	39. 55	21. 47	39. 36	21. 82	39. 17	22. 16	38. 97	22. 50
50	43. 94	23. 86	43. 73	24. 24	43. 52	24. 62	43. 30	25. 00
55	48. 33	26. 24	48. 10	26. 66	47. 87	27. 08	47. 63	27. 50
60	52. 73	28. 63	52. 48	29. 09	52. 22	29. 55	51. 96	30. 00
65	57. 12	31. 02	56. 85	31. 51	56. 57	32. 01	56. 29	32. 50
70	61. 52	33. 40	61. 22	33. 94	60. 92	34. 47	60. 62	35. 00
75	65. 91	35. 79	65. 60	36. 36	65. 28	36. 93	64. 95	37. 50
80	70. 31	38. 17	69. 97	38. 78	69. 63	39. 39	69. 28	40. 00
85	74. 70	40. 56	74. 34	41. 21	73. 98	41. 86	73. 61	42. 50
90	79. 09	42. 94	78. 72	43. 63	78. 33	44. 32	77. 94	45. 00
95	83. 49	45. 33	83. 09	46. 06	82. 68	46. 78	82. 27	47. 50
100	87. 88	47. 72	87. 46	48. 48	87. 04	49. 24	86. 60	50. 00
	Dep.	Lat.	Dep.	Lat.	Dep.	Lat.	Dep.	Lat.
	61½ Deg.		61 Deg.		60½ Deg.		60 Deg.	

Distance.	30½ Deg.		31 Deg.		31½ Deg.		32 Deg.	
	Lat.	Dep.	Lat.	Dep.	Lat.	Dep.	Lat.	Dep.
1	0. 86	0. 51	0. 86	0. 51	0. 85	0. 52	0. 85	0. 53
2	1. 72	1. 02	1. 71	1. 03	1. 71	1. 04	1. 70	1. 06
3	2. 58	1. 52	2. 57	1. 55	2. 56	1. 57	2. 54	1. 59
4	3. 45	2. 03	3. 43	2. 06	3. 41	2. 09	3. 39	2. 12
5	4. 31	2. 54	4. 29	2. 58	4. 26	2. 61	4. 24	2. 65
6	5. 17	3. 05	5. 14	3. 09	5. 12	3. 13	5. 09	3. 18
7	6. 03	3. 55	6. 00	3. 61	5. 97	3. 66	5. 94	3. 71
8	6. 89	4. 06	6. 86	4. 12	6. 82	4. 18	6. 78	4. 24
9	7. 75	4. 57	7. 71	4. 64	7. 67	4. 70	7. 63	4. 77
10	8. 62	5. 08	8. 57	5. 15	8. 53	5. 22	8. 48	5. 30
11	9. 48	5. 58	9. 43	5. 67	9. 38	5. 75	9. 33	5. 83
12	10. 34	6. 09	10. 29	6. 18	10. 23	6. 27	10. 18	6. 36
13	11. 20	6. 60	11. 14	6. 70	11. 08	6. 79	11. 02	6. 89
14	12. 06	7. 11	12. 00	7. 21	11. 94	7. 31	11. 87	7. 42
15	12. 92	7. 61	12. 86	7. 73	12. 79	7. 84	12. 72	7. 95
16	13. 79	8. 12	13. 71	8. 24	13. 64	8. 36	13. 57	8. 48
17	14. 65	8. 63	14. 57	8. 77	14. 49	8. 88	14. 42	9. 01
18	15. 51	9. 14	15. 43	9. 27	15. 35	9. 40	15. 26	9. 54
19	16. 37	9. 64	16. 29	9. 79	16. 20	9. 93	16. 11	10. 07
20	17. 23	10. 15	17. 14	10. 30	17. 05	10. 45	16. 96	10. 60
21	18. 09	10. 66	18. 00	10. 82	17. 91	10. 97	17. 81	11. 13
22	18. 96	11. 17	18. 86	11. 33	18. 76	11. 49	18. 66	11. 66
23	19. 82	11. 67	19. 71	11. 85	19. 61	12. 02	19. 51	12. 19
24	20. 68	12. 18	20. 57	12. 36	20. 46	12. 54	20. 35	12. 72
25	21. 54	12. 69	21. 43	12. 88	21. 32	13. 06	21. 20	13. 25
26	22. 40	13. 20	22. 29	13. 39	22. 17	13. 58	22. 05	13. 78
27	23. 26	13. 70	23. 14	13. 91	23. 02	14. 11	22. 90	14. 31
28	24. 13	14. 21	24. 00	14. 42	23. 87	14. 63	23. 75	14. 84
29	24. 99	14. 72	42. 86	14. 94	24. 73	15. 15	24. 59	15. 37
30	25. 85	15. 23	25. 71	15. 45	25. 58	15. 67	25. 44	15. 90
35	30. 16	17. 76	30. 00	18. 03	29. 84	18. 29	29. 68	18. 55
40	34. 47	20. 30	34. 29	20. 60	34. 11	20. 90	33. 92	21. 20
45	38. 77	22. 84	38. 57	23. 18	38. 37	23. 51	38. 16	23. 85
50	43. 08	25. 38	42. 86	25. 75	42. 63	26. 12	42. 40	26. 50
55	47. 39	27. 91	47. 14	28. 33	46. 90	28. 74	46. 64	29. 15
60	51. 70	30. 45	51. 53	30. 90	51. 16	31. 35	50. 88	31. 80
65	56. 01	32. 99	55. 72	33. 48	55. 42	33. 96	55. 12	34. 44
70	60. 31	35. 53	60. 00	36. 05	59. 68	36. 57	59. 36	37. 09
75	64. 62	38. 07	64. 29	38. 63	63. 95	39. 19	63. 60	39. 74
80	68. 93	40. 60	68. 57	41. 20	68. 21	41. 80	67. 84	42. 39
85	73. 24	43. 14	72. 86	43. 78	72. 47	44. 41	72. 08	45. 04
90	77. 55	45. 68	77. 15	46. 35	76. 74	47. 02	76. 32	47. 69
95	81. 85	48. 22	81. 43	48. 93	81. 00	49. 64	80. 56	50. 34
100	86. 16	50. 75	85. 72	51. 50	85. 26	52. 25	84. 80	52. 59
	Dep.	Lat.	Dep.	Lat.	Dep.	Lat.	Dep.	Lat.
	59½ Deg.		59 Deg.		58½ Deg.		58 Deg.	

Distance.	32½ Deg.		33 Deg.		33½ Deg.		34 Deg.	
	Lat.	Dep.	Lat.	Dep.	Lat.	Dep.	Lat.	Dep.
1	0.84	0.54	0.84	0.54	0.83	0.55	0.83	0.56
2	1.69	1.07	1.68	1.09	1.67	1.10	1.66	1.12
3	2.53	1.61	2.52	1.63	2.50	1.66	2.49	1.68
4	3.37	2.15	3.35	2.18	3.34	2.21	3.32	2.24
5	4.22	2.69	4.19	2.72	4.17	2.76	4.15	2.80
6	5.06	3.22	5.03	3.27	5.00	3.31	4.97	3.36
7	5.90	3.76	5.87	3.81	5.84	3.86	5.80	3.91
8	6.75	4.30	6.71	4.36	6.67	4.42	6.63	4.47
9	7.59	4.84	7.55	4.90	7.50	4.97	7.46	5.03
10	8.43	5.37	8.39	5.45	8.34	5.52	8.29	5.59
11	9.28	5.91	9.23	5.99	9.17	6.07	9.12	6.15
12	10.12	6.45	10.06	6.54	10.01	6.62	9.95	6.71
13	10.96	6.98	10.90	7.08	10.84	7.18	10.78	7.27
14	11.81	7.52	11.74	7.62	11.67	7.73	11.61	7.83
15	12.65	8.06	12.58	8.17	12.51	8.28	12.44	8.39
16	13.49	8.60	13.42	8.71	13.34	8.83	13.26	8.95
17	14.34	9.13	14.26	9.26	14.18	9.38	14.09	9.51
18	15.18	9.67	15.10	9.80	15.01	9.93	14.92	10.07
19	16.02	10.21	15.93	10.35	15.84	10.49	15.75	10.62
20	16.87	10.75	16.77	10.89	16.68	11.04	16.58	11.18
21	17.71	11.28	17.61	11.44	17.51	11.59	17.41	11.74
22	18.55	11.82	18.45	11.98	18.35	12.14	18.24	12.30
23	19.40	12.36	19.29	12.53	19.18	12.69	19.07	12.86
24	20.28	12.90	20.13	13.07	20.01	13.25	19.90	13.42
25	21.08	13.43	20.97	13.62	20.85	13.80	20.73	13.98
26	21.93	13.97	21.81	14.16	21.68	14.35	21.55	14.54
27	22.77	14.51	22.64	14.71	22.51	14.90	22.38	15.10
28	23.61	15.04	23.48	15.25	23.35	15.45	23.21	15.66
29	24.46	15.58	24.32	15.97	24.18	16.01	24.04	16.22
30	25.30	16.12	25.16	16.34	25.02	16.56	24.87	16.78
35	29.52	18.81	29.35	19.06	29.19	19.32	29.02	19.57
40	33.74	21.49	33.55	21.79	33.36	22.08	33.16	22.37
45	37.95	24.18	37.74	24.51	37.52	24.84	37.31	25.16
50	42.17	26.86	41.93	27.23	41.69	27.60	41.45	27.96
55	46.39	29.55	46.13	29.96	45.86	30.36	45.60	30.76
60	50.60	32.24	50.32	32.68	50.08	33.12	49.74	33.55
65	54.82	34.92	54.51	35.40	54.20	35.88	53.89	36.35
70	59.04	37.61	58.72	38.12	58.37	38.64	58.03	39.14
75	63.25	40.30	62.90	40.85	62.54	41.40	62.18	41.94
80	67.47	42.98	67.09	43.57	66.71	44.15	66.32	44.74
85	71.69	45.67	71 29	46.29	70.88	46.91	70.47	47.53
90	75.91	48.36	75.48	49.02	75.05	49.67	74.61	50.33
95	80.12	51.04	79.67	51.74	79.22	54.43	78.76	53.12
100	84.34	53.73	83.87	54.46	83.39	55.19	82.90	55.92
	Dep.	Lat.	Dep.	Lat.	Dep.	Lat.	Dep.	Lat.
	57½ Deg.		57 Deg.		56½ Deg.		56 Deg.	

Distance.	34½ Deg.		35 Deg.		35½ Deg.		36 Deg.	
	Lat.	Dep.	Lat.	Dep.	Lat.	Dep.	Lat.	Dep.
1	0. 82	0. 57	0. 82	0. 57	0. 81	0. 58	0. 81	0. 59
2	1. 65	1. 13	1. 64	1. 15	1. 63	1. 16	1. 62	1. 18
3	2. 47	1. 70	2. 46	1. 72	2. 44	1. 74	2. 43	1. 76
4	3. 30	2. 27	3. 28	2. 29	3. 26	2. 32	3. 24	2. 35
5	4. 12	2. 83	4. 10	2. 87	4. 07	2. 90	4. 05	2. 94
6	4. 94	3. 40	4. 91	3. 44	4. 88	3. 48	4. 85	3. 53
7	5. 77	3. 96	5. 73	4. 01	5. 70	4. 06	5. 66	4. 11
8	6. 59	4. 53	6. 55	4. 59	6. 51	4. 65	6. 47	4. 70
9	7. 42	5. 10	7. 37	5. 16	7. 33	5. 23	7. 28	5. 29
10	8. 24	5. 66	8. 19	5. 74	8. 14	5. 81	8. 09	5. 88
11	9. 07	6. 23	9. 01	6. 31	8. 96	6. 39	8. 90	6. 47
12	9. 89	6. 80	9. 83	6. 88	9. 77	6. 97	9. 71	7. 05
13	10. 71	7. 36	10. 65	7. 46	10. 58	7. 55	10. 52	7. 64
14	11. 54	7. 93	11. 47	8. 03	11. 40	8. 13	11. 33	8. 23
15	12. 36	8. 50	12. 29	8. 60	12. 21	8. 71	12. 14	8. 82
16	13. 19	9. 06	13. 11	9. 18	13. 03	9. 29	12. 94	9. 40
17	14. 01	9. 63	13. 93	9. 75	13. 84	9. 87	13. 75	9. 99
18	14. 83	10. 20	14. 74	10. 32	14. 65	10. 45	14. 56	10. 58
19	15. 66	10. 76	15. 56	10. 90	15. 47	11. 03	15. 37	11. 17
20	16. 48	11. 33	16. 38	11. 47	16. 28	11. 61	16. 18	11. 76
21	17. 31	11. 89	17. 20	12. 05	17. 10	12. 19	16. 99	12. 34
22	18. 13	12. 46	18. 02	12. 62	17. 91	12. 78	17. 80	12. 93
23	18. 95	13. 03	18. 84	13. 19	18. 72	13. 36	18. 61	13. 52
24	19. 78	13. 59	19. 66	13. 77	19. 54	13. 94	19. 42	14. 11
25	20. 60	14. 16	20. 48	14. 34	20. 35	14. 52	20. 23	14. 69
26	21. 43	14. 73	21. 30	14. 91	21. 17	15. 10	21. 03	15. 28
27	22. 25	15. 29	22. 12	15. 49	21. 98	15. 68	21. 84	15. 87
28	23. 08	15. 86	22. 94	16. 06	22. 80	16. 26	22. 65	16. 46
29	23. 90	16. 43	23. 76	16. 63	23. 61	16. 84	23. 46	17. 05
30	24. 72	16. 99	24. 57	17. 21	24. 42	17. 42	24. 27	17. 63
35	28. 84	19. 82	28. 67	20. 08	28. 49	20. 32	28. 32	20. 57
40	32. 97	22. 66	32. 77	22. 94	32. 56	23. 23	32. 36	23. 51
45	37. 09	25. 49	36. 86	25. 81	36. 64	26. 13	36. 41	26. 45
50	41. 21	28. 32	40. 96	28. 68	40. 71	29. 04	40. 45	29. 39
55	45. 33	31. 15	45. 05	31. 55	44. 78	31. 94	44. 50	32. 23
60	49. 45	33. 98	49. 15	34. 41	48. 85	34. 84	48. 54	35. 27
65	53. 57	36. 82	53. 24	37. 28	52. 92	37. 75	52. 59	38. 21
70	57. 69	39. 65	57. 34	40. 15	56. 99	40. 65	56. 63	41. 14
75	61. 81	42. 48	61. 44	43. 02	61. 06	43. 55	60. 68	44. 08
80	65. 93	45. 31	65. 53	45. 89	65. 13	46. 46	64. 72	47. 02
85	70. 05	48. 14	69. 63	48. 75	69. 20	49. 36	68. 77	49. 96
90	74. 17	50. 98	73. 72	51. 62	73. 27	52. 26	72. 81	52. 90
95	78. 29	53. 81	77. 82	54. 49	77. 34	55. 17	76. 86	55. 84
100	82. 41	56. 64	81. 92	57. 36	81. 41	58. 07	80. 90	58. 78
	Dep.	Lat.	Dep.	Lat.	Dep.	Lat.	Dep.	Lat.
	55½ Deg.		55 Deg.		54½ Deg.		54 Deg.	

Distance.	36½ Deg.		37 Deg.		37½ Deg.		38 Deg.	
	Lat.	Dep.	Lat.	Dep.	Lat.	Dep.	Lat.	Dep.
1	0. 80	0. 59	0. 80	0. 60	0. 79	0. 61	0. 79	0. 62
2	1. 61	1. 19	1. 60	1. 20	1. 59	1. 22	1. 58	1. 23
3	2. 41	1. 78	2. 40	1. 81	2. 38	1. 83	2. 36	1. 85
4	3. 22	2. 38	3. 19	2. 41	3. 17	2. 43	3. 15	2. 46
5	4. 02	2. 97	3. 99	3. 01	3. 97	3. 04	3. 94	3. 08
6	4. 82	3. 57	4. 79	3. 61	4. 76	3. 65	4. 73	3. 69
7	5. 63	4. 16	5. 59	4. 21	5. 55	4. 20	5. 52	4. 31
8	6. 43	4. 76	6. 39	4. 81	6. 35	4. 87	6. 30	4. 93
9	7. 23	5. 35	7. 19	5. 42	7. 14	5. 48	7. 09	5. 54
10	8. 04	5. 95	7. 99	6. 02	7. 93	6. 09	7. 88	6. 16
11	8. 84	6. 54	8. 78	6. 62	8. 73	6. 70	8. 67	6. 77
12	9. 65	7. 14	9. 58	7. 22	9. 52	7. 31	9. 46	7. 39
13	10. 45	7. 73	10. 38	7. 82	10. 31	7. 91	10. 24	8. 00
14	11. 25	8. 33	11. 18	8. 43	11. 11	8. 52	11. 03	8. 62
15	12. 06	8. 92	11. 98	9. 03	11. 90	9. 13	11. 82	9. 23
16	12. 86	9. 52	12. 78	9. 63	12. 69	9. 74	12. 61	9. 85
17	13. 67	10. 11	13. 58	10. 23	13. 49	10. 35	13. 40	10. 47
18	14. 47	10. 71	14. 38	10. 83	14. 28	10. 96	14. 18	11. 08
19	15. 27	11. 30	15. 17	11. 43	15. 07	11. 57	14. 97	11. 70
20	16. 08	11. 90	15. 97	12. 04	15. 87	12. 18	15. 76	12. 31
21	16. 88	12. 49	16. 77	12. 64	16. 66	12. 78	16. 55	12. 93
22	17. 68	13. 09	17. 57	13. 24	17. 45	13. 39	17. 34	13. 54
23	18. 49	13. 68	18. 37	13. 84	18. 25	14. 00	18. 12	14. 16
24	19. 29	14. 28	19. 17	14. 44	19. 04	14. 61	18. 91	14. 78
25	20. 10	14. 87	19. 97	15. 05	19. 83	15. 22	19. 70	15. 39
26	20. 90	15. 47	20. 76	15. 65	20. 63	15. 83	20. 49	16. 01
27	21. 70	16. 06	21. 56	16. 25	21. 42	16. 44	21. 28	16. 62
28	22. 51	16. 65	22. 36	16. 85	22. 21	17. 05	22. 06	17. 24
29	23. 31	17. 25	23. 16	17. 45	23. 01	17. 65	22. 85	17. 85
30	24. 12	17. 84	23. 96	18. 05	23. 80	18. 26	23. 64	18. 47
35	28. 13	20. 82	27. 95	21. 06	27. 77	21. 31	27. 58	21. 55
40	32. 15	23. 79	31. 95	24. 07	31. 73	24. 35	31. 52	24. 63
45	36. 17	26. 77	35. 94	27. 08	35. 70	27. 39	35. 46	27. 70
50	40. 19	29. 74	39. 93	30. 09	39. 67	30. 44	39. 40	30. 78
55	44. 21	32. 72	43. 92	33. 10	43. 63	33. 48	43. 34	33. 86
60	48. 23	35. 69	47. 92	36. 11	47. 60	36. 53	47. 28	36. 94
65	52. 25	38. 66	51 91	39. 12	51. 57	39. 57	51. 22	40. 02
70	56. 27	41. 64	55. 90	42. 13	55. 53	42. 61	55. 16	43. 10
75	60. 29	44. 61	59. 90	45. 14	59. 50	45. 66	59. 10	46. 17
80	64. 31	47. 59	63. 89	48. 15	63. 47	48. 70	63. 04	49. 25
85	68. 33	50. 56	67. 48	51. 15	67. 43	51. 74	66. 98	52. 33
90	72. 35	53. 53	71. 88	54. 16	71. 40	54. 79	70. 92	55. 41
95	76. 37	56. 51	75. 87	57. 17	75. 37	57. 83	74. 86	58. 49
100	80. 39	59. 48	79. 86	60. 18	79. 34	60. 88	78. 80	61. 57
	Dep.	Lat.	Dep.	Lat.	Dep.	Lat.	Dep.	Lat.
	53½ Deg.		53 Deg.		52½ Deg.		52 Deg.	

Distance.	38½ Deg.		39 Deg.		39½ Deg.		40 Deg.	
	Lat.	Dep.	Lat.	Dep.	Lat.	Dep.	Lat.	Dep.
1	0. 78	0. 62	0. 78	0. 63	0. 77	0. 64	0. 77	0. 64
2	1. 57	1. 24	1. 55	1. 26	1. 54	1. 27	1. 53	1. 29
3	2. 35	1. 87	2. 33	1. 89	2. 31	1. 91	2. 30	1. 93
4	3. 13	2. 49	3. 11	2. 52	3. 09	2. 54	3. 06	2. 57
5	3. 91	3. 11	3. 89	3. 15	3. 86	3. 18	3. 83	3. 21
6	4. 70	3. 74	4. 66	3. 98	4. 63	3. 82	4. 60	3. 86
7	5. 48	4. 36	5. 44	4. 41	5. 40	4. 45	5. 36	4. 50
8	6. 26	4. 98	6. 22	5. 03	6. 17	5. 09	6. 13	5. 14
9	7. 04	5. 60	6. 99	5. 66	6. 94	5. 72	6. 89	5. 79
10	7. 83	6. 23	7. 77	6. 29	7. 72	6. 36	7. 66	6. 43
11	8. 61	6. 85	8. 55	6. 92	8. 49	7. 00	8. 43	7. 07
12	9. 39	7. 47	9. 33	7. 55	9. 26	7. 63	9. 19	7. 71
13	10. 17	8. 09	10. 10	8. 18	10. 03	8. 27	9. 96	8. 36
14	10. 96	8. 72	10. 88	8. 81	10. 80	8. 91	10. 72	9. 00
15	11. 74	9. 34	11. 66	9. 44	11. 57	9. 54	11. 49	9. 64
16	12. 52	9. 96	12. 43	10. 07	12. 35	10. 18	12. 26	10. 28
17	13. 30	10. 58	13. 21	10. 70	13. 12	10. 81	13. 02	10. 93
18	14. 09	11. 21	13. 99	11. 33	13. 89	11. 45	13. 79	11. 57
19	14. 87	11. 83	14. 77	11. 96	14. 66	12. 09	14. 55	12. 21
20	15. 65	12. 45	15. 54	12. 59	15. 43	12. 72	15. 32	12. 86
21	16. 43	13. 07	16. 32	13. 22	16. 20	13. 36	16. 09	13. 50
22	17. 22	13. 70	17. 10	13. 84	16. 98	13. 99	16. 85	14. 14
23	18. 00	14. 32	17. 87	14. 47	17. 75	14. 63	17. 62	14. 78
24	18. 78	14. 94	18. 65	15. 10	18. 52	15. 27	18. 39	15. 43
25	19. 57	15. 56	19. 43	15. 73	19. 29	15. 90	19. 15	16. 07
26	20. 35	16. 19	20. 21	16. 36	20. 06	16. 54	19. 92	16. 71
27	21. 13	16. 81	20. 98	16. 99	20. 83	17. 17	20. 68	17. 36
28	21. 91	17. 43	21. 76	17. 62	21. 61	17. 81	21. 45	18. 00
29	22. 70	18. 05	22. 54	18. 25	22. 38	18. 45	22. 22	18. 64
30	23. 48	18. 68	23. 31	18. 88	23. 15	19. 08	22. 98	19. 28
35	27. 39	21. 79	27. 20	22. 03	27. 01	22. 26	26. 81	22. 50
40	31. 30	24. 90	31. 09	25. 17	30. 86	25. 44	30. 64	25. 71
45	35. 22	28. 01	34. 97	28. 32	34. 72	28. 62	34. 47	28. 93
50	39. 13	31. 13	38. 86	31. 47	38. 58	31. 80	38. 30	32. 14
55	43. 04	34. 24	42. 74	34. 61	42. 44	34. 98	42. 13	35. 35
60	46. 96	37. 35	46. 63	37. 76	46. 30	38. 16	45. 96	38. 57
65	50. 87	40. 46	50. 51	40. 91	50. 16	41. 35	49. 79	41. 78
70	54. 78	43. 58	54. 40	44. 05	54. 01	44. 53	53. 62	45. 00
75	58. 70	46. 69	58. 29	47. 20	57. 87	47. 71	57. 45	48. 21
80	62. 61	49. 80	62. 17	50. 35	61. 73	50. 89	61. 28	51. 42
85	66. 52	52. 91	66. 06	53. 49	65. 59	54. 07	65. 11	54. 64
90	70. 43	56. 03	69. 94	56. 64	69. 45	57. 25	68. 94	57. 85
95	74. 35	59. 14	73. 83	59. 79	73. 30	60. 43	72. 77	61. 06
100	78. 26	62. 25	77. 71	62. 93	77. 16	63. 61	76. 60	64. 28
	Dep.	Lat.	Dep.	Lat.	Dep.	Lat.	Dep.	Lat.
	51½ Deg.		51 Deg.		50½ Deg.		50 Deg.	

Distance.	40½ Deg. Lat.	40½ Deg. Dep.	41 Deg. Lat.	41 Deg. Dep.	41½ Deg. Lat.	41½ Deg. Dep.	42 Deg. Lat.	42 Deg. Dep.
1	0. 76	0. 65	0. 75	0 66	0. 75	0. 66	0. 74	0. 67
2	1. 52	1. 30	1. 51	1. 31	1. 50	1. 33	1. 49	1. 34
3	2. 28	1. 95	2. 26	1. 97	2. 25	1. 99	2. 23	2. 01
4	3. 04	2. 60	3. 02	2. 62	3. 00	2. 65	2. 97	2. 68
5	3. 80	3. 25	3. 77	3. 28	3. 74	3. 31	3. 72	3. 35
6	4. 56	3. 90	4. 53	3. 94	4. 49	3. 98	4. 46	4. 01
7	5. 32	4. 55	5. 28	4. 59	5. 24	4. 64	5. 20	4. 68
8	6. 08	5. 20	6. 04	5. 25	5. 99	5. 30	5. 95	5. 35
9	6. 84	5. 84	6. 79	5. 90	6. 74	5. 96	6. 69	6. 02
10	7. 60	6. 49	7. 55	6. 56	7. 49	6. 63	7. 43	6. 69
11	8. 36	7. 14	8. 30	7. 22	8. 24	7. 29	8. 17	7. 36
12	9. 12	7. 79	9. 06	7. 87	8. 99	7. 95	8. 92	8. 03
13	9. 89	8. 44	9. 81	8. 53	9. 74	8. 61	9. 66	8. 70
14	10. 65	9. 09	10. 57	9. 18	10. 49	9. 28	10. 40	9. 37
15	11. 41	9. 74	11. 32	9. 84	11. 23	9. 94	11. 15	10. 04
16	12. 17	10. 39	12. 08	10. 50	11. 98	10. 60	11. 89	10. 71
17	12. 93	11. 04	12. 83	11. 15	12. 73	11. 26	12. 63	11. 38
18	13. 69	11. 69	13. 58	11. 81	13. 48	11. 93	13. 38	12. 04
19	14. 45	12. 34	14. 34	12. 47	14. 23	12. 59	14. 12	12. 71
20	15. 21	12. 99	15. 09	13. 12	14. 98	13. 25	14. 86	13. 38
21	15. 97	13. 64	15. 85	13. 78	15. 73	13. 91	15. 61	14. 05
22	16. 73	14. 29	16. 60	14. 43	16. 48	14. 58	16. 35	14. 72
23	17. 49	14. 94	17. 36	15. 09	17. 23	15. 24	17. 09	15. 39
24	18. 25	15. 59	18. 11	15. 75	17. 97	15. 90	17. 84	16. 06
25	19. 01	16. 24	18. 87	16. 40	18. 72	16. 57	18. 58	16. 73
26	19. 77	16. 89	19. 62	17. 06	19. 47	17. 23	19. 32	17. 40
27	20. 53	17. 54	20. 38	17. 71	20. 22	17. 89	20. 06	18. 07
28	21. 29	18. 18	21. 13	18. 37	20. 97	18. 55	20. 81	18. 74
29	22. 05	18. 83	21. 89	19. 03	21. 72	19. 22	21. 55	19. 40
30	22. 81	19. 48	22. 64	19. 68	22. 47	19. 88	22. 29	20. 07
35	26. 61	22. 73	26. 41	22. 96	26. 21	23. 19	26. 01	23. 42
40	30. 42	25. 98	30. 19	26. 24	29. 96	26. 50	29. 73	26. 77
45	34. 22	29. 23	33. 96	29. 52	33. 70	29. 82	33. 44	30. 11
50	38. 02	32. 47	37. 74	32. 80	37. 45	33. 13	37. 16	33. 46
55	41. 82	35. 72	41. 51	36. 08	41. 19	36. 44	40. 87	36. 80
60	45. 62	38. 97	45. 28	39. 36	44. 94	39. 76	44. 59	40. 15
65	49. 43	42. 21	49. 06	42. 64	48. 68	43. 07	48. 30	43. 49
70	53. 23	45. 46	52. 83	45. 92	52. 43	46. 38	52. 02	46. 84
75	57. 03	48. 71	56. 60	49. 20	56. 17	49. 70	55. 74	50. 18
80	60. 83	51. 96	60. 38	52. 48	59. 92	53. 01	59. 45	53. 53
85	64. 63	55. 20	64. 15	55. 76	63. 66	56. 32	63. 17	56. 88
90	68. 44	58. 45	67. 92	59. 05	67. 41	59. 64	66. 88	60. 22
95	72. 24	61. 70	71. 70	62. 33	71. 15	62. 95	70. 60	63. 57
100	76. 04	64. 98	75. 47	65. 61	74. 90	66. 26	74. 31	66. 91
	Dep.	Lat.	Dep.	Lat.	Dep.	Lat.	Dep.	Lat.
	49½ Deg.		49 Deg.		48½ Deg.		48 Deg.	

Distance.	42½ Deg.		43 Deg.		43½ Deg.		44 Deg.	
	Lat.	Dep.	Lat.	Dep.	Lat.	Dep.	Lat.	Dep.
1	0. 74	0. 68	0. 73	0. 68	0. 73	0. 69	0. 72	0. 69
2	1. 47	1. 35	1. 46	1. 36	1. 45	1. 38	1. 44	1. 39
3	2. 21	2. 03	2. 19	2. 05	2. 18	2. 07	2. 16	2. 08
4	2. 95	2. 70	2. 93	2. 73	2. 90	2. 75	2. 88	2. 78
5	3. 69	3. 38	3. 66	3. 41	3. 63	3. 44	3. 60	3. 47
6	4. 42	4. 05	4. 39	4. 09	4. 35	4. 13	4. 32	4. 17
7	5. 16	4. 73	5. 12	4. 77	5. 08	4. 82	5. 04	4. 86
8	5. 90	5. 40	5. 85	5. 46	5. 80	5. 51	5. 75	5. 56
9	6. 64	6. 08	6. 58	6. 14	6. 53	6. 20	6. 47	6. 25
10	7. 37	6. 76	7. 31	6. 82	7. 25	6. 88	7. 19	6. 95
11	8. 11	7. 43	8. 04	7. 50	7. 98	7. 57	7. 91	7. 64
12	8. 85	8. 11	8. 78	8. 18	8. 70	8. 26	8. 63	8. 34
13	9. 58	8. 78	9. 51	8. 87	9. 43	8. 95	9. 35	9. 03
14	10. 32	9. 46	10. 24	9. 55	10. 16	9. 64	10. 07	9. 73
15	11. 06	10. 13	10. 97	10. 23	10. 88	10. 33	10. 79	10. 42
16	11. 80	10. 81	11. 70	10. 91	11. 61	11. 01	11. 51	11. 11
17	12. 53	11. 48	12. 43	11. 59	12. 33	11. 70	12. 23	11. 81
18	13. 27	12. 16	13. 16	12. 28	13. 06	12. 39	12. 95	12. 50
19	14. 01	12. 84	13. 90	12. 96	13. 78	13. 08	13. 67	13. 20
20	14. 75	13. 51	14. 63	13. 64	14. 51	13. 77	14. 39	13. 89
21	15. 48	14. 19	15. 36	14. 32	15. 23	14. 46	15. 11	14. 59
22	16. 22	14. 86	16. 09	15. 00	15. 96	15. 14	15. 83	15. 28
23	16. 96	15. 54	16. 82	15. 69	16. 88	15. 83	16. 54	15. 98
24	17. 69	16. 21	17. 55	16. 37	17. 41	16. 52	17. 26	16. 67
25	18. 43	16. 89	18. 28	17. 05	18. 13	17. 21	17. 98	17. 37
26	19. 17	17. 57	19. 02	17. 73	18. 86	17. 90	18. 70	18. 06
27	19. 91	18. 24	19. 75	18. 41	19. 59	18. 59	19. 42	18. 76
28	20. 64	18. 92	20. 48	19. 10	20. 31	19. 27	20. 14	19. 45
29	21. 38	19. 59	21. 21	19. 78	21. 04	19. 96	20. 86	20. 15
30	22. 12	20. 27	21. 94	20. 46	21. 76	20. 65	21. 58	20. 84
35	25. 80	23. 65	25. 60	23. 87	25. 39	24. 09	25. 18	24. 31
40	29. 49	27. 02	29. 25	27. 28	29. 01	27. 53	28. 77	27. 79
45	33. 18	30. 40	32. 91	30. 69	32. 64	30. 98	32. 37	31. 26
50	36. 86	33. 78	36. 57	34. 10	36. 27	34. 42	35. 57	34. 73
55	40. 55	37. 16	40. 22	37. 51	39. 90	37. 86	39. 96	38. 21
60	44. 24	40. 54	43. 88	40. 92	43. 52	41. 30	43. 16	41. 68
65	47. 92	43. 91	47. 54	44. 33	47. 15	44. 74	46. 76	45. 15
70	51. 61	47. 29	51. 19	47. 74	50. 78	48. 18	50. 35	48. 63
75	55. 30	50. 67	54. 85	51. 15	54. 40	51. 63	53. 95	52. 10
80	58. 98	54. 05	58. 51	54. 56	58. 03	55. 07	57. 55	55. 57
85	62. 67	57. 43	62. 17	57. 97	61. 66	58. 51	61. 14	59. 05
90	66. 35	60. 80	65. 82	61. 38	65. 28	61. 95	64. 74	62. 52
95	70. 04	64. 18	69. 48	64. 79	68. 91	65. 39	68. 34	65. 99
100	73. 73	67. 56	73. 14	68. 20	72. 54	68. 84	71. 93	69. 47
	Dep.	Lat.	Dep.	Lat.	Dep.	Lat.	Dep.	Lat.
	47½ Deg.		47 Deg.		46½ Deg.		46 Deg.	

Distance.	44½ Deg.		45 Deg.	
	Lat.	Dep.	Lat.	Dep.
1	0. 71	0. 70	0. 71	0 71
2	1. 43	1. 40	1. 41	1. 41
3	2. 14	2. 10	2. 12	2. 12
4	2. 85	2. 80	2. 83	2. 83
5	3. 57	3. 50	3. 54	3. 54
6	4. 28	4. 21	4. 24	4. 24
7	4. 99	4. 91	4. 95	4. 95
8	5. 71	5. 61	5. 66	5. 66
9	6. 42	6. 31	6. 36	6. 36
10	7. 13	7. 01	7. 07	7. 07
11	7. 85	7. 71	7. 78	7. 78
12	8. 56	8. 41	8. 49	8. 49
13	9. 27	9. 11	9. 19	9. 19
14	9. 99	9. 81	9. 90	9. 90
15	10. 70	10. 51	10. 61	10. 61
16	11. 41	11. 21	11. 31	11. 31
17	12. 13	11. 92	12. 02	12. 02
18	12. 84	12. 62	12. 73	12. 73
19	13. 55	13. 32	13. 43	13. 43
20	14. 26	14. 02	14. 14	14. 14
21	14. 98	14. 72	14. 85	14. 85
22	15. 69	15. 42	15. 56	15. 56
23	16. 40	16. 12	16. 26	16. 26
24	17. 12	16. 82	16. 97	16. 97
25	17. 83	17. 52	17. 68	17. 68
26	18. 54	18. 22	18. 38	18. 38
27	19. 26	18. 92	19. 09	19. 09
28	19. 97	19. 63	19. 80	19. 80
29	20. 68	20. 33	20. 51	20. 51
30	21. 40	21. 03	21. 21	21. 21
35	24. 96	24. 53	24. 75	24. 75
40	28. 53	28. 04	28. 28	28. 28
45	32. 10	31. 54	31. 82	31. 82
50	35. 66	35. 05	35. 36	35. 36
55	39. 23	38. 55	38. 89	38. 89
60	42. 79	42. 05	42. 43	42. 43
65	46. 36	45. 56	45. 96	45. 96
70	49. 93	49. 06	49. 50	49. 50
75	53. 49	52. 57	53. 03	53. 03
80	57. 06	56. 07	56. 57	56. 57
85	60. 63	59. 58	60. 10	60. 10
90	64. 19	63. 08	63. 64	63. 64
95	67. 76	66. 59	67. 18	67. 18
100	71. 33	70. 09	70. 71	70. 71
	Dep.	Lat.	Dep.	Lat.
	45½ Deg.		45 Deg.	

′	0°	1	2°	3°	4°	5°	6°	7°	8°	9°	10°	11°	12°	13°	14°	15°
0	0	60	120	180	240	300	361	421	482	542	603	664	725	787	848	910
1	1	61	121	181	241	301	362	422	483	543	604	665	726	788	850	911
2	2	62	122	182	242	302	363	423	484	544	605	666	727	789	851	913
3	3	63	123	183	243	303	364	424	485	545	606	667	728	790	852	914
4	4	64	124	184	244	304	365	425	486	546	607	668	729	791	853	915
5	5	65	125	185	245	305	366	426	487	547	608	669	730	792	854	916
6	6	66	126	186	246	306	367	427	488	548	609	670	731	793	855	917
7	7	67	127	187	247	307	368	428	489	549	610	671	732	794	856	918
8	8	68	128	188	248	308	369	429	490	550	611	672	734	795	857	919
9	9	69	129	189	249	309	370	430	491	551	612	673	735	796	858	920
10	10	70	130	190	250	310	371	431	492	552	613	674	736	797	859	921
11	11	71	131	191	251	311	372	432	493	553	614	675	737	798	860	922
12	12	72	132	192	252	312	373	433	494	554	615	676	738	799	861	923
13	13	73	133	193	253	313	374	434	495	555	616	677	739	800	862	924
14	14	74	134	194	254	314	375	435	496	556	617	678	740	801	863	925
15	15	75	135	195	255	315	376	436	497	557	618	679	741	802	864	926
16	16	76	136	196	256	316	377	437	468	558	619	680	742	803	865	927
17	17	77	137	197	257	317	378	438	499	559	620	681	743	804	366	928
18	18	78	138	198	258	318	379	439	500	560	621	682	744	805	867	929
19	19	79	139	199	259	319	380	440	501	561	622	683	745	806	868	930
20	20	80	140	200	260	320	381	441	502	562	623	684	746	807	869	931
21	21	81	141	201	261	321	382	442	503	564	624	685	747	808	870	932
22	22	82	142	202	262	322	383	443	504	565	625	687	748	809	871	933
23	23	83	143	203	263	323	384	444	505	566	626	688	749	810	872	934
24	24	84	144	204	264	824	385	445	506	567	627	689	750	811	873	935
25	25	85	145	205	265	325	386	446	507	568	628	690	751	812	874	936
26	26	86	146	206	266	326	387	447	508	569	629	691	752	313	875	937
27	27	87	147	207	267	327	388	448	509	570	631	692	753	815	876	938
28	28	88	148	208	268	328	389	449	510	571	632	693	754	816	877	939
29	29	89	149	209	269	330	390	450	511	572	633	694	755	817	878	941
30	30	90	150	210	270	331	391	451	512	573	634	695	756	818	879	942
31	31	91	151	211	271	332	392	452	513	574	655	696	757	819	880	943
32	32	92	152	212	272	333	393	453	514	575	636	697	758	820	882	944
33	33	93	153	213	273	334	394	454	515	576	637	698	759	821	883	945
34	34	94	154	214	274	335	395	455	516	577	638	699	760	822	884	946
35	35	95	155	215	275	336	396	456	517	578	639	700	761	823	885	947
36	36	96	156	216	276	337	397	457	518	579	640	701	762	824	886	948
37	37	97	157	217	277	338	398	458	519	580	941	702	763	825	887	949
38	38	98	158	218	278	339	399	459	520	581	642	703	764	826	888	950
39	39	99	159	219	279	340	400	460	521	582	643	704	765	827	889	951
40	40	100	160	220	280	341	401	461	522	583	644	705	766	828	890	952
41	41	101	161	221	281	342	402	462	523	584	645	706	767	829	891	953
42	42	102	162	222	282	343	403	463	524	585	646	707	768	830	892	954
43	43	103	163	223	283	344	404	464	525	586	647	708	769	831	893	955
44	44	104	164	224	284	345	405	465	526	587	648	709	770	832	894	956
45	45	105	165	225	285	346	406	466	527	588	649	710	771	833	895	957
46	46	106	166	226	286	347	407	467	528	589	650	711	772	834	896	958
47	47	107	167	227	287	348	408	468	529	590	651	712	773	835	897	959
48	48	108	168	228	288	349	409	469	530	591	652	713	774	836	898	960
49	49	109	169	229	289	350	410	470	531	592	653	714	775	837	899	961
50	50	110	170	230	290	351	411	471	532	593	654	715	777	838	900	962
51	51	111	171	231	291	352	412	472	533	594	655	716	778	839	901	963
52	52	112	172	232	292	353	413	473	534	595	656	717	779	840	902	964
53	53	113	173	233	293	354	414	474	535	596	657	718	780	841	903	965
54	54	114	174	234	294	355	415	476	536	597	658	719	781	842	904	966
55	55	115	175	235	295	356	416	477	537	598	659	720	782	843	905	968
56	56	116	176	236	296	357	417	478	538	599	660	721	783	844	606	969
57	57	117	177	237	297	358	418	479	539	600	661	722	784	845	907	970
58	58	118	178	238	298	359	419	480	540	601	662	723	785	846	908	971
59	59	119	179	239	299	360	420	481	541	602	663	724	786	847	909	972

'	16°	17°	18°	19°	20°	21°	22°	23°	24°	25°	26°	27°	28°
0	973	1035	1098	1161	1225	1289	1354	1419	1484	1550	1616	1684	1751
1	974	1036	1099	1163	1226	1290	1355	1420	1485	1551	1618	1685	1752
2	975	1037	1100	1164	1227	1291	1356	1421	1486	1552	1619	1686	1753
3	976	1038	1101	1165	1228	1292	1357	1422	1487	1553	1620	1687	1755
4	977	1039	1102	1166	1229	1293	1358	1423	1488	1554	1621	1688	1756
5	978	1041	1103	1167	1230	1295	1359	1424	1490	1556	1622	1689	1757
6	979	1042	1105	1168	1232	1296	1360	1425	1491	1557	1623	1690	1758
7	980	1043	1106	1169	1233	1297	1361	1426	1492	1558	1624	1692	1759
8	981	1044	1107	1170	1234	1298	1362	1427	1493	1559	1625	1693	1760
9	982	1045	1108	1171	1235	1299	1363	1428	1494	1560	1626	1694	1761
10	983	1046	1109	1172	1236	1300	1364	1430	1495	1561	1628	1695	1762
11	984	1047	1110	1173	1237	1301	1366	1431	1496	1562	1629	1696	1764
12	985	1048	1111	1174	1238	1302	1367	1432	1497	1563	1630	1697	1765
13	986	1049	1112	1175	1239	1303	1368	1433	1498	1564	1631	1698	1766
14	987	1050	1113	1176	1240	1304	1369	1434	1499	1565	1632	1699	1767
15	988	1051	1114	1177	1241	1305	1370	1435	1500	1567	1633	1700	1768
16	989	1052	1115	1178	1242	1306	1371	1436	1502	1568	1634	1701	1769
17	990	1053	1116	1179	1243	1307	1372	1437	1503	1569	1635	1703	1770
18	991	1054	1117	1181	1244	1308	1373	1438	1504	1570	1637	1704	1772
19	993	1055	1118	1182	1245	1310	1374	1439	1505	1571	1638	1705	1773
20	994	1056	1119	1183	1246	1311	1375	1440	1506	1572	1639	1706	1774
21	995	1057	1120	1184	1248	1312	1376	1441	1507	1573	1640	1707	1775
22	996	1058	1121	1185	1249	1313	1377	1443	1508	1574	1641	1708	1776
23	997	1059	1122	1186	1250	1314	1379	1444	1509	1575	1642	1709	1777
24	998	1060	1123	1187	1251	1315	1380	1445	1510	1577	1643	1711	1778
25	999	1161	1125	1188	1252	1316	1381	1446	1511	1578	1644	1712	1780
26	1000	1063	1126	1189	1253	1317	1382	1447	1513	1579	1645	1713	1781
27	1001	1064	1127	1190	1254	1318	1383	1448	1514	1580	1647	1714	1782
28	1002	1065	1128	1191	1255	1319	1384	1449	1515	1581	1648	1715	1783
29	1003	1066	1129	1192	1256	1320	1385	1450	1516	1582	1649	1716	1784
30	1004	1067	1130	1193	1257	1321	1386	1451	1517	1583	1650	1717	1785
31	1005	1068	1131	1194	1258	1322	1387	1452	1518	1584	1651	1718	1786
32	1006	1069	1132	1195	1259	1324	1388	1453	1519	1585	1652	1720	1787
33	1007	1070	1133	1196	1260	1325	1389	1455	1520	1586	1653	1721	1789
34	1008	1071	1134	1198	1261	1326	1390	1456	1521	1588	1654	1722	1790
35	1009	1072	1135	1199	1262	1327	1392	1457	1522	1589	1656	1723	1791
36	1010	1073	1136	1200	1264	1328	1393	1458	1524	1590	1657	1724	1792
37	1011	1074	1137	1201	1265	1329	1394	1459	1525	1591	1658	1725	1793
38	1012	1075	1138	1202	1266	1330	1395	1460	1526	1592	1659	1726	1794
39	1013	1076	1139	1203	1267	1331	1396	1461	1527	1593	1660	1727	1795
40	1014	1077	1140	1204	1268	1332	1397	1462	1528	1594	1661	1729	1797
41	1015	1078	1141	1205	1269	1333	1398	1463	1529	1595	1662	1730	1798
42	1016	1079	1142	1206	1270	1334	1399	1464	1530	1596	1663	1731	1799
43	1018	1080	1144	1207	1271	1335	1400	1465	1531	1598	1664	1732	1800
44	1019	1081	1145	1208	1272	1336	1401	1467	1532	1599	1666	1733	1801
45	1020	1082	1146	1209	1273	1338	1402	1468	1533	1600	1667	1734	1802
46	1021	1084	1147	1210	1274	1339	1403	1469	1535	1601	1668	1735	1803
47	1022	1085	1148	1211	1275	1340	1405	1470	1536	1602	1669	1736	1805
48	1023	1086	1149	1212	1276	1341	1406	1471	1537	1603	1670	1737	1806
49	1024	1087	1150	0213	1277	1342	1407	1472	1538	1604	1671	1739	1807
50	1025	1088	1151	1215	1278	1343	1408	1473	1539	1605	1672	1740	1808
51	1026	1089	1152	1216	1280	1344	1409	1474	1540	1606	1673	1741	1809
52	1027	1090	1153	1217	1281	1345	1410	1475	1541	1608	1675	1742	1810
53	1028	1091	1154	1218	1282	1346	1411	1476	1542	1609	1676	1743	1811
54	1029	1092	1155	1219	1283	1347	1412	1477	1543	1610	1677	1744	1813
55	1030	1093	1156	1220	1284	1348	1413	1479	1544	1611	1678	1746	1814
56	1031	1094	1157	1221	1285	1349	1414	1480	1546	1612	1679	1747	1815
57	1032	1095	1158	1222	1286	1350	1415	1481	1547	1613	1680	1748	1816
58	1033	1096	1159	1223	1287	1352	1416	1482	1548	1614	1681	1749	1817
59	1034	1097	1160	1224	1288	1353	1418	1483	1549	1615	1682	1750	1818

′	29°	30°	31°	32°	33°	34°	35°	36°	37°	38°	39°	40°	41°
0	1819	1888	1958	2028	2100	2171	2244	2318	2393	2468	2545	2623	2702
1	1821	1890	1959	2030	2101	2173	2246	2319	2394	2470	2546	2624	2703
2	1822	1891	1960	2031	2102	2174	2247	2320	2395	2471	2648	2625	2704
3	1823	1892	1962	2032	2103	2175	2248	2322	2396	2472	2549	2627	2706
4	1924	1893	1963	2033	2104	2176	2249	2323	2398	2473	2550	2628	2707
5	1825	1894	1964	2034	2105	2178	2250	2324	2399	2475	2551	2629	2708
6	1826	1895	1965	2035	2107	2179	2252	2325	2400	2479	2553	2631	2710
7	1827	1896	1966	2037	2108	2180	2253	2327	2401	2477	2554	2632	2711
8	1829	1898	1967	2038	2109	2181	2254	2328	2403	2478	2555	2633	2712
9	1830	1899	1969	2039	2110	2182	2255	2329	2404	2480	2557	2634	2714
10	1831	1900	1970	2040	2111	2184	2257	2330	2405	2481	2558	2636	2715
11	1832	1901	1971	2041	2113	2185	2258	2332	2406	2482	2559	2637	2716
12	1833	1902	1972	2043	2114	2186	2259	2333	2408	2484	2560	2638	2718
13	1834	1903	1973	2044	2115	2187	2260	2334	2409	2485	2562	2640	2719
14	1835	1905	1974	2045	2116	2188	2261	2335	2410	2486	2563	2641	2720
15	1837	1906	1976	2046	2117	2190	2263	2337	2411	2487	2564	2642	2722
16	1838	1907	1977	2047	2119	2191	2264	2338	2413	2489	2566	2644	2723
17	1839	1908	1978	2048	2120	2192	2265	2339	2414	2490	2567	2645	2724
18	1840	1909	1679	2050	2121	2193	2266	2340	2415	2491	2568	2646	2726
19	1841	1910	1980	2051	2122	2194	2268	2342	2416	2492	2569	2648	2727
20	1842	1912	1981	2052	2123	2196	2269	2343	2418	2494	2571	2649	2728
21	1843	1913	1983	2053	2125	2197	2270	2344	2419	2495	2572	2650	2729
22	1845	1914	1984	2054	2126	2198	2271	2345	2420	2496	2573	2551	2731
23	1846	1915	1985	2056	2127	2199	2272	2346	2422	2498	2575	2653	2732
24	1847	1916	1986	2057	2128	2200	2274	2348	2423	2499	2576	2654	2733
25	1848	1917	1987	2058	2129	2202	2275	2349	2424	2500	2577	2655	2735
29	1849	1918	1988	2059	2131	2203	2266	2350	2425	2501	2578	2657	2736
27	1850	1920	1990	2060	2132	2204	2267	2351	2427	2503	2580	2658	2737
28	1852	1921	1991	2061	2133	2205	2279	2353	2428	2504	2581	2659	2739
29	1853	1922	1992	2063	2134	2207	2280	2354	2429	2505	2582	2661	2740
30	1854	1923	1993	2064	2135	2208	2281	2355	2430	2506	2584	2662	2742
31	1855	1924	1994	2065	2137	2209	2232	2356	2432	2508	2585	2663	2743
32	1856	1925	1995	2066	2138	2210	2283	2358	2433	2509	2586	2665	2744
33	1857	1927	1997	2067	2139	2211	2285	2359	2434	2510	2588	2666	2746
34	1858	1928	1998	2069	2140	2213	2286	2360	2435	2512	2589	2667	2747
35	1360	1929	1999	2070	2141	2214	2287	2361	2437	2513	2590	2669	2748
36	1861	1930	2000	2071	2143	2215	2288	2363	2438	2514	2591	2670	2750
37	1892	1931	2001	2072	2144	2216	2290	2364	2439	2515	2593	2671	2751
38	1863	1932	2002	2073	2145	2217	2291	2365	2440	2517	2594	2673	2752
39	1864	1934	2004	2075	2146	2219	2292	2366	2442	2518	2595	2674	2754
40	1865	1935	2005	2076	2147	2220	2293	2368	2443	2519	2597	2675	2755
41	1866	1936	2006	2077	2149	2221	2295	2369	2444	2521	2598	2676	2756
42	1868	1937	2077	2078	2150	2222	2296	2370	2445	2522	2599	2678	2758
43	1869	1938	2008	2079	2151	2224	2297	2371	2447	2523	2601	2679	2759
44	1870	1939	2010	2080	2152	2225	2298	2373	2448	2524	2602	2680	2760
45	1871	1941	2011	2082	2153	2226	2299	2374	2449	2526	2603	2682	2762
46	1872	1942	2012	2083	2155	2227	2301	2375	2451	2527	2604	2683	2763
47	1873	1943	2013	2084	2156	2228	2302	2376	2452	2528	2606	2684	2764
48	1875	1944	2014	2085	2157	2230	2303	2378	2453	2530	2907	2636	2766
49	1876	1945	2015	2086	2158	2231	2304	2379	2454	2531	2608	2687	2767
50	1877	1946	2017	2088	2159	2232	2306	2380	2456	2532	2610	2688	2768
51	1878	1948	2018	2089	2161	2233	2307	2381	2457	2533	2611	2690	2770
52	1879	1949	2019	2090	2162	2235	2308	2383	2458	2535	2612	2691	2771
53	1880	1950	2020	2091	2163	2236	2309	2384	2459	2536	2614	2692	2772
54	1881	1951	2021	2092	2164	2237	2311	2385	2461	2537	2615	2694	2774
55	1883	1952	2022	2094	2165	2238	2312	2386	2462	2538	2616	2695	2775
56	1884	1953	2024	2095	2167	2239	2313	2388	2463	2540	2617	2696	2776
57	1885	1955	2025	2096	2168	2241	2314	2389	2464	2541	2619	2698	2778
58	1886	1956	2026	2097	2169	2242	2316	2390	2466	2542	2620	2699	2779
59	1887	1957	2027	2098	2170	2243	2317	2391	2467	2544	2621	2700	2780

′	42°	43°	44°	45°	46°	47°	48°	49°	50°	51°	52°	53°	54°
0	2782	2863	2946	3030	3116	3203	3292	3382	3474	3569	3665	3764	3865
1	2783	2864	2947	3031	3117	3204	3293	3384	3476	3570	3667	3765	3866
2	2734	2866	2949	3033	3118	3206	3295	3385	3478	3572	3668	3767	3868
3	2786	2867	2950	3034	2120	3207	3296	3387	3479	3573	3670	3769	3870
4	2787	2869	2951	3036	3121	3209	3298	3388	3481	3575	3672	3770	3871
5	2788	2870	2953	3037	3123	3210	3299	3390	3482	3577	3673	3772	3873
6	2790	2871	2954	3038	3123	3212	3301	3391	3484	3578	3675	3774	3375
7	2791	2873	2956	3040	3126	3213	3302	3393	3485	3580	3677	3775	3877
8	2792	2874	2957	3041	3127	3214	3303	3394	3487	3582	3678	3777	3878
9	2794	2875	2958	3043	3129	3216	3305	3396	3488	3583	3680	3779	3880
10	2795	2877	2960	3044	3130	3217	3306	3397	3490	3585	3681	3780	3882
11	2797	2878	2961	3046	3131	3219	3308	3399	3492	3586	3683	3782	3883
12	2798	2880	2963	3047	3133	3220	3309	3400	3493	3588	3685	3784	3885
13	2799	2881	2964	3048	3134	3222	3311	3402	3495	3590	3686	3785	3887
14	2801	2882	2965	3050	3136	3224	3312	3403	3496	3591	3688	3787	3889
15	2802	2884	2967	3051	3137	3225	3314	3405	3498	3593	3690	3799	3890
16	2803	2885	2968	3053	3139	3226	3316	3407	3499	3594	3691	3790	3892
17	2805	2886	2970	3054	3140	3228	3317	3408	3501	3596	3693	3792	3894
18	2806	2888	2971	3055	3142	3229	3319	3410	3503	3598	3695	3794	3895
19	2807	2889	2972	3057	3143	3231	3320	3411	3504	3599	3696	3795	3897
20	2809	2891	2974	3058	3144	3232	3322	3413	3506	3601	3698	3797	3899
21	2810	2892	2975	3060	3146	3234	3323	3414	3507	3602	3699	3799	3901
22	2811	2893	2976	3061	3147	3235	3325	3416	3509	3604	3701	3800	3902
23	2813	2895	2978	3063	3149	3237	3326	3417	3510	3606	3703	3802	3904
24	2814	2896	2979	3064	3150	3238	3328	3419	3512	3607	3704	3804	3906
25	2815	2897	2981	3065	3152	3240	3329	3420	3514	3609	3706	3806	3907
26	2817	2899	2982	3067	3153	3241	3331	3422	3515	3610	3708	3807	3909
27	2818	2900	2983	3068	3155	3242	3332	3423	3517	3612	3709	3809	3911
28	2820	2902	2985	3070	3156	3244	3334	3425	3518	3614	3711	3811	3913
29	2821	2903	2986	3071	3157	3245	3335	3427	3520	3615	3713	3812	3914
30	2822	2904	2988	3073	3159	3247	3337	3428	3521	3617	3714	3814	3916
31	2824	2906	2989	3074	3160	3248	3338	3430	3523	3618	3716	3816	3918
32	2825	2907	2991	3075	3162	3250	3340	3431	3525	3620	3717	3817	3919
33	2826	2908	2992	3077	3163	3251	3341	3433	3526	3622	3719	3819	3921
34	2828	2910	2993	3078	3165	3253	3343	3434	3528	3623	3721	3821	3923
35	2829	2911	2995	3080	3166	3254	3344	3436	3529	3625	3722	3822	3925
36	2830	2913	2996	3081	3168	3256	3346	3437	3531	3626	3724	3824	3926
37	2832	2914	2998	3083	3169	3257	3347	3439	3532	3628	3726	3826	3928
38	2833	2915	2999	3084	3171	3259	3349	3440	3534	3630	3727	3827	3930
39	2834	2917	3000	3085	3172	3260	3350	3442	3536	3631	3729	3829	3932
40	2836	2918	3002	3087	3173	3262	3352	3543	3537	3633	3731	3831	3933
41	2837	2919	3003	3088	3175	3263	3353	3445	3539	3634	3732	3832	3935
42	2839	2921	3005	3090	3176	3265	3355	3447	3540	3636	3734	3834	3937
43	2840	2922	2006	3091	3178	3266	3356	3448	3542	3638	3736	3836	3938
44	2841	2924	3007	3093	3179	3268	3358	3450	3543	3639	3737	3838	3940
45	2843	2925	3009	3094	3181	3269	3359	3551	3545	3641	3739	3839	3942
46	2844	2926	3010	3095	3182	3271	3361	3453	3547	3643	3741	3841	3944
47	2845	2928	3012	3097	3184	3272	3362	3454	3548	3644	3742	3843	3945
48	2847	2929	3013	3098	3185	3274	3364	3456	3550	3646	3744	3844	3947
49	2848	2931	3014	3100	3187	3275	3365	3457	3551	3647	3746	3846	3949
50	2849	2932	3016	3101	3188	3277	3367	3459	3553	3649	3747	3848	3951
51	2851	2933	3017	3103	3190	3278	3368	3460	3555	3651	3749	3849	3952
52	2852	2935	3019	3104	3191	3280	3370	3462	3556	3652	3750	3851	3954
53	2854	2936	3020	3105	3192	3281	3371	3464	3558	3654	3752	3853	3956
54	2855	2937	3021	3107	3194	3283	3373	3465	3559	3655	3754	3854	3958
55	2856	2939	3023	3108	3195	3284	3374	3467	3561	3657	3755	3856	3959
56	2858	2940	3024	3110	3197	3286	3376	3468	3562	3659	3757	3858	3961
57	2859	2942	3026	3111	3198	3287	3378	3470	3564	3660	3759	3860	3963
58	2860	2943	3027	3113	3200	3289	3379	3471	3566	3662	3760	3861	3964
59	2862	2944	3029	3114	3201	3290	3381	3473	3567	3664	3762	3863	3966

′	55°	56°	57°	58°	59°	60°	61°	62°	63°	64°	65°	66°	67°
0	3968	4074	4183	4294	4409	4527	4649	4775	4905	5039	5179	5324	5474
1	3970	4076	4184	4296	4411	4529	4651	4777	4907	5042	5181	5326	5477
2	3971	4077	4186	4298	4413	4531	4653	4779	4909	5044	5184	5328	5479
3	3973	4079	4188	4300	4415	4533	4655	4781	4912	5046	5186	5331	5482
4	3975	4081	4190	4302	4417	4535	4657	4784	4914	5049	5188	5333	5484
5	3977	4083	4192	4304	4419	4537	4660	4786	4916	5051	5191	5336	5487
6	3978	4085	4194	4306	4421	4539	4662	4788	4918	5053	5193	5338	5489
7	3980	4086	4195	4308	4423	4541	4664	4790	4920	5055	5195	5341	5492
8	3982	4058	4197	4309	4425	4543	4666	4792	4923	5058	5198	5343	5495
9	3984	4080	4199	4311	4427	4545	4668	4794	4925	5060	5200	5346	5497
10	3985	4092	4202	4313	4429	4547	4670	4796	4927	5062	5203	5348	5500
11	3987	4094	4203	4315	4431	4549	4672	4798	4929	5065	5205	5351	5502
12	3989	4095	4205	4317	4433	4551	4674	4801	4931	5067	5207	5353	5505
13	3991	4097	4207	4319	4434	4553	4676	4805	4934	5069	5210	5356	5507
14	3992	4099	4208	4321	4436	4555	4678	4808	4936	5071	5212	5358	5510
15	3994	4101	4210	4323	4438	4557	4680	4807	4938	5074	5214	5361	5513
16	3996	4103	4212	4325	4440	4559	4682	4809	4940	5076	5217	5363	5515
17	3998	4104	4214	4327	4442	4562	4684	4811	4943	5078	5219	5366	5518
18	3999	4106	4216	4328	4444	4564	4687	4814	4945	5081	5222	5368	5520
19	4001	4108	4218	4330	4446	4566	4689	4816	4947	5083	5224	5371	5523
20	4003	4110	4220	4332	4448	4568	4691	4818	4949	5085	5226	5373	5526
21	4005	4112	4221	4334	4450	4570	4693	4820	4951	5088	5229	5376	5528
22	4006	4113	4223	4336	4452	4572	4695	4822	4954	5090	5231	5378	5531
23	4008	4115	4225	4338	4454	4574	4697	4824	4956	5092	5234	5380	5533
24	4010	4117	4227	4340	4456	4576	4699	4826	4958	5095	5236	5383	5536
25	4012	4119	4229	4342	4458	4578	4701	4829	4960	5097	5238	5385	5539
26	4014	4121	4231	4344	4460	4580	4703	4831	4963	5099	5241	5388	5541
27	4015	4122	4232	4346	4462	4582	4705	4833	4965	5102	5243	5390	5544
28	4017	4124	4234	4347	4464	4584	4707	4835	4967	5104	5246	5393	5546
29	4019	4126	4236	4349	4466	4586	4710	4837	4969	5106	5248	5395	5549
30	4021	4128	4238	4351	4468	4588	4712	4839	4972	5108	5250	5398	5552
31	4022	4130	4240	4353	4470	4590	4714	4842	4974	5111	5253	5401	5554
32	4024	4132	4242	4355	4472	4592	4716	4844	4976	5113	5255	5403	5557
33	4026	4133	4244	4357	4474	4594	4718	4846	4978	5115	5258	5406	5559
34	4028	4135	4246	4359	4476	4596	4720	4848	4981	5118	5260	5408	5562
35	4029	4137	4247	4361	4478	4598	4722	4850	4983	5120	5263	5411	5565
36	4031	4139	4249	4363	4480	4600	4724	4852	4985	5122	5265	5413	5567
37	4033	4141	4251	4365	4482	4602	4726	4855	4987	5125	5267	5416	5570
38	4035	4142	4253	4367	4484	4604	4728	4857	4990	5127	5270	5418	5573
39	4037	4144	4255	4369	4486	4606	4731	4859	4992	5129	5272	5421	5575
40	4038	4146	4257	4370	4488	4608	4733	4861	4994	5132	5275	5423	5578
41	4040	4148	4259	4372	4490	4610	4735	4863	4996	5134	5277	5426	5580
42	4042	4150	4250	4374	4492	4612	4736	4865	4999	5136	5280	5428	5583
43	4044	4152	4262	4376	4494	4614	4739	4868	5001	5139	5282	5431	5586
44	4045	4153	4264	4378	4495	4616	4741	4870	5003	5141	5284	5433	5588
45	4047	4155	4266	4380	4497	4618	4743	4872	5005	5143	5287	5436	5591
46	4049	4157	4268	4382	4499	4620	4745	4874	5008	5146	5289	5438	5594
47	4051	4159	4270	4384	4501	4623	4747	4876	5010	5148	5292	5441	5596
48	4052	4161	4272	4386	4503	4625	4750	4879	5012	5151	5294	5443	5599
49	4054	4162	4274	4388	4505	4627	4752	4881	5014	5153	5297	5446	5602
50	4056	4164	4275	4390	4507	4629	4754	4883	5017	5155	5299	5448	5604
51	4058	4166	4277	4392	4509	4631	4756	4885	5019	5158	5301	5451	5607
52	4060	4168	4279	4394	4511	4633	4758	4887	5021	5160	5304	5454	5610
53	4061	4170	4281	4396	4513	4635	4760	4890	5023	5162	5306	5456	5612
54	4063	4172	4283	4398	4515	4637	4762	4892	5026	5165	5309	5458	5615
55	4065	4173	4285	4399	4517	4639	4764	4894	5028	5167	5311	5461	5617
56	4067	4175	4287	4401	4519	4641	4766	4896	5030	5169	5314	5464	5620
57	4069	4177	4289	4403	4521	4643	4769	4898	5033	5172	5316	5466	5623
58	4070	4179	4291	4405	4523	4645	4771	4901	5035	5174	5319	5469	5625
59	4072	4181	4292	4407	4525	4647	4773	4903	5037	5176	5321	5471	5628

′	68°	69°	70°	71°	72°	73°	74°	75°	76°	77°	78°	79°	80°
0	5631	5795	5966	6146	6335	6534	6746	6970	7210	7467	7745	8046	8375
1	5633	5797	5969	6149	6338	6538	6749	6974	7214	7472	7749	8051	8381
2	5636	5800	5972	6152	6341	6541	6753	6978	7218	7476	7754	8056	8387
3	5639	5803	5975	6155	6345	6545	6757	6982	7222	7481	7759	8061	8393
4	5642	5806	5978	6158	6348	6548	6760	6986	7227	7485	7764	8067	8398
5	5644	5809	5981	6161	6351	6552	6764	6990	7231	7490	7769	8072	8404
6	5646	5811	5984	6164	6354	6555	6768	6994	7235	7494	7774	8077	8410
7	5650	5814	5986	6167	6358	6558	6771	6997	7239	7498	7778	8083	8416
8	5652	5817	5989	6170	6361	6562	6775	7001	7243	7503	7783	8088	8422
9	5655	5820	5992	6173	6364	6565	6779	7005	7247	7507	7788	8093	8427
10	5658	5823	5995	6177	6367	6569	6782	7009	7252	7512	7793	8099	8433
11	5660	5825	5998	6180	6371	6572	6786	7013	7256	7516	7798	8104	8439
12	5663	5828	6001	6183	6374	6576	6790	7017	7260	7521	7803	8109	8445
13	5666	5831	6004	6186	6377	6579	6793	7021	7264	7525	7808	8115	8451
14	5668	5834	6007	6189	6380	9583	6797	7025	7268	7530	7813	8120	8457
15	5671	5837	6010	6192	6384	6586	6801	7029	7273	7535	7817	8125	8463
16	5674	5839	6013	6195	6387	6590	6804	7033	7277	7439	7821	8131	8469
17	5676	5842	6016	6198	6390	6593	6808	7027	7281	7544	7827	8136	8474
18	5679	5845	6019	6201	6394	6597	6812	7041	7285	7548	7832	8141	8480
19	5682	5848	6022	6205	6397	6600	6815	7045	7289	7553	7837	8147	8486
20	5685	5851	6025	6208	6400	6603	6819	7048	7294	7557	7842	8152	8492
21	5687	5854	6028	6211	6403	5607	6823	7052	7298	7562	7847	8158	8498
22	5690	5856	6031	6214	6407	6610	6826	7056	7302	7566	7852	8163	8504
23	5693	5859	6034	6217	6410	6614	6830	7060	7306	7571	7857	8168	8510
24	5695	5862	6037	6220	6413	6617	6834	7064	7311	7576	7862	8174	8516
25	5698	5865	6040	6223	6417	6621	6838	7068	7315	7580	7867	9179	8522
26	5701	5868	6043	6226	6420	6624	6841	7072	7319	7585	7872	8185	8528
27	5704	5871	6046	6230	6423	6628	6845	7076	7323	7589	7877	8190	8534
28	5706	5874	6049	6233	6427	6631	6849	7080	7328	7594	7882	8196	8540
29	5709	5876	6052	6236	6430	6635	6853	7084	7332	7599	7887	8201	8546
30	5712	5879	6055	6239	6433	6639	6856	7088	7336	7603	7892	8207	8552
31	5715	5882	6058	6242	6437	6642	6860	7092	7341	7608	7897	8212	8558
32	5717	5885	6061	6245	6440	6646	6864	7096	7345	7612	7902	8218	8565
33	5720	5888	6064	6249	6443	6649	6868	7100	7349	7617	7907	8223	8571
34	5723	5891	6067	6252	6447	6653	6871	7104	7353	7622	7912	8229	8577
35	5725	5894	6070	6255	6450	6656	6875	7108	7358	7626	7917	8234	8583
36	5728	5896	6073	6258	6453	6660	6879	7112	7362	7631	7922	8240	8589
37	5731	5899	6076	6261	6457	6663	6883	7116	7366	7636	7927	8245	8595
38	5734	5902	6079	6264	6460	6667	6886	7128	7371	7640	7932	8251	8601
39	5736	5905	6082	6268	6463	6670	6890	7124	7375	7645	7937	8256	8607
40	5739	5908	6085	6271	6467	6674	6894	7128	7379	7650	7942	8262	8614
41	5742	5911	6088	6274	6470	6677	6898	7132	7384	7654	7948	8267	8620
42	5745	5914	6091	6277	6473	6681	6901	7136	7388	7659	7953	8273	8626
43	5747	5917	6094	6280	6477	6685	6905	7140	7392	7664	7958	8279	8632
44	5750	5919	6097	6283	6480	6688	6909	7145	7397	7668	7963	8284	8638
45	5753	5922	6100	6287	6483	6692	6913	7149	7401	7673	7968	8290	8644
46	5756	5925	6103	6290	6487	6695	6917	7153	7406	7678	7973	8295	8651
47	5758	5928	6106	9293	6490	6699	6920	7157	7410	7683	7978	8301	8657
48	5761	5931	6109	6296	6494	6702	6924	7161	7414	7687	7983	8307	8663
49	5764	5934	6112	6299	6497	6706	6928	7165	7419	7692	7989	8312	8669
50	5767	5937	6115	6303	6500	6710	6932	7169	7423	7697	7994	8318	8676
51	5770	5940	6118	6306	6504	6713	6936	7173	7427	7702	7999	8324	8682
52	5772	5943	6121	6309	6507	6717	6940	7177	7432	7706	8004	8329	8688
53	5775	5946	6124	6312	6511	6720	6943	7181	7436	7711	8009	8335	8695
54	5778	5948	6127	6315	6514	6724	6947	7185	7441	7716	8914	8341	8701
55	5781	5951	6130	6319	6517	6728	6951	7189	7445	7721	8020	8347	8707
56	5783	5954	6133	6322	6521	6731	6955	7194	7449	7725	8025	8352	8714
57	5786	5957	6136	6325	6524	6735	6959	7198	7454	7730	8030	8358	8720
58	5789	5960	6140	6328	6528	6738	6963	7202	7458	7735	8035	8364	8726
59	5792	5963	6143	6332	6531	6742	6966	7206	7463	7740	8040	8369	8733

′	81°	82°	83°	84°	85°
0	8739	9145	9606	10137	10765
1	8745	9153	9614	10146	10776
2	8752	9160	9622	10156	10788
3	8758	9167	9631	10166	10799
4	8765	9174	9639	10175	10811
5	8771	9182	9647	10185	10822
6	8778	9189	9655	10195	10834
7	8784	9197	9664	10205	10846
8	8791	9203	9672	10214	10858
9	8797	9211	9683	10224	10869
10	8804	9218	9689	10234	10881
11	8810	9225	9697	10244	10893
12	8817	9233	9706	10254	10905
13	8823	9240	9714	10364	10917
14	8830	9248	9723	10273	10929
15	8836	9255	9731	10283	10941
16	8843	9262	9740	10293	10953
17	8849	9270	9748	10303	10965
18	8856	9277	9757	10314	10978
19	8863	9285	9765	10324	10990
20	8869	9292	9774	10334	11002
21	8876	9300	9783	10344	11014
22	8883	9307	9791	10354	11027
23	8889	9315	9800	10364	11039
24	8896	9322	9809	10374	11052
25	8903	9330	9817	10385	11064
26	8909	9337	9826	10395	11077
27	8916	9345	9835	10405	11089
28	8923	9353	9844	10416	11102
29	8930	9360	9852	10426	11115
30	8936	9368	9861	10437	11127
31	8943	9376	9870	10447	11140
32	8950	9383	9879	10457	11153
33	8957	9391	9888	10468	11166
34	8963	9399	9897	10479	11179
35	8970	9407	9906	10489	11192
36	8977	9414	9915	10500	11205
37	8984	9422	9924	10510	11218
38	8991	9430	9933	10521	11231
39	8998	9438	9942	10532	11244
40	9005	9445	9951	10542	11257
41	9012	9453	9960	10553	11270
42	9018	9461	9969	10564	11284
43	9025	9469	9978	10575	11297
44	9032	9477	9987	10586	11310
45	9039	9485	9996	10597	11324
46	9046	9493	10005	10608	11337
47	9053	9501	10015	10619	11351
48	9060	9509	10024	10630	11365
49	9067	9517	10033	10641	11378
50	9074	9525	10043	10652	11392
51	9081	9533	10052	10663	11406
52	9088	9541	10061	10674	11420
53	9096	9549	10071	10685	11434
54	9103	9557	10080	10696	11448
55	9110	9565	10089	10708	11462
56	9117	9573	10099	10719	11476
57	9124	9581	10108	10730	11490
58	9131	9589	10118	10742	11504
59	9138	9598	10127	10753	11518

TABLE V.
Dip of the Sea Horizon.

Hight of Eye in Ft.	Dip of the Horizon. ′ ″	Hight of Eye in Ft.	Dip of the Horizon. ′ ″
1	0 59	38	6 4
2	1 24	41	6 18
3	1 42	44	6 32
4	1 58	47	6 45
5	2 12	50	6 58
6	2 25	53	7 10
7	2 36	56	7 12
8	2 47	59	7 24
9	2 57	62	7 45
10	3 07	65	7 56
11	3 16	68	8 07
12	3 25	71	8 18
13	3 33	74	8 28
14	3 41	77	8 38
15	3 49	80	8 48
16	3 56	83	8 58
17	4 04	86	9 08
18	4 11	89	9 17
19	4 17	92	9 26
20	4 24	95	9 36
21	4 31	98	9 45
22	4 37	101	9 54
23	4 43	104	10 02
24	4 49	107	10 11
25	4 55	110	10 19
26	5 01	113	10 28
27	5 07	116	10 36
28	5 13	119	10 44
29	5 18	122	10 52
30	5 24	125	11 00
31	5 29	128	11 08
32	5 34	131	11 16
33	5 39	134	11 24
34	5 44	137	11 31
35	5 49	140	11 39

TABLE VI.
Dip of the Sea Horizon at different Distances from it.

Dist. in Miles.	Hight of Eye in Ft. 5 ′	10 ′	15 ′	20 ′	25 ′	30 ′
¼	11	22	34	45	56	68
½	6	11	17	22	28	34
¾	4	8	12	15	19	23
1	4	6	9	12	15	17
1¼	3	5	7	9	12	14
1½	3	4	6	8	9	12
2	2	3	5	6	8	10
2½	2	3	5	6	7	8
3	2	3	4	5	6	7
3½	2	3	4	5	6	6
4	2	3	4	4	5	6
5	2	3	4	4	5	5
6	2	3	4	4	5	5

TABLE VII.
Mean Refraction of Celestial Objects.

Alt. ° ′	Refr. ′ ″	Alt. ° ′	Refr. ′ ″	Alt. ° ′	Refr. ′ ″	Alt. ° ′	Refr. ′ ″	Alt. °	Refr ″
0 0	33 0	10 0	5 15	20 0	2 35	32 0	1 30	67	24
10	31 32	10	5 10	10	2 24	40	1 29	68	23
20	29 50	20	5 05	20	2 22	33 0	1 28	69	22
30	28 23	30	5 00	30	2 21	20	1 26	70	21
40	27 00	40	4 56	40	2 29	40	1 25	71	19
50	25 42	50	4 51	50	2 28	34 0	1 24	72	18
1 0	24 29	11 0	4 47	21 0	2 27	20	1 23	73	17
10	23 20	10	4 43	10	2 26	40	1 22	74	16
20	22 15	20	4 39	20	2 25	35 0	1 21	75	15
30	21 15	30	4 34	30	2 24	20	1 20	76	14
40	20 18	40	4 31	40	2 23	40	1 19	77	13
50	19 25	50	4 27	50	2 21	36 0	1 18	78	12
2 0	18 35	12 0	4 23	22 0	2 20	30	1 17	79	11
10	17 48	10	4 20	10	2 19	37 0	1 16	80	10
20	17 04	20	4 16	20	2 18	30	1 14	81	9
30	16 24	30	4 13	30	2 17	38 0	1 13	82	8
40	15 45	40	4 09	40	2 16	30	1 11	83	7
50	15 09	50	4 06	50	2 15	39 0	1 10	34	6
3 0	14 34	13 0	4 03	23 0	2 14	30	1 09	85	5
10	14 04	10	4 00	10	2 13	40 0	1 08	86	4
20	13 34	20	3 57	20	2 12	30	1 07	87	3
30	13 06	30	3 54	30	2 11	41 0	1 05	88	2
40	12 40	49	3 51	40	2 10	30	1 04	89	1
50	12 15	50	3 48	50	2 09	42 0	1 03	90	0
4 0	11 51	14 0	3 45	24 0	2 08	30	1 02		
10	11 29	10	3 43	10	2 07	43 0	1 01		
20	11 08	20	3 40	20	2 06	30	1 00		
30	10 48	30	3 38	30	2 05	44 0	0 59		
40	10 29	40	3 35	40	2 04	30	0 58		
50	10 11	50	3 33	50	2 03	45 0	0 57		
5 0	9 54	15 0	3 30	25 0	2 02	30	0 56		
10	9 38	10	3 28	10	2 01	46 0	0 55		
20	9 23	20	3 26	20	2 00	30	0 54		
30	9 08	30	3 24	30	1 59	47 0	0 53		
40	8 54	40	3 21	40	1 58	30	0 52		
50	8 41	50	3 19	50	1 57	48 0	0 51		
6 0	8 28	16 0	3 17	26 0	1 56	30	0 50		
10	8 15	10	3 15	10	1 55	49 0	0 49		
20	8 03	20	3 12	20	1 55	30	0 49		
30	7 15	30	3 10	30	1 54	50 0	0 48		
40	7 40	40	3 08	40	1 53	30	0 47		
50	7 30	50	3 06	50	1 52	51 0	0 46		
7 0	7 20	17 0	3 04	27 0	1 51	30	0 45		
10	7 11	10	3 03	15	1 50	52 0	0 44		
20	7 02	20	3 01	30	1 49	30	0 44		
30	6 53	30	2 59	45	1 48	53 0	0 43		
40	6 45	40	2 57	28 0	1 47	30	0 42		
50	6 37	50	2 55	15	1 46	54 0	0 41		
8 0	6 29	18 0	2 54	30	1 45	55 0	0 40		
10	6 22	10	2 52	45	1 44	56 0	0 38		
20	6 15	20	2 51	29 0	1 42	57 0	0 37		
30	6 08	30	2 49	20	1 41	58 0	0 35		
40	6 01	40	2 47	40	1 40	59 0	0 34		
50	5 55	50	2 46	30 0	1 38	60 0	0 33		
9 0	5 98	19 0	2 44	20	1 37	61 0	0 32		
10	5 42	10	2 43	40	1 36	62 0	0 30		
20	5 46	20	2 41	31 0	1 35	63 0	0 29		
30	5 41	30	2 40	20	1 33	64 0	0 28		
40	5 25	40	2 38	40	1 32	65 0	0 26		
50	5 20	50	2 37	32 0	1 31	66 0	0 25		

www.ingramcontent.com/pod-product-compliance
Lightning Source LLC
LaVergne TN
LVHW020217110826
845151LV00003B/744

* 9 7 8 1 4 2 5 5 3 3 6 2 5 *